ENRICHMENT MATHEMATICS for the GRADES

TWENTY-SEVENTH YEARBOOK

18
23
11

National Council of Teachers of Mathematics
WASHINGTON, D.C., 1963

HISTORY AND ACKNOWLEDGEMENTS

This, the Twenty-Seventh Yearbook, and its companion volume, the Twenty-Eighth Yearbook, have been long in the making. Into these two books have gone the combined contributions, suggestions, and criticisms of innumerable individuals, committees, and subcommittees. To these individuals and groups the Editorial Committee extends its deep appreciation.

In its April 1959 report the National Council of Teachers of Mathematics (NCTM) Committee on Mathematics for the Talented recommended to the Board of Directors that the Board consider the feasibility of a yearbook containing materials for the talented. The members of that committee were Mary Lee Foster Henderson, Frances Johnson, Glen D. Vannatta, Robert S. Fouch, Joseph N. Payne, Harry D. Ruderman, and Julius H. Hlavaty (chairman).

The Board of Directors at its summer 1959 meeting acted favorably on the suggestion and named a committee, consisting of Albert A. Blank, Vincent J. Glennon, Bruce E. Meserve, Harry D. Ruderman, and Julius H. Hlavaty (chairman), to prepare an outline of a possible yearbook. This committee, with the encouragement and support of NCTM President Harold P. Fawcett, prepared such an outline and submitted it to the Board of Directors in April 1960. The Board thereupon authorized the preparation and publication of the present yearbook.

The new president of the NCTM, Phillip S. Jones, then appointed the undersigned Editorial Committee. President Jones, fresh from his successful experience as editor of the Twenty-Fourth Yearbook, gave the Editorial Committee a detailed, helpful, and explicit charge. The Editorial Committee wishes to express its gratitude to him. Others who gave invaluable advice to the Committee at this stage were Henry Van Engen and Bruce E. Meserve.

The Editorial Committee also extends its thanks to Edward G. Begle, Director of the School Mathematics Study Group (SMSG), for his interest and support. Not only did Dr. Begle put at the disposal of the Committee the rich files of the SMSG Writing Groups, but he also generously gave the Committee permission to adapt and publish some materials originally prepared for SMSG. The individual acknowledgements are noted in the appropriate places in this yearbook.

Gordon L. Walker, editor, *Notices of the American Mathematical Society;* E. Glenadine Gibb, editor, the *Arithmetic Teacher;* and Robert

E. Pingry, editor, the *Mathematics Teacher*, graciously printed in their respective journals requests by the Committee for contributions.

In response to these appeals, many mathematicians and teachers sent in suggestions and contributions. While it has proved impossible to publish all of the contributions, the Editorial Committee is most grateful to all the respondents, for their suggestions and contributions—even when not explicitly used—guided the Committee in its planning.

The following, in addition to those whose names appear elsewhere in the book, made significant contributions to the yearbook: Robert L. Armstrong, Casa Grande Union High School, Casa Grande, Ariz.; I. C. Barker, Lowell High School, San Francisco, Calif.; Douglas M. Beamish, Los Angeles City Unified School District, Los Angeles, Calif.; Richard H. Beyerle, Edmund Scientific Co., Barrington, N. J.; R. H. Bing, University of Wisconsin, Madison, Wis.; Robert L. Burch, Sharon, Mass.; George K. Black, Newhall, Calif.; Florence W. Borgeson, Roosevelt Junior High School, Westfield, N. J.; Brother T. Brendan, FSC, Saint Mary's College of California, St. Mary's, Calif.; John Brown, University of Delaware, Newark, Del.; C. Stuart Brewster, Addison-Wesley Publishing Co., Reading, Mass.; Bancroft H. Brown, Dartmouth College, Hanover, N. H.; Paul C. Burns, University of Kansas, Lawrence, Kans.; Leonard C. Cahen, Stanford University, Stanford, Calif.; Caroline H. Clark, New Hope, Pa.; John R. Clark, New Hope, Pa.; Wagner Collins, Robbinsdale, Minn.; Elizabeth M. Cooper, Syracuse, N. Y.; Dwight Daugherty, State Teachers College, Kutztown, Pa.; Homer R. DeGraff, Dewitt Junior High School, Ithaca, N. Y.; John J. Fiumaro, Darien, Conn.; Herbert Fremont, Queens College, Flushing, N. Y.; Abe Gelbart, Yeshiva University, New York, N. Y.; Harry M. Gelder, Western Washington College of Education, Bellingham, Wash.; Harriet Griffith, Brooklyn College, Brooklyn, N. Y.; Clarence P. Hammarlund, Junior High School, Janesville, Wis.; Carl E. Heilman, Department of Public Education, Harrisburg, Pa.; Irving Hollingshead Jr., Rancocas, N. J.; Rodney T. Hood, Franklin College, Franklin, Ind.; W. Robert Houston, University of Texas, Austin, Tex.; Lawrence Hyman, Board of Education, Cleveland, Ohio; Humphrey C. Jackson, Grosse Pointe Public School System, Grosse Pointe, Mich.; Paul B. Johnson, University of California, Los Angeles, Calif.; John J. Kinsella, New York University, New York, N. Y.; Mrs. A. C. Kruer, Saint Louis, Mo.; Ernestine M. Lawson, Hattiesburg, Miss.; Norbert Lerner, State University College, Cortland, N. Y.; Eleanor Lazansky, Oakland Public Schools, Oakland, Calif.; Daniel B. Lloyd, District of Columbia Teachers College, Washington, D. C.; John Lukas, Northbrook Elementary Schools, Northbrook, Ill.; Mary E. McDermott, Mt. Diablo Unified School District,

Concord, Calif.; George W. McLaughlin, Machias, Maine; George McMeen, California State Polytechnic College, San Luis Obispo, Calif.; William H. Meyer, University of Chicago, Chicago, Ill.; Paul M. Nemecek, Riverside-Brookfield High School, Riverside, Ill.; Claire M. Newman, Edgemont High School, Scarsdale, N. Y.; Monte S. Norton, Lincoln Public Schools, Lincoln, Nebr.; Paul Olum, Cornell University, Ithaca, N. Y.; Cynthia Parsons, Cos Cob, Conn.; Francis Regan, St. Louis University, St. Louis, Mo.; L. A. Ringenberg, Eastern Illinois University, Charleston, Ill.; Gerald R. Rising, Norwalk High School, Norwalk, Conn.; Tillie D. Sadowsky, Philadelphia, Pa.; Helen C. Sears, Ketchikan, Alaska; Sister Anne Agnes von Steiger, CSJ, St. Thomas Aquinas Parish High School, Florissant, Mo.; Sister M. Edith, Sacred Heart Academy, Cullman, Ala.; Sister M. Seraphine, OP, Saint Catherine High School, Racine, Wis.; Lehi T. Smith, Arizona State University, Temple, Ariz.; Stanley A. Smith, Dumbarton Junior High School, Baltimore, Md.; Max A. Sobel, Montclair State College, Upper Montclair, N. J.; Nancy Soverign, Central Michigan University, Mount Pleasant, Mich.; Jay E. Strum, New York University, N. Y.; Dmitri Thoro, San Jose State College, San Jose, Calif.; James F. Ulrich, Arlington High School, Arlington Heights, Ill.; Lowell Van Tassel, Hoover Senior High School, San Diego, Calif.; Stacey Wahl, Ridgefield, Conn.; William C. Weber, Derry Area Joint High School, Derry, Pa.; Lester Weinstein, Carey Junior High School, Cheyenne, Wyo.; Karl H. West, Jr., Needham Public Schools, Needham, Mass.; B. T. Whittington, Chilton County Training School, Clanton, Ala.; Gayle W. Wolff, Southwest High School, Minneapolis, Minn.; and Edward J. Zoll, Newark State College, Newark, N. J.

Many devoted people have contributed to the technical preparation of the manuscript. We express our appreciation to Muriel Bitensky and Esther Unkel of Syracuse University; Giselle Hendel and Mia C. Fienemann of the College Entrance Examination Board; Josephine and Thomas Joyce of New York City; Paul Ferber and Jerry Neu of De Witt Clinton High School, New York, N. Y.; and Fancille H. Hlavaty, New Rochelle, N. Y.

Finally, the Editorial Committee is most grateful to Myrl H. Ahrendt, NCTM Executive Secretary, and Doris Wigglesworth, NCTM Editorial Associate, for the dedicated, detail work which made final publication possible.

The Editorial Committee

ALBERT A. BLANK   RICHARD S. PIETERS
VINCENT J. GLENNON   HARRY D. RUDERMAN
JOSEPH N. PAYNE   HENRY W. SYER
             JULIUS H. HLAVATY, *Chairman*

CONTENTS

vii

GENERAL INTRODUCTION

JULIUS H. HLAVATY

Editor, Twenty-Seventh Yearbook

De Witt Clinton High School
New York, New York

OVERVIEW

The academically talented student in mathematics has always been with us, and most teachers have always attempted to provide for his special needs in one way or another. The principal purpose of this yearbook is to give teachers one more resource in order to provide for the talented more efficiently and more quickly.

In the last ten years there has been a great increase in interest on the part of teachers and parents in the treatment of the academically talented student. One result of this increased interest on the national scale was the Conference on the Academically Talented Student held in Washington, D. C., in February 1958, under the chairmanship of James B. Conant. Another result was the organization of the Academically Talented Student Project by the National Education Association, under the chairmanship of Charles E. Bish. This project, with the cooperation of various subject matter organizations, has resulted in a whole series of pamphlets, covering the various disciplines. One of these pamphlets, *Mathematics for the Academically Talented Student*, might well be considered as an introduction to the Twenty-Seventh and the Twenty-Eighth Yearbooks.

The literature on various aspects of the problem, in mathematics alone, is huge. It is not our intention to review the literature in any detailed fashion. We append to this introduction a bibliography for the use of those teachers who wish to study any aspect of the problem in detail.

Most of the articles emphasize enrichment materials rather than ways of using these materials. However, an exception is found in Chapter 1 of the Twenty-Seventh Yearbook where Glennon devotes his whole discussion to a consideration of the broad problems in education which must be kept in focus in order to make most effective provision for the talented. Provision for the gifted in mathematics has in the past taken

1

the form of either enrichment or acceleration. Glennon discusses these two approaches. In essence his conclusions are applicable to all levels of instruction and of interest to readers of both yearbooks. It is important to consider these fundamentals of approach as well as of content since a proper solution determines the effectiveness of teaching.

The purpose of this General Introduction is to give a brief outline of the book, and to suggest possible uses of the book. The purpose of the individual introductions to the various chapters and sections of the yearbook is similar to that of the General Introduction.

CONTENTS OF THE YEARBOOK

The principal objective of this yearbook is to present content-material for use as enrichment for talented students. Enrichment has usually been achieved by the following: (a) guiding students to deeper considerations of standard topics in a course of study; (b) encouraging individual research; and (c) organizing extra-curricular activities in the form of mathematics clubs, or mathematics teams (for intra- or extra-mural mathematical competitions).

The content of the yearbook, therefore, divides naturally into the following categories:

1. Enrichment units to be taken up by a class or a smaller group of students under the guidance of a teacher,
2. Articles that can be read and studied by individual students,
3. Articles and activities that can be pursued as extra-curricular activities by individual students or by small groups of students.

ORGANIZATION OF THE YEARBOOK

The contributions to the Twenty-Seventh and Twenty-Eighth Yearbooks have been divided into four levels as follows:

(Twenty-Seventh Yearbook)
1. The elementary school years
2. The junior high school years
(Twenty-Eighth Yearbook)
3. The high school years
4. The transition to college.

It is clear that this grouping is quite arbitrary and the suitability of any one article may overlap two or more levels. As a matter of fact, the overlaps are even more extensive than may appear on the surface, and may extend from one yearbook to the other. This is so particularly because of the great ferment in mathematical education at the present time. For example, many topics which are here included at the elementary level may be genuine experiences in enrichment for a high school senior who has had no exposure to some of the newer ideas in current curricula. On the other hand some of the topics, even in the senior high school section or the transition to college section, may be wonderful and accessible activities for a bright eighth grader who has already had an enriched elementary school experience.

If we refer to the three categories of activities suggested above, it will be observed that the activities of the first category predominate in the elementary section, and that there is a gradual shifting in the remaining sections of the book, until, in the last section of the book, the articles are almost wholly of the individual reading and research type.

USES OF THIS YEARBOOK

This book is addressed to teachers of mathematically gifted students. Some of the individual articles, however, are addressed only to the teachers of bright students, while others are addressed to the students and to their teachers.

Some of the articles—from any of the levels—may be utilized as enrichment units to be inserted between standard units in a standard curriculum, as the time and the interest of the students and of the teacher may make feasible. Such units may profitably take several days—or even some weeks—for instruction with a given class.

There are other articles which may be given to individual students for study and investigation. In such cases the student must have the opportunity to consult other materials, his own teacher, or some other interested adult about the progress of his investigation.

Finally, some of the articles make ideal materials for small groups of students, along with their teacher, to use as a basis for investigations in depth into some mathematical idea.

The last two categories of articles raise an important question in teaching—and not only in the teaching of gifted students. No teacher of mathematics, whether at the elementary level or at the senior high school level, in fact no professional mathematician is or can be expected to know fully and completely *all* the material covered in this book. Some of the articles present very novel broachings of questions usually arising as quite incidental asides in advanced discussions of mathematical topics. Therefore, no teacher need hesitate to undertake, with an individual student or with a small group of students, a joint study of a topic that is entirely novel, both to the teacher and to the students. Some most worthwhile results can come from just such experiences. For the student it may be the beginning of a deep interest in mathematical investigation.

For the teacher, it may provide a fresh insight into the learning process, a deeper understanding of the characteristics of the gifted student, and an additional resource to be added to the teacher's arsenal of materials.

Since the various grade levels suggested by the organization of the materials in this book are quite arbitrary, teachers and students are advised to browse through this book beyond the confines of their basic grade-placement interests. Materials in all sections of the book may be helpful, interesting, and appropriate for individual students at any given grade level. It bears repetition that many of the articles in the elementary school section are first-rate enrichment possibilities for high school students who have not had the opportunity to encounter them in their earlier schooling. Also, some of the articles in the more advanced sections are quite accessible to students who have done a lot of individual reading on their own, or who have had the benefits of an enriched elementary curriculum.

Moreover, alert teachers and students will detect that some topics (e.g. numeration systems, number theory, geometric concepts, abstract algebraic notions, problem solving, etc.) crop up in different sections of the book with increasingly broader treatments. A student, or a teacher, may be interested in following through the threads of a particular topic across the various grade levels. A guide for doing this may be suggested by glancing through the brief introductions to the two sections of the book, and by reading the introductions given at the beginning of each article. This injunction applies with equal force to the companion yearbook.

CONCLUSION

Among the many characteristics of the talented and interested student in mathematics, we may select four for special emphasis. He is interested in the relevance of mathematics to life—its applications. He is excited by his ability to raise questions and his ability to answer his own questions or those raised by others. He is delighted with the organization of abstract concepts into patterns and structures and with his perception and understanding of such patterns and structures. Finally he wants to follow through on any of these to see where they lead.

In selecting articles for this book, the following have been among the considerations we have kept in mind. We have tried to illustrate the applications of mathematics—either in mathematics or elsewhere; we have tried to provide materials for doing things—problem solving specifically (and we have included the solutions); we have tried to give some novel and simple samples of mathematical structures; and we have, in most articles, suggested further investigations and readings.

One final point needs stressing. Though we have been so overwhelmed with contributions and materials that two volumes are required to contain these materials, yet the content of the Twenty-Seventh and the Twenty-Eighth Yearbooks is in no sense exhaustive or prescriptive. The articles in this book and the many other possible ones that we could not include for lack of space are still only a sampling of the many types of enrichment that can be given to talented students. We are confident not only that teachers and students will hit on subsidiary investigations suggested by the contents of the yearbook, but that they will find many others not even hinted at here.

MATHEMATICS FOR THE GIFTED:
A Bibliography for the Grades

WILLIAM L. SCHAAF

Brooklyn College
Brooklyn, New York

GENERAL EDUCATION DISCUSSION

BARBOUR, RICHMOND. *Educational Offerings for Exceptional Children*. San Diego: City Schools, 1952.

BOARD OF SCHOOL COMMISSIONERS. *The Superior Child in the Baltimore Public Schools*. Baltimore: Public Schools, 1953. pp. 9, 22–23.

BRIGGS, THOMAS, and OTHERS. "Some Issues and Problems Raised by the Conference on Education for the Gifted, Seminar V, Education of Gifted Pupils in Secondary Schools." *Teachers College Record* 43: 44–51; February 1941.

CONANT, JAMES B. *The Identification and Education of the Academically Talented Student in the American Secondary School*. National Education Association. Washington, D. C.: The Association; February 1958. pp. 97–103.

CORNOG, W. H. "The High School Can Educate the Exceptionally Able Student." *Bulletin, National Association of Secondary School Principals* 39: 380–86; April 1955.

CRUICKSHANK, WILLIAM. *Psychology of Exceptional Children and Youth*. Englewood Cliffs, N.J.: Prentice-Hall, 1955. pp. 475–529.

CUTTS, NORMAN, and MOSELY, NICHOLAS. *Teaching the Bright and Gifted*. Englewood Cliffs, N. J.: Prentice-Hall, 1957.

DAVIS, F. B., and OTHERS. "Identification and Classroom Behavior of Gifted Elementary School Children." *The Gifted Student*. U. S. Department of Health, Education, and Welfare, Office of Education, Cooperative Research Monograph No. 2. Washington, D. C.: Superintendent of Documents, Government Printing Office, 1960. pp. 19–32.

FLIEGLER, LOUIS, and BISH, CHARLES. "The Gifted and Talented." *Review of Educational Research* 29: 408–50; December 1959.

HAVIGHURST, R. J., STEVERS, E., and DE HAAN, R. F. *A Survey of the Education of Gifted Children. The Committee on Human Development*. Chicago: University of Chicago Press, No. 83, November 1955.

HILDRETH, GERTRUDE. *Educating Gifted Children*. New York: Harper & Brothers, 1952.

HILDRETH, GERTRUDE. "School-Wide Planning for the Gifted." *Educational Administration and Supervision* 41: 1–10; January 1955.

JANSEN, WILLIAM. *Our Public Schools, Part I; Gifted Children and Slow Learners in the Junior High School.* New York: Board of Education, 1953. pp. 3–11.

JUSTMAN, J., and WRIGHTSTONE, J. W. "Opinions of Junior High School Principals Concerning the Organization of Special Classes for Gifted Children." *Educational Administration and Supervision* 37: 396–404; November 1951.

KOUGH, JACK. "Administrative Provisions for the Gifted". *Working with Superior Students: Theories and Practices.* (Edited by Bruce Shertzer.) Chicago: Science Research Associates, 1960. Chapter 13, pp. 142–60.

MAKOVIC, SISTER MARY VERNICE. *The Gifted Child. Special Education of the Exceptional Child.* Washington, D.C.: Catholic University of America Press, 1953. pp. 48–69.

MCWILLIAMS, E. M. "The Gifted Pupil in the High School." *Bulletin, National Association of Secondary School Principals* 39: 1–9; May 1955.

MOSKOWITZ, DAVID. "Educating Superior Students." *High Points* 28: 5–9; June 1946.

NATIONAL EDUCATION ASSOCIATION, Research Division. "High-School Methods with Superior Students." *Research Bulletin* 19: 155–97; September 1941.

NATIONAL SOCIETY FOR THE STUDY OF EDUCATION. *Education of Exceptional Children.* Forty-Ninth Yearbook, Part II. Chicago: University of Chicago Press, 1950. pp. 259–80.

NEW YORK, UNIVERSITY OF THE STATE OF. *Fifty-Six Practices for the Gifted for Secondary Schools of New York State.* (With Selected Bibliography.) Albany: State Education Department, Bureau of Secondary School Curriculum Development, 1959.

OHIO, BOARD OF EDUCATION. *Selected and Annotated Bibliography on the Gifted.* Columbus, Ohio: The Board, 1960.

OLIVER, ALBERT. "Administrative Problems in Educating the Gifted." *The Nation's Schools* 48: 44–46; November 1951.

PARKER, J. C., and RUSSELL, D. H. "Ways of Providing for Individual Differences." *Educational Leadership* 11: 168–74; December 1953.

PASCHAL, ELIZABETH. *Encouraging the Excellent.* New York: Fund for the Advancement of Education, 1960. 79 pp.

PRITCHARD, MIRIAM. "Total School Planning for the Gifted Child." *Exceptional Children* 18: 107–10, January 1952; 143–47, February 1952; 174–80, March 1952.

SIMPSON, ROY, and MARTINSON, RUTH. *Educational Programs for Gifted Pupils.* Sacramento: California State Department of Education, 1961. 274 pp.

STRANG, RUTH. "Inner World of Gifted Adolescents." *Exceptional Children* 16: 97–101, 125; January 1950.

STRANG, RUTH (chairman). "The Gifted Child." *Journal of Teacher Education* 53: 210–32; September 1954.

SUBURBAN SCHOOL STUDY COUNCIL. *Guiding Your Gifted.* Philadelphia: Educational Service Bureau, School of Education, University of Pennsylvania, 1950.

U.S. DEPARTMENT OF HEALTH, EDUCATION, AND WELFARE, Office of Education. *Teaching Rapid and Slow Learners in High Schools.* Bulletin 1954, No. 5. Washington, D.C.: Government Printing Office, 1954.

WEITZ, LEO, and OTHERS. "The Rapid Learner in Our High Schools: A Report." *High Points*, February 1956. pp. 5–36.

WILSON, FRANK T. "The Evidence About Acceleration of Gifted Youth." *School and Society* 73: 409–10; June 1951.

WILSON, FRANK T. "Preparation for Teachers of Gifted Children in the United States." *Exceptional Children* 20: 78–80; November 1953.
WITTY, PAUL, editor. *The Gifted Child.* Boston: D. C. Heath and Co., 1951. pp. 215–22; 261–62.

GENERAL MATHEMATICAL DISCUSSION

AHRENDT, M. H. "Education for the Mathematically Gifted." *Phi Delta Kappan* 34: 285–87; April 1953.
AIKEN, D. J. "Some Comment on Accelerated Mathematics." *Mathematics Teachers* 51: 292–93; April 1958.
BERGER, E. "Enriching Instruction via Television." *Mathematics Teacher* 51: 550–52; November 1958.
BRANDWEIN, PAUL. "New Patterns in the Education of Gifted Children in Mathematics and Science." *Kentucky School Journal* 35: 20–21+; November 1956.
BRANDWEIN, PAUL. "Mathematics and Science." *Education for the Gifted.* Fifty-Seventh Yearbook, National Society for the Study of Education. Chicago: University of Chicago Press, 1958. Chapter 12, pp. 290–93.
BROWN, KENNETH, and JOHNSON, PHILIP. *Education for the Talented in Mathematics and Science.* U.S. Office of Education, Federal Security Agency. Bulletin 1952, No. 15. Washington, D.C.: Government Printing Office, 1952.
BUTCHART, J. H. "High-Ability Students' Institute at Flagstaff." *Mathematics Teacher* 54: 245–47; April 1961.
√ CLARK, N. "Challenge to the Gifted." *Mathematics Teacher* 48: 434–35; October 1955.
√ COE, BURR. "Ungraded Classes for Superior Pupils." *Mathematics Teacher* 37: 81–83; 1944.
COLE, CHARLES C., JR. *Encouraging Scientific Talent.* New York: College Entrance Examination Board, 1956. 259 pp.
DEVINE, D. F. "Mathematics for Gifted Students." *Illinois Education* 46: 166–67; January 1958.
DOUGLASS, EARL R. "Issues in Elementary and Secondary School Mathematics." *Mathematics Teacher* 46: 290–94; May 1954.
DRENCKHAHN, FRIEDRICH. "Der Mathematische Unterricht im Lichte der Differentiellen Psychologie". *Der Mathematische Unterricht für die 6-bis 15-jährige Jugend in der Bundesrepublik Deutschland.* Göttingen: Vandenhoeck & Ruprecht, 1958. pp. 75–86.
ECKELBERRY, R. H. "Accelerated Study in Mathematics and Science." *Educational Research Bulletin* 35: 47–48; February 1956.
ELDER, F. L. "Providing for the Student with High Mathematical Potential." *Mathematics Teacher* 50: 502–06; November 1957.
FEHR, HOWARD. "General Ways to Identify Students with Scientific and Mathematical Potential." *Mathematics Teacher* 46: 230–34; April 1953.
FEHR, HOWARD. "Mathematics for the Gifted." *Bulletin, National Association of Secondary School Principals* 38: 103–10; May 1954.
FEHR, HOWARD. "Mathematics Instruction and Scientific Manpower." *School Science and Mathematics* 54: 169–72; March 1954.
FULLERTON, G. "Summer-School Science and Mathematics." *Journal, National Education Association* 49: 35–36; April 1960.

GORDON, GARFORD. *Providing for Outstanding Science and Mathematics Students.* Los Angeles: University of Southern California Press, 1955. 100 pp.

HANKINS, DONALD. *Realignment of High School Mathematics Programs.* San Diego: San Diego, California City Schools, 1956.

HETLAND, M. J., and GLENN, H. "Program for the Mathematically Gifted." *California Journal of Secondary Education* 32: 334–37; October 1957.

JACKSON, H. O. "Superior Pupil in Mathematics." *Mathematics Teacher* 52: 202–04; March 1959.

JACKSON, H. O. "Superior Pupil in Mathematics." *Mathematics Teacher* 52: 389–417; May 1959.

KELLY, INEZ. "Challenging the Gifted Student." *School Life* 35: 27–28, November 1952.

KENNEDY, J. W., AND MAYOR, J. R. "Individual Differences and Mathematics." *Phi Delta Kappan* 37: 222–23; February 1956.

KRAFT, ONA. "Providing a Challenging Program in Science and Mathematics for Pupils of Superior Mental Ability." *School Science and Mathematics* 52: 143–47; February 1952.

LANGER, R. E. "Time Is Running Out." *Mathematics Teacher* 49: 418–24; October 1956.

LEE, D. M. "Study of Specific Ability and Attainment in Mathematics." *British Journal of EducationalPsychology* 25: 178–89; November 1955. (Bibliography)

LEVY, N. "Toward Discovery and Creativity." *Mathematics Teacher* 50: 19–22; January 1957.

LLOYD, DANIEL B. "Ultra-curricular Stimulation for the Superior Students." *Mathematics Teacher* 46: 487–89; November 1953.

LONDON TIMES. "Maths in the Cradle; Fostering Ability." *London Times Educational Supplement* 2191: 695; May 17, 1957.

MALLINSON, G. G. "Creativity in Science and Mathematics." *Educational Leadership* 18: 24–27; October 1960. (Bibliography)

MAYOR, JOHN R., and BROWN, JOHN A., editors. "The Mathematics Program at the Phillips Exeter Academy." *American Mathematical Monthly* 65: 705–07; November 1958.

METZNER, J., and REINER, W. B. "Provisions for the Academically Talented Student in Science and Mathematics." *Review of Educational Research* 31: 323–30; June 1961. (Bibliography)

NICHOLS, E. D. "Summer Mathematics Program for the Mathematically Talented." *Mathematics Teacher* 53: 235–40; April 1960.

NORTON, M. S. "Educating the Gifted Pupil in Mathematics and Science; an Important Project for Our Nation." *School Science and Mathematics* 56: 665–67; November 1956. (Bibliography)

NORTON, M. S. "Enrichment as a Provision for the Gifted in Mathematics." *School Science and Mathematics* 57: 339–45; May 1957.

NORTON, M. S. "What Are Some of the Factors to Consider in a Program of Identifying the Gifted Pupil in Science and Mathematics?" *School Science and Mathematics* 57: 103–08; February 1957.

NORTON, M. S. "Study of Practices and Provisions for the Gifted Pupil in Mathematics." *Journal of Educational Research* 53: 316–17; April 1960.

PECKMAN, EUGENE. "Providing a Challenging Program in Mathematics and Science for Pupils of Superior Mental Ability." *School Science and Mathematics* 52: 187–92; March 1952.

PENK, G. L. "St. Paul Vitalizes Science and Mathematics for the Gifted." *American School Board Journal* 138: 19–21; March 1959.

PRICE, G. BALEY. "A Mathematics Program for the Able." *Mathematics Teacher* 44: 369–76; October 1951.

REEVE, WILLIAM D. "The Problem of Varying Abilities among Students in Mathematics." *Mathematics Teacher* 49: 70–78; 1956.

SMITH, ROLLAND R. "Provisions for Individual Differences." *The Learning of Mathematics: Its Theory and Practice.* Twenty-First Yearbook. Washington, D.C.: National Council of Teachers of Mathematics, a department of the National Education Association, 1953. pp. 271–302.

VANNATTA, GLEN D. *Background, Choices, and Opinions of Superior Mathematics Students as a Basis for an Attack On the Scientific Manpower Shortage.* Doctor's thesis. Bloomington: Indiana University, 1958. Abstracted in U.S. Office of Education Bulletin 1958, No. 4. Washington, D.C.: Government Printing Office, 1958, p. 70.

WATSON, BERNARD B. "The AAAS Cooperative Committee on the Teaching of Science and Mathematics." *School Science and Mathematics* 54: 116–18; February 1954.

WEST, J. "Science Gifted Work with Local Scientists." *School Executive* 78: 74–75; September 1958.

WOLFE, DAEL. "Future Supply of Science and Mathematics Students." *Mathematics Teacher* 46: 227–29; May 1953.

KINDERGARTEN—GRADE 8

BANKS, L. E.; SHEEHAN, N. L.; and WILLWERTH, R. F. *An Evaluation of Self-Administering Arithmetic Enrichment Activities for Grades Four, Five and Six.* Master's thesis (unpublished). Boston: Boston University School of Education, 1959.

BARRETT, D.; LIGHTMAN, P. M.; MINDEL, H. B.; and SHATZ, F. *Self-Administering Arithmetic Enrichment Activities for the More Rapid Learner.* Master's thesis (unpublished). Boston: Boston University School of Education, 1958.

BRUECKNER, L. J., and GROSSNICKLE, F. E. *Making Arithmetic Meaningful.* Philadelphia: John C. Winston Co., 1953. Pp. 99–100, 453–66 deal with "Activities Suitable for Gifted Children," "Variations in Major Areas in Arithmetic," and "Providing for Individual Differences in Ability and in Rate of Learning."

BRUECKNER, L. J.; GROSSNICKLE, F. E.; and RECKZEH, J. *Developing Mathematical Understandings in the Upper Grades.* Philadelphia: John C. Winston Co., 1957. Pp. 502–51: "Mathematics for the Gifted Student."

BRYDEGAARD, MARGUERITE. "Creative Teaching Points the Way to Help the Brighter Child in Mathematics." *Arithmetic Teacher* 1: 21–24; February 1954.

BUREAU OF ELEMENTARY CURRICULUM DEVELOPMENT. New York State Education Department. *Curriculum Adaptations for the Gifted.* Albany: The Department, 1958. 52 pp.
Bibliography. Arithmetic, pp. 26–27.

BURNS, P. C. "Research for the Library-Minded Mentally Advanced Arithmetic Pupils in Grades 4, 5 and 6." *School Science and Mathematics* 61. 694–96; December 1961.

CARLSON, R. K., and TYLDSLEY, C. H. "Bibliography of Books for Enrichment in Arithmetic." *Arithmetic Teacher* 7: 189–93; April 1960.

CLARK, JOHN R. "A Promising Approach to Provision for Individual Differences." *Journal of Education* 136: 94–96; December 1954.

CLARK, JOHN R., and EADS, LAURA K. *Guiding Arithmetic Learning.* Yonkers, N. Y.: World Book Co., 1954. Pp. 245–52: "Teaching Children in Groups."

COBURN, MAUDE. "Flexibility in the Arithmetic Program: To Promote Maximum Pupil Growth." *Arithmetic Teacher* 2: 48–54; April 1955.

CRUMLEY, R. D. "Unifying Ideas in the Arithmetic Curriculum." *School Science and Mathematics* 58: 341–46; May 1958.

DELONG, A. R., and CLARK, R. M. "Developing Creativity through Arithmetic." *Arithmetic Teacher* 6: 206–08; October 1959.

DENNY, ROBERT. *How to Challenge the Gifted in Arithmetic.* Second edition. Des Moines, Iowa: The Author, 1959.

DUMAS, E.; KITTELL, J.; and GRANT, B. *How to Meet Individual Differences in Teaching Arithmetic.* San Francisco: Fearon Publishers, 1957.

DURR, W. K. "Use of Arithmetic Workbooks in Relation to Mental Abilities and Selected Achievement Levels." *Journal of Educational Research* 51: 561–71; April 1958.

ERHART, M. "Arithmetic for the Academically Talented." *Arithmetic Teacher* 7: 53–60; February 1960.

FLOURNOY, F. "Meeting Individual Differences in Arithmetic." *Arithmetic Teacher* 7: 80–86; February 1960.

FOOTE, E. W. "Using Teachers' Manuals for Deeper Learning." *Arithmetic Teacher* 6: 17–22; February 1959. (Bibliography)

GASKILL, A. R. "Stimulating the Better Arithmetic Pupil with Insight Producing Activities." *Arithmetic Teacher* 4: 33–34; February 1957.

GLENNON, VINCENT J. "Arithmetic for the Gifted Child." *Elementary School Journal* 58: 91–96; November 1957. (Bibliography)

GLENNON, VINCENT J. "Arithmetic for the Gifted Child." *Frontiers of Elementary Education.* Syracuse: Syracuse University, School of Education, 1956. Vol. 3, pp. 39–46.

GLENNON, VINCENT. J. "Evaluation in Arithmetic and Talented Students." *Bulletin, National Association of Secondary School Principals* 43: 134–36; May 1959.

GROSSNICKLE, FOSTER E. "Arithmetic for Those Who Excel." *Arithmetic Teacher* 3: 41–48; March 1956.

GROSSNICKLE, FOSTER E. "Individualizing Instruction in Arithmetic." *Instructor* 63: 65, 72; May 1954.

GROSSNICKLE, FOSTER E., and BRUECKNER, L. J. *Discovering Meanings in Arithmetic.* Philadelphia: John C. Winston Co., 1959. Pp. 373–77: "Individual and Trait Differences"; pp. 398–421: "Enrichment of Learning in Arithmetic."

HANNAN, HERBERT. " 'Sets' Aid in Adding Fractions." *Arithmetic Teacher* 6: 35–38; February 1959.

HESS, A. L. "Bibliography of Books for Enrichment in Arithmetic." *Arithmetic Teacher* 6: 12–16; February 1959.

HESS, A. L. "Bibliography of Mathematics Books for Elementary School Libraries." *Arithmetic Teacher* 4: 15–20; February 1957.

HILLMAN, G. D. "Horizontally, Vertically, and Deeper Work for the Fast-Moving Class." *Arithmetic Teacher* 5: 34–37; February 1958.

HUTCHESON, RUTH; MANTOR, EDNA; and HOLMBERG, MARJORIE. "The Elementary School Mathematics Library." *Arithmetic Teacher* 3: 8–16; February 1956.

ISAACS, A. F. "Gifted Underachiever in Arithmetic: A Case Study." *Arithmetic Teacher* 6: 257–61; November 1959.

IVIE, C.; GUNN, L.; and HOLLADAY, I. "Grouping in Arithmetic in the Normal Classroom." *Arithmetic Teacher* 4: 219–21; November 1957.

JOHNSON, CHARLES E. "Grouping Children for Arithmetic Instruction." *Arithmetic Teacher* 1: 16–20; February 1954.

JUNGE, C. W. "Arithmetic for the Able Learner." *Proceedings: Pennsylvania University Schoolmen's Week* 44: 169–75.

JUNGE, C. W. "Gifted Ones: How Shall We Know Them?" *Arithmetic Teacher* 4: 141–46; October 1957.

KLING, P. T., and TRIEBS, P. M. *A Collection of Arithmetic Exercises Designed Particularly for the Superior Child in Grade Three or Four to Supplement the Standard Program at these Grade Levels.* Master's thesis (unpublished). Boston: Boston University School of Education, 1957.

LARSEN, HAROLD D. *Enrichment Program for Arithmetic.* Evanston, Ill.: Row, Peterson & Co., 1956. A series of thirty-two 16-page booklets for Grades 3–6.

LERNER, N., and SOBEL, MAX. " 'Sets' and Elementary School Mathematics." *Arithmetic Teacher* 5: 239–46; November 1958.

MARKS, J. L.; PURDY, C. R.; and KINNEY, L. B. *Teaching Arithmetic for Understanding.* New York: McGraw-Hill Book Co., 1958. Pp. 373–409: "Adjusting to Individuals."

McMEEN, GEORGE. "Differentiating Arithmetic Instruction for Various Levels of Achievement." *Arithmetic Teacher* 6: 113–20; April 1959.

McSWAIN, E. T., and COOKE, R. J. *Understanding and Teaching Arithmetic.* New York: Henry Holt and Co., 1958. Pp. 353–58: "Meeting Individual Differences."

MORTON, R. L. "Providing for Individual Differences by Grouping in Depth." *Bulletin, National Association of Secondary School Principals* 43: 119–22; May 1959.

MOSER, HAROLD. "Levels of Learning (Planning in Depth)." *Arithmetic Teacher* 3: 221–25; December 1956.

NIES, RUTH. "Classroom Experiences with Recreational Arithmetic." *Arithmetic Teacher* 3: 90–93; April 1956.

PANEK, A. "Providing for the Gifted Child." *Arithmetic Teacher* 6: 246–50; November 1959. (Bibliography)

PARSONS, CYNTHIA. "Arithmetic for the Gifted." *Arithmetic Teacher* 6: 84–86; March 1959.

PAUKNER, LILLIAN. "Milwaukee's In-Service Arithmetic Education Program." *Arithmetic Teacher* 4: 222–23; November 1957.

REZSEK, A. N., and NORRIS, R. B. "Enrichment in Arithmetic in the Primary Grades." *Grade Teacher* 73: 39+; October 1955.

SAWYER, W. W. "Why Is Arithmetic Not the End?" *Arithmetic Teacher* 6: 95–96, 99; March 1959.

SCHWARTZ, A. N. "Challenging the Rapid Learner." *Arithmetic Teacher* 6: 311–13+; December 1959.

SHEPARD, JOHN P. "The Pythagorean Theorem in the Fifth Grade." *Elementary School Journal* 58: 398–400; April 1958.

SPITZER, HERBERT F. *Enriching the Teaching of Arithmetic.* Boston: Houghton Mifflin Co., 1954.

SPITZER, HERBERT. F. *Practical Classroom Procedures for Enriching Mathematics.* St. Louis: Webster Publishing Co., 1956. pp. 33, 144.

Sprague, C. "Individual Projects in Creative Arithmetic." *Instructor* 68: 34+; September 1958. (Bibliography)

SWENSON, ESTHER, J. "Rate of Progress in Learning Arithmetic." *Mathematics Teacher* 48: 70–76; February 1955.

WEAVER, J. FRED. "Big Dividends from Little Interviews." *Arithmetic Teacher* 2: 40–47; April 1955.

WEAVER, J. FRED. "Differentiated Instruction in Arithmetic: An Overview and a Promising Trend." *Education* 74: 300–05; January 1954.

WEAVER, J. FRED. "Recurring Mathematical Ideas in the Arithmetic Program." *Frontiers of Elementary Education.* Syracuse: Syracuse University Press, 1958. Vol. 5, 43–54.

WEAVER, J. FRED, and BRAWLEY, C. F. "Enriching the Elementary School Mathematics Program for More Capable Children." *Journal of Education* 142: 1–40; October 1959. (Bibliography)

WILLERDING, MARGARET. "Codes for Boys and Girls." *Arithmetic Teacher* 2: 23–24; February 1955.

WILLERDING, MARGARET. "History of Mathematics in Teaching Arithmetic." *Arithmetic Teacher* 1: 24–25; April 1954.

YOOD, ETHEL. *An Evaluation of Enrichment Material to Accompany the Teaching of Two- and Three-Figure Multipliers in the Fourth Grade.* Master's thesis (unpublished). Boston: Boston University School of Education, 1956.

GRADES 7–10: JUNIOR HIGH SCHOOL

BRYAN, EUNICE R. *A Mathematics Program for Superior Upper Grade Pupils.* Master's thesis. Normal, Ill.: Illinois State University, 1958. Abstracted in the U.S. Office of Education Bulletin 1960, No. 8, Washington, D. C.: Government Printing Office, 1960. p. 25.

COMMITTEE FOR THE VERY SUPERIOR PUPIL. *The Very Superior Pupil, a Tentative Plan.* Junior High School edition. Long Beach: Long Beach California Public Schools, 1952. pp. 5–6.

ELDER, F. L. "Junior High School Seminar for Talented Students." *Bulletin, National Association of Secondary School Principals* 43: 95–98, May 1959.

GORDON, D. X. "Problem-Solving in Ninth-Year Algebra." *High Points* 36: 19–31; February 1954.

HEELY, R. B. "Using Seventh and Eighth Grade Mathematics as a Pattern of Creative Ability." *Education* 66: 183–89; November 1945.

IRVIN, LEE. "The Organization of Instruction in Arithmetic and Basic Mathematics in Selected Secondary Schools." *Mathematics Teacher* 46: 235–40; April 1953.

KEEDY, M. L. "University of Maryland Project and Talented Students." *Bulletin, National Association of Secondary School Principals* 43: 84–85; May 1959.

LATINO, J. J. "An Algebra Program for the Bright Ninth-Grader." *Mathematics Teacher* 49: 179–84; March 1956. (Bibliography)

LICHTENSTEIN, S. "Enrichment of Mathematics in Junior High School." *High Points* 30: 79–80; April 1948.

MARCANTONIO, EDWARD. "There's Life in Ninth-Year Mathematics." *High Points* 35: 60–62; February 1953.

McWILLIAMS, E. M., AND BROWN, KENNETH. *The Superior Pupil in Junior High School Mathematics.* Office of Education, Bulletin 1955, No. 4. Washington, D.C.: Government Printing Office, 1955.

NIELSEN, R. A. AND GOHMAN, W. "Junior High School Seminar in Science and Mathematics." *Mathematics Teacher* 52: 295–98; April 1959.

NORTON, M. S. "Enrichment Units in Junior High School Grades." *Arithmetic Teacher* 4: 260–61; December 1957.

NORTON, M. S. "Successful Practices and Provisions for Enriching the Educational Program for Gifted Students in Junior High School Mathematics and Science." *School Science and Mathematics.* 59: 101–06; February 1959.

TWISS, R. M. "Providing for the Fast Learner in Business Arithmetic." *American Business Education* 14: 209–13; May 1958.

VAN ENGEN, H. "Concepts Pervading Elementary and Secondary Mathematics." *Bulletin, National Association of Secondary School Principals* 43: 116–18; May 1959.

SECTION I | THE ELEMENTARY SCHOOL YEARS

INTRODUCTION

VINCENT J. GLENNON

Syracuse University
Syracuse, New York

SUGGESTED USES FOR SECTION 1

Section I deals with specific strands of mathematical subject matter which can be drawn upon by the teacher who is working with talented children in whatever manner they are grouped for teaching. The content of the strands is primarily abstract mathematics rather than applied mathematics. That is to say, it is derived from the content of mathematics, not from its applications in daily life. Today, it seems that teachers are still more concerned with social utility than with the mathematical meanings of the present program.

At the end of Section I is appended a graded list of books, which the teacher can use to supplement the enrichment program.

In the following pages are articles on many topics from which elementary school personnel can draw for ideas for enriching the program. Not all the topics will appeal to all talented children or all teachers. Nor need all the articles be used in the same way. In some instances an article might be given to the talented to read on their own; in other instances the teacher might study the content and prepare his own lesson plans and exercise material. At the request of the chairman of this subcommittee, two contributors, Rudd and Johnson, kindly prepared a set of lesson plans to illustrate ways to adapt the content of their articles (Chapters 2 and 10 respectively) to the elementary grades.

ELEMENTARY SCHOOL YEARS SUBCOMMITTEE

The following subcommittee gathered, appraised, and organized the materials for the Elementary School Years Section.
WILLIAM A. BROWNELL, University of California, Berkeley, California
CHARLOTTE W. JUNGE, Wayne State University, Detroit, Michigan

17

HAROLD L. LARSEN†, Albion College, Albion, Michigan
ESTHER SWENSON, University of Alabama, University, Alabama
J. FRED WEAVER, Boston University, Boston, Massachusetts
JOHN W. WILSON, Syracuse University, Syracuse, New York
                              VINCENT J. GLENNON, *Chairman*
                              Syracuse University, Syracuse, New York

† Deceased.

SOME PERSPECTIVES
IN EDUCATION

VINCENT J. GLENNON

Syracuse University
Syracuse, New York

The purpose of this article is to discuss the problem of providing learning experiences in mathematics for talented children in the early school years (K-6) and to treat the several related educational questions that must be kept in focus as we seek solutions for our major problem. These related problems cluster around the following questions.

1. What is a balanced program in elementary school mathematics?
2. What mathematics should be provided for the talented children?
3. How can we identify the mathematically talented children?
4. How shall we group mathematically talented children for instructional purposes?
5. What teaching methods should we use with mathematically talented children?

WHAT IS A BALANCED PROGRAM?

One of the educator's easiest tasks is to *change* the curriculum; one of his most difficult tasks is to *improve* the curriculum. Change requires little professional training in educational philosophy, educational psychology, or in the subject matter to be taught. Improvement, however, demands a high degree of each. Change is concerned only with answering the question: *"Can* the child learn a given topic?" Improvement is concerned with the infinitely more profound question: *"Ought* the child learn a particular topic?" Change can be implemented through a monolithic, authoritarian decision; improvement must call upon the combined judgment of the best minds in the several disciplines that impinge upon the school curriculum.

At the outset let it be agreed that the mathematics program which is good, appropriate, and necessary for all children is also, and by that fact, equally good, appropriate, and necessary for the talented. The program for the talented should, in general terms, be a set of well selected and well ordered learning experiences that go beyond, yet are continuous with, those for the average child. For this reason it is of primary importance to first ask the question: "What mathematics is most appropriate, of most worth, *for the average child?*" Having answered this question, we can then know the nature of the foundation upon which we are to build the superstructure for the talented child.

Informed curriculum theorists are in general agreement that the primary criterion for determining the school curriculum in general and of the mathematics curriculum in particular for the average child is that of *balance.* They are also in agreement that balance in the curriculum is difficult to define, obtain, and maintain. This is especially true during periods of rapid social change such as the one in which we are now living.

Herbert Spencer's question of one hundred years ago, "What knowledge is of most worth?," is still the phoenix of today. Evidence of the recurring nature of the problem is the fact that the 1961 Yearbook of the Association for Supervision and Curriculum Development, *Balance in the Curriculum,* is addressed to this problem. Foshay (8: iii)[1] discusses the magnitude of the problem this way:

> A conception of what balance means in the curriculum is a necessity in any time. In these days of upheaval in education, however, such a conception is an urgent necessity. It is possible that the new curriculum patterns, when they have emerged, will prove to be in better balance than anything we have known. However, taken as a whole, it could be that the new curriculum will imply a distorted version of our culture, or our ideals as a people, even what we want an American to be. This has happened in the past, at those times when it has become apparent that the existing curriculum no longer fits the times. The changes have not always proved to be improvements; sometimes, despite the best efforts of wise men, the result has been only to substitute one distortion for another.

In a word then, the search for balance in the curriculum is an unending one and is very much with us today.

The Problem of Balance in Elementary School Mathematics

If we assume some 15 to 30 minutes of arithmetic instruction time per day in the primary grades and some 40 to 50 minutes in the upper elementary grades, which amounts to the rather few clock hours per

---

[1] The first number indicates the reference in the bibliography at the end of this article, the second number indicates the page.

year of 45 to 90 and of 120 to 150 respectively (with 180 school days per year), our problem becomes: what content is of sufficient importance to merit part of these very few hours out of each whole year in the life of a child? Viewed thus, society can hardly afford the luxury of assuming that all knowledge is of equal worth, and that consequently it matters not *what* one teaches young children.

As has so often happened in the past, when the need for change has been recognized the tendency has been to overcompensate. A recent case in point was the shift away from the mental discipline point of view and toward a narrow interpretation of the social utility point of view. This happened in the 1920's, 30's and 40's. It required the Herculean efforts of such men as Brownell, Clark, Judd, and many others to counteract narrow utilitarianism and restore balance to the program.

Today's problem of balance, of what we *ought* to teach, is not new; it is many hundreds of years old. It is one that each society has had to solve for itself. Yesterday's solution does not fit today's problem. The need for thoughtful, mature, systematic, and unbiased efforts in maintaining balance in all of our school programs (here specifically in the elementary school mathematics program through evolutionary curriculum adjustments) is unending. It is our purpose to aid this effort by presenting next a discussion of balance as a function of the interrelationship of theories of curriculum organization.

THE THREE THEORIES OF CURRICULUM ORGANIZATION

Many educators have presented the problem of balance as that of finding a middle ground between two curriculum theories: the mathematical and the social utility (the pure and the applied) with the mathematical at one end of a line segment and the social utility at the other end. There are in fact *three* curriculum theories—the third one being the psychological theory. Hence the interrelationships among the three theories are more accurately represented by a triangle. All three theories when viewed together constitute a curriculum model for assessing and evaluating the worth of new proposals.

Each is commonly known by other names. The psychological theory is known as the needs-of-the-individual theory, the theory of felt-needs, and the expressed-needs theory. The social theory is known as the needs-of-society theory, the sociological theory, the social-utility theory, and instrumentalism. The theory which stresses the structure of the subject is known generally as the needs-of-the-subject theory, or the logical organization theory; and in mathematics in particular it is known as the pure-game or structural, or meaning, theory of arithmetic. In this

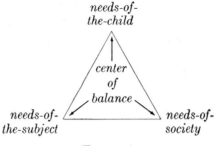

FIGURE 1

discussion we will refer to these theories as the needs-of-the-individual theory, the needs-of-society theory, and the needs-of-the-subject theory.

As part of all three theories, those concerned with elementary school mathematics will continue to expect computational competence and problem solving ability; hence, both of these socially desirable learnings are presupposed in this discussion. When all three theories are influencing the program in *appropriate but perhaps not equal amounts* we say the program is a balanced one. Each theory has merits until it becomes the *only* theory.

We can use the triangle in Figure 1 to represent the three theories of curriculum organization which have influenced the elementary school mathematics program in this century. When the effect of any one of the theories exerts a pull or a force that is either disproportionately or inappropriately greater than the other two, the balance is disturbed, distorted, or destroyed.

During this century, these three theories have influenced in varying amounts the content of the elementary school mathematics program. In some schools the effect of the newly created knowledge in mental health, in clinical psychology, and in developmental psychology resulted in arithmetic being taught only when the children expressed a need for it. This approach tended to distort balance in some such manner as shown in Figure 2. The new center of balance has shifted from the region near point A to the region near point X.

In other schools, a rigorous application of the theory of social utility caused certain topics to be deferred to later grades, and other topics to be discarded. The effect on balance is shown in Figure 3, with the new center of balance in the region near point Y.

From the mid-1920's to the mid-1950's many mathematics educators directed their efforts toward correcting the excesses of these two trends by emphasizing the meaning (needs-of-the-subject) theory. And since

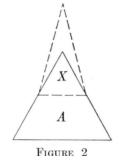

FIGURE 2

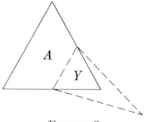

FIGURE 3

the mid-1950's, this trend has received new impetus from individuals and groups, committees and commissions, foundations and the federal government. From this new activity have come several projects and programs which vary from the moderate—such as suggesting better ways to teach the present program—to the extreme such as advocating content which has traditionally been a part of the high school algebra and geometry programs. Somewhere between the moderate and the extreme would be that program which not only proposes better ways to teach the present program but also includes that part of the new mathematics that is appropriate to the elementary school level.

An extreme application of the needs-of-the-subject theory would cause the center of balance to shift in some such manner as shown in Figure 4. Little regard would be given to either the ability of the child to learn the mathematics or to the usefulness of the content. Such a shift would be unfortunate.

Just as unfortunate as *too great* a shift in the balance of the program would be a reactionary effort which would cause the program to be frozen to the *status quo* position. Over the past few years several individuals and groups have expressed their concern about both the possibility of too much change and of too much rigidity—but particularly about the former; see Brownell (2:44), Fehr (6:34), and Jones (13:65).

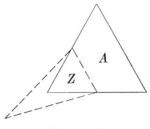

FIGURE 4

Perhaps the most significant statement on the need for progress, but *balanced* progress, is the recent one endorsed by some 65 mathematicians and mathematics educators (18:191). They readily admit that: "the teaching of mathematics in the elementary and secondary schools lags far behind present-day requirements and highly needs essential improvement." But they also expressed this caution against too rapid a shift: "Mathematicians, reacting to the dominance of education by professional educators who may have stressed pedagogy at the expense of content, may now stress content at the expense of pedagogy and be equally ineffective."

The Newly Emerging Concept of Balance

There is little question among the majority of mathematics educators that reform is needed in the elementary school mathematics program. Further, there is little question but that out of the efforts of the many who are working on the problem, and whose efforts are to be applauded, a new concept of balance will emerge. However, it is very unlikely that the new program will include *for all children* in the primary grades, say, the construction of perpendicular bisectors to given line segments; and for those in the middle grades quadratic equations, even though these topics *can* be taught and some sort of learning does take place. The fact that a given topic can be taught and learned is in itself no logical argument that it *ought* to be learned. Determining "oughtness" requires as Clark stated (4:388) "the combined judgment of teachers, psychologists, mathematicians and educational philosophers."

The newly emerging concept of balance can be illustrated by the triangle in Figure 5 in which the shaded area indicates that (a) more new socially useful content will be skillfully integrated into the present program, and (b) all will be taught with greater emphasis on the mathematical nature of the content, by (c) teachers who have an increased understanding of the teaching-learning process.

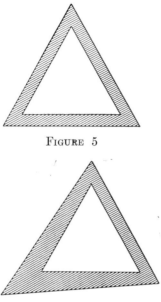

FIGURE 5

FIGURE 6

WHAT MATHEMATICS SHOULD BE PROVIDED FOR THE TALENTED?

As was noted above, the program that is good and necessary for the average child is not a sufficient program for the talented child. Further the program that is a balanced program for the average child may indeed be an *unbalanced* program for the talented. Or stated otherwise, it may well be that the program that leans heavily toward the mathematical point of view, and would therefore be an unbalanced one for the average child, may be very appropriate for the talented child. Since it is characteristic for talented children to see relationships and make generalizations more quickly than average children and since these characteristics are important in mathematics learning, it would seem reasonable to provide a program that capitalizes these abilities.

Returning to our triangle symbol, it would seem that a balanced program for the talented child could be shown in some such manner as in Figure 6. That is, the gifted child would learn all that the average child learns in a balanced program, and in addition would go far beyond this in the direction of abstract mathematics. Therefore, it might well be that making simple geometric constructions using a compass would be quite appropriate for him.

Acceleration and Enrichment

There are two major approaches to selecting mathematics for the talented child—*academic acceleration* and *enrichment*. *Academic* acceleration, not to be confused with *administrative* acceleration or grade-skipping, simply makes available to a rapidly learning fourth grade child the fifth grade textbook upon his completion of the work in the fourth grade book. Whenever the child completes the work in the fifth grade book, whether he is still in the fourth grade or has moved to the fifth grade, he is provided with the sixth grade book and proceeds to work through that one. The shortcoming of academic acceleration is readily seen when one recalls that any textbook in a standard arithmetic series is written for the average child on that grade level. Hence, when a child is academically accelerated through a series of standard textbooks, he is in effect learning that content, and only that content, which the average child learns, but he learns it somewhat sooner. In a word, this approach views the talented child merely as a fast learner of average content. Such a perception of the talented child and his mathematics program is wholly insufficient and inadequate.

Enrichment is the most widely accepted practice for providing learning experiences for the talented child. This approach is equally desirable whether the children are grouped in a self-contained classroom (sometimes called a coordinated classroom since no classroom except a one-room rural school is fully self-contained) or whether they are grouped in some form of departmentalized organization.

Inherent in the idea of enrichment are the concepts of *breadth* and *depth*. Breadth implies that the enrichment material will be broader in scope than that which occurs in the standard textbook program but that it will be *related to* and *continuous with* this program, and *appropriate for* elementary school children. There is, of course, no single set of concepts or topics that constitute the only appropriate set, but those presented in the following chapters are representative of topics that can be used as reservoirs to be drawn upon by the teacher for classroom lessons.

Sources for Enrichment Ideas

Several exploratory programs being developed for use in the elementary grade levels contain material which, although not written specifically for the talented, is sufficiently different from the standard textbook content to serve well as enrichment material. Other exploratory programs that can be culled for reservoir material are listed by Weaver (25).

Still another fruitful source of ideas for enriching the elementary school mathematics program can be found in the more complete teacher's editions accompanying several textbook series. Here the teacher can find suggestions for differentiating and extending the textbook content for the above-average child and for the slower-learning child.[2]

Developing Depth

As mentioned above, a second way to enrich the program is through providing for greater *depth* in the learning. Whereas the concept of breadth is concerned with the introduction of *new but related* topics, the concept of depth is concerned with developing new insights into what is *presently* taught. Junge (14:342) says that depth in the learning, or fuller utilization of the content, involves emphasis upon development of inventiveness and reflective thinking. It is best developed by

1. Confronting the student with challenging problems—but problems within his power of comprehension
2. Leading him from the very beginning to see the futility of thought without dependable data
3. Maturing him in those methods of disciplined thought which have been found to facilitate the work of mathematicians
4. Providing opportunity to discover and to find original solutions
5. Steadily encouraging him to new levels of creative thinking.

SUMMARY. It should be noted that the concepts of breadth and depth are not mutually exclusive. On the contrary, the outcomes of depth-in-learning can be achieved as well through topics that are new, but appropriate and related, as through topics that are part of the present programs. Enrichment, whether through breadth within appropriate topics, or through depth, or through both, is the richest reservoir of content and experiences for talented children in elementary school mathematics.

The reader should keep in mind that the several topics developed in Section I are suggestive not prescriptive of the sources from which the teacher can obtain ideas for enriching the program. Furthermore, when

---

[2] Some materials currently being made available to accompany textbook programs are as follows: Ginn & Company, has published a set of self-teaching "write-in" texts for grades 4 to 6, and other supplementary materials for grades 7 and 8 under the general title *Ginn Arithmetic Enrichment Program*; Row, Peterson & Company has published sets of 16-page pamphlets for grades 3 to 6; Harcourt, Brace and World, Inc. has produced a set of five programmed textbooks under the general title *Mathematics Enrichment*; Webster Publishing Company has published a set of 12 booklets on as many different topics under the general series titled, *Exploring Mathematics On Your Own*. Other companies will undoubtedly soon publish materials similar in purpose to the above.

enrichment is viewed as acquiring new insights and greater depth in thought processes, rather than new content, the textbook alone could conceivably be a thoroughly adequate source of content for enrichment.

IDENTIFYING MATHEMATICALLY TALENTED CHILDREN

Some persons seem to think that we can select the mathematically talented rather easily by administering intelligence tests and choosing those with IQ's above some selected cut-off point. The evidence does not warrant such a complacent view. Too, the problem of identifying the mathematically talented elementary school child is considerably more difficult than identifying the mathematically talented learner on the high school or college level. We have only a little research evidence to help us, since it is only recently that we have become concerned with this problem.

Because not enough is yet known of the cognitive and affective components of talent in mathematics, we must assume on the basis of what we do know that this talent is not identical with measured intelligence. That is, we cannot assume that the child who is generally talented is also equally talented in mathematics; he may be either more or less so. However, it is reasonable to assume that there is a substantial overlap in these two traits. Support for this assumption can be found in the overlap of many characteristics of the generally talented and the mathematically talented child.

Weaver and Brawley (24:6) have listed the following as characteristics of the *mathematically* talented child:

1. Sensitivity to, awareness of, and curiosity regarding quantity and the quantitative aspects of things within the environment
2. Quickness in perceiving, comprehending, understanding, and dealing effectively with quantity and the quantitative aspects of things within the environment
3. Ability to think and work abstractly and symbolically when dealing with quantity and quantitative ideas
4. Ability to communicate quantitative ideas effectively to others, both orally and in writing; and to readily receive and assimilate quantitative ideas in the same way
5. Ability to perceive mathematical patterns, structures, relationships, and inter-relationships
6. Ability to think and perform in quantitative situations in a flexible rather than in a stereotyped manner: with insight, imagination, creativity, originality, self-direction, independence, eagerness, concentration, and persistence

7. Ability to think and reason analytically and deductively; ability to think and reason inductively, and to generalize
8. Ability to transfer learning to new or novel "untaught" quantitative situations
9. Ability to apply mathematical learning to social situations, to other curriculum areas, and the like
10. Ability to remember and retain that which has been learned.

For other lists of characteristics of the mathematically talented child similar to the above see: Hlavaty (12: 11), Junge (15: 142), Kough and DeHaan (17), and Woolcock (26:45).

Stumbling Blocks in Identification

More difficult than identifying the talented child is the problem of identifying the person who, like Albert Einstein or Winston Churchill, demonstrates only average ability as a child but who will, when adult, be a genius in the sciences, humanities, or social sciences. At present there is no known way of identifying these "late bloomers."

Rendering still more difficult the problem of identifying the mathematically talented child is the fact that the exercises making up the group and individual tests of intelligence tend to be more oriented to the life of the middle and upper class child than to the life of the lower class child. To the degree the tests are thus oriented, they tend to discriminate against the child from the lower socio-economic class. Hence the teacher needs to use extra care to make sure that he does not exclude the child who is talented but whose measured intelligence score and achievement scores do not clearly indicate his talent.

Further, the teacher must keep in mind that creative thinking ability is both sufficiently different from measured intelligence and at the same time so closely related to subject matter achievement as to be a significant factor in identifying the talented. (The interested reader should study the recent, 1962, contribution of Getzels and Jackson, *Creativity and Intelligence*, published by John Wiley and Sons.)

A Research Report

One of the most intensive studies of the personality traits of talented (high achieving) children in *arithmetic* is that of d'Heurle, Mellinger and Haggard (5:14). In this longitudinal investigation of a group of highly gifted children over the span of grades 3 through 9, the researchers compared the personality traits of high arithmetic achievers, high reading achievers, high spelling achievers, and high general achievers. For our purposes here it is interesting to note that: (a) The high arithmetic achievers tend to be more at peace with the world and with themselves. (b) In their relations with their parents and other authority figures, they show less strain than the high general achievers and the high reading achievers, and greater independence than the high spelling achievers. (c) They show the greatest degree of ego integration as well as greater maturity in dealing with the outside world of people and things. (d) They are able to express their feelings freely, but at the same time they are emotionally controlled and flexible. (e) Their intellectual processes also tend to be spontaneous and creative, and they are the most skilled in the manipulation of abstract symbols.

SUMMARY. While we have as yet no highly reliable single instrument available to us for identifying the mathematically talented children, the teacher is not without help. By using a combination of such traits as measured intelligence, general achievement, creative ability, and ego integration the teacher should have a high degree of success in identifying the mathematically talented child.

HOW SHALL WE GROUP TALENTED CHILDREN?

Having decided what content to teach and how to identify talented children, the teacher must then decide how to group the children to best advantage for instruction.[3] Here we encounter an educational problem that is often more heavily weighted with opinion than with fact. Opinion alone is not undesirable as a basis for action if the opinion is based upon mature judgement. However, some facts are available from research studies, and when used in conjunction with informed opinion they provide a reasonable basis for making decisions. The following discussion draws upon both informed opinion and fact.

The Self-Contained Classroom

One of the leading advocates of the self-contained classroom is the Association for Supervision and Curriculum Development (ASCD) of the National Education Association (NEA). In its recently published pamphlet, *The Self-Contained Classroom* (7:v), the ASCD reaffirms its belief in and support of this way of grouping children for instructional purposes. In the pamphlet's Foreword Foshay says:

... the (writing) committee hopes that the values this kind of organization permits will not be casually overlooked when questions of reorganization are considered. Too many educational innovators (or re-innovators, as is often the case these days) have undertaken change in ignorance.

---

[3] For a first-hand report of grouping procedures used in the English elementary schools see Pace's article, Chapter 12, Section I.

If the authors plead the case of the self-contained classroom, it is because a strong case exists. Problems can be met through this organization that must inevitably be met if one is to have a good school.

We should note here that the ASCD concept of a self-contained classroom is a flexible one. Continuing, Foshay states:

The self-contained classroom plan does not negate the need that children and youth have for experiences with more than one teacher. It is not intended that boys and girls live on a 'secluded island' day after day with one teacher. The teacher's resources are supplemented by specialists in the areas in which the students are learning.

Another ardent supporter of the self-contained classroom concept is the Department of Elementary School Principals (DESP). This department of the NEA holds a position similar to that of the ASCD quoted above, and in a recent Resolution (19:56) declared:

The findings of research and experience indicate that children need the security and stability that comes from being taught by one teacher in a group which does not basically change in its membership during the school day. A teacher also needs the close contact with the individual child in order to understand him better and provide for his needs. The program may be enriched and enhanced by providing supplementary teaching services in specialized subject areas, such as physical education, music and art.

From these statements we may reasonably conclude that there is little likelihood the strong support and widespread acceptance of the self-contained classroom idea presently held by the vast majority of elementary school personnel will be modified in any substantial degree in the immediate years ahead. But it is important to note well that the statements of both of these associations express recognition of the need for the services of some specialist teachers—if the self-contained classroom is to operate at its optimum level.

Support from Mathematics Educators

Nor is support for the self-contained classroom idea limited to the generalists in elementary education. Support is also expressed by mathematicians and mathematics educators. Fehr (6:35) stated his position recently this way:

I believe that all theory of child development and child growth is against specialist teaching during the ages 5 to 10 years. I doubt that a specialist teacher would achieve more than the regular teacher except perhaps for a few bright children. Of course I would like to see every elementary school teacher a capable teacher of arithmetic. Platoon teaching below the fifth grade could be as detrimental to total educational growth as it could be beneficial in growth in just the one subject matter area of arithmetic.

Approaches to Grouping

While opposing total departmentalization or segregation and complete teaching by specialists in the elementary school, almost all students at the same time do recognize the urgent need for some solution to the problem of improving instruction in elementary school mathematics. Already widely known are several programs concerned with talented children generally—the Hunter College (New York) elementary school program; the Cleveland (Ohio) Major Work classes; the Colfax (Pittsburgh, Pennsylvania) plan; and the University City (Missouri) program. With the exception of the Hunter College elementary school program, in which there is a complete segregation of the talented children in a separate building, these programs have varying amounts of time during the week in which the talented children are segregated for teaching by specialists.

THE DUAL PROGRESS PLAN. Very recently Stoddard (21:2) has proposed and experimented with the Dual Progress Plan. Under this plan mathematics, science, art, and music are taught by specialist teachers. The children are assigned for these subjects to classes on the basis of their maturity in each subject rather than by age or grade. Stoddard points out that: "Thus bright third graders will be brigaded with older pupils who are at about the same level, let us say, in mathematics or science."

After two year's work with the plan Stoddard reported that: (a) Even with the major dislocations caused by the new plan, the expected rate of academic growth is maintained. (b) The majority of parents favor the plan. (c) The majority of pupils like the plan and enjoy working under it. (d) The teachers are divided in their acceptance or approval of the plan, but with experience in it and the correction of certain defects, they increasingly register approval.

THE SPECIALIST TEACHER CONCEPT. Acceptance of the specialist teacher within a broad interpretation of the self-contained classroom concept is receiving increasing support. In its recent report, *Contemporary Issues in Elementary Education*, the Educational Policies Commission stated:

In the past, fully departmentalized types of organization have been tried in some elementary schools, but research and experience have generally caused these experiments to be discarded. Research and experimentation continue with forms of organization lying between the completely departmentalized and the completely nondepartmentalized. Many of these forms appear to offer significant advantages.

Use of special teachers, team teaching, and teacher aids, for example, can be of benefit as long as they do not hamper the close contact of the classroom teacher and pupil. Pupils require stable personal relationships, but this need not mean that a child must remain in the same classroom all day with one teacher. Special teachers may enter and children may leave. The essential condition is that some one teacher have major responsibility for curriculum and guidance of a group of pupils.

And Woolcock (26: 50) made the same point when he said, "The self-contained classroom organization of the typical elementary school can be retained, but it should be modified in terms of permitting some periods of instruction in subjects of strength and high promise by the best teacher available in the school or school system."

Most enrichment programs operating within the self-contained classroom that are without the help of specialists fall short of their goal. The general elementary classroom teacher has neither the time nor the resources available to become well enough informed on all subject matter areas to make the enrichment program work. With the rapid accretion of subject matter in elementary school mathematics, elementary school science, and all the other content subjects, the teacher would be rare indeed who could carry out good enrichment programs in all areas without the aid of resource persons.

Nor can the teacher expect the building principal, the elementary supervisor, or the curriculum director to be all things to all teachers. These persons have neither the time nor the resources at hand to become deeply informed in the subject matter and professional knowledge of all subjects in the elementary school program.

The Present Dilemma and a Reasonable Solution

As mentioned above, the overwhelming majority of the elementary classrooms in the country are operating on the self-contained classroom concept and will continue to do so in the foreseeable future. At the same time, these teachers are being told to enrich the programs for the talented children—in mathematics, in science, in the language arts, etc. How, then, can we maintain the values of the self-contained classroom and at the same time provide the specialized help that the classroom teacher must have if he is to enrich the program for the mathematically talented child? In a word, how can we have our cake and eat it too?

In a recent critical analysis of the theory of supervision under which the elementary school of today operates, Glennon (10:48) concludes that this theory is no longer adequate. He points out that this present-day theory, which is now well over 100 years old, "places the responsibility for the improvement of instruction on the principal of each building unit," who, as noted above, has neither the time nor the resources available to carry out effectively this part of a many-faceted job.

Glennon proposes a solution to the problem of providing enrichment in the self-contained classroom. He suggests that the superior teachers be developed through graduate study into specialists while they remain self-contained classroom teachers. In the beginning, there may be only a few teachers in a school building unit who have the interest and time to devote to becoming well-informed in one subject matter area. But over a few years the principal can select several who are not only superior classroom teachers, but who are also highly professionally motivated to study deeply one subject matter area. Studying a subject deeply would mean the equivalent of obtaining a master's degree, or beyond, in that one area and its closely related professional areas.

In the future, it may be that Thelen's suggestion (23: 131) will serve as a basis for grouping children with teachers; that is—giving teachers teachable classes, classes containing children who appeal to teachers. As Thelen stated: "We do not know at present how to compose groups for instructional purposes, but we do know a good deal about what is going to be involved, and there is a growing belief that the composition of the group may be at least as important as the method of the teacher in determining the quality of the educational product."

WHAT TEACHING METHODS SHOULD WE USE?

A study of the history of methodology shows a variety of practices that have ranged from imitation and memoriter activities at one end of a continuum to complete freedom to learn through unguided discovery at the other end.

The greatest contribution to methodology among early cultures was that of Socrates. Through a questioning-telling method Socrates led his students to discover knowledge, or at least to arrive at generalizations not previously known to the learner. For our purposes in this yearbook one of the most pertinent dialogues was the one held between

Socrates and Meno. In this dialogue, Socrates led a boy, Meno, who was quite naïve about geometry, to demonstrate a geometrical theorem.

Unguided Discovery

In our time the major problem of method is that of striking some sort of balance between *telling* (authoritative identification), at one end of a continuum, and *unguided discovery* at the other end. Gagné (9:55) expressed two doubts about the use of an extreme form of discovery:

1. Carried to a logical extreme, the principle is obviously false. An individual cannot solve a truly original problem without having acquired a considerable background of concepts and values.
2. There are many instances of learning other types of behavior in which the dictum 'the best way to learn is to practice the ultimate task' has been shown to be false. For example, the best way to learn a motor performance is not always to practice the task in its final form.

The problem of "what method *ought* I use?" is greatly complicated by four variables: the nature of the cognitive material, the personality traits of the learner, the personality traits of the teacher, and the subject matter competence of both the learner and the teacher.

Fortunately, we are today able to subject hypotheses about method to investigation. Studies of some aspects of this general problem have been carried out. Ausubel (1:56) has provided us with one of the most thorough summary discussions of the self-discovery method. While recognizing that learning by discovery has indeed some value when used to achieve certain objectives under certain conditions, he points out that on the whole the claims made for the method are not supported in the research literature. On the contrary, "research on meaningful verbal and symbolic problem solving indicates, on the whole, that providing guidance to the learner, in the form of verbal explanation of the underlying principles, almost invariably facilitates learning and retention, and sometimes transfer as well. Self-discovery methods, in contrast, are relatively much less effective."

Guided Discovery and the Nature of the Cognitive Material

There is little question but that most students of the teaching-learning process today favor increased emphasis on guiding the learner to discover new (to him) cognitive material. But cognitive material (subject matter) is not all of a kind. For convenience here the subject matter can be categorized into three groups—arbitrary associations, concepts, and understandings. Understandings are sometimes called meanings, generalizations, number relationships, structures, or integrating ideas.

The use of pure discovery methods for learning arbitrary associations would be wasteful of school time. By way of an illustration, why have a group of children take all the time that would be needed to discover that the symbol < means "is less than"? Perhaps direct telling or at most guided discovery would be the appropriate method for this type of cognitive material.

The general ambiguity of professional knowledge in this area is pointed out by Bruner (3:21). In quoting the University of Illinois Committee on School Mathematics, Bruner states that "the method of discovery would be too time-consuming for presenting all of what a student must cover in mathematics. The proper balance between the two is anything but plain. . . ."

But some work has been done. Sobel (20) found the discovery method superior for the teaching of algebraic concepts *but only* for pupils with an IQ of 100 to 115. Kersh (16), working with small groups of college students, compared independent discovery with directed (guided) discovery. He concluded that directed discovery results in maximum learning, but that independent discovery results in maximum motivation to continue learning. This conclusion forces the teacher into a value choice of either maximum *learning* or maximum *motivation*.

Another dimension of the whole problem of method is what Hendrix refers to as "nonverbal awareness." Hendrix claims that the learning which a child acquires but which he does not verbalize is superior to the learning he is required to verbalize. In one study Hendrix (11) found that discovery and non-verbalized awareness of a principle were superior to discovery with verbalization of that principle, and that direct presentation of a principle with illustrations was inferior to both of the discovery methods.

Relationship of Guided Discovery and Personality Traits

The other three variables mentioned above which impinge upon the value of the guided discovery method are the personality traits of the children and of the teachers, and the subject matter competence of both children and teachers. Too little is presently known about the interaction among these variables to support the statement that guided discovery is the best method for all children in all teaching situations. Such a dogmatic statement would presume that we have knowledge which we simply do not have at this time.

It may well be that guided discovery is helpful to some children and harmful to others. Specifically it may be that it is helpful to the bright child who has a high tolerance for frustration and ambiguity, but harmful to the bright child who has low tolerance in these traits. In the same way, it may be that teacher *A* can easily accept the idea of using guided discovery methods, while teacher *B* becomes highly anxious and insecure when it is suggested that she change her methods of teaching to discovery methods.

SUMMARY

Elementary school education is different from all other levels above it in that the entire program is *general* education, the kind of education which has for its purpose enabling a person *to live*—live better, live more fully, live more wisely and prudently. On every grade level above grade 6, some part of the program and some part of the school day are devoted to *special or vocational* education—that kind of education that has for its purpose enabling a person to make a living. On increasingly higher grade levels the proportion of the program and of the school day devoted to general education decreases, while that devoted to special education increases, until, on the graduate levels of education, engineering, law, medicine, and other professional fields the student spends 100 per cent of his school day on his vocational area.

On the elementary school level it is not the intended purpose of the educational curriculum to make the child into a mathematician, a scientist, a physician, a lawyer, a poet, or any other kind of professional person; rather it is intended that he acquire some general learnings in many areas of knowledge. The mathematician (scientist, musician, or whatever) who views the elementary school as an opportunity to make all children into his own likeness shows a profound unawareness of the function of elementary education in our society. Thoughtful people want today's child to be more "mathematically minded," not a mathematician; more "musically minded," not a musician; more "scientifically minded," not a scientist, etc. Thoughtful people have always wanted these aims for elementary education. But this is not at all the same thing as making the child into a mathematician for 45 minutes per day; then into a scientist for 45 minutes per day, three days per week. Such a view of the elementary school program would quickly result in a completely departmentalized elementary school program with each group of children being taught by some eight or more teachers per day.

The elementary schools of America are presently committed to the self-contained classroom concept with flexibility that allows for varying amounts of specialist teaching. It is for this reason that the first question posed here is concerned with the need for and the nature of a balanced program in elementary school mathematics. Mature leaders in elementary school mathematics are in general agreement on the need for balance, and in fairly substantial agreement on the nature of balance.

In a few years we can expect that the dust created by what Phillip S. Jones (13: 65) called "the denunciators and innovators" will have settled. Also we can expect that the National Council of Teachers of Mathematics will have heeded the concern expressed by Sueltz, (22: 279) and by others, to the effect that the decisions on the elementary school mathematics program will not be "formed by those who have shouted the loudest" nor will they be made by people who "happen, for the moment, to be placed in positions of control . . ." but rather by those who "have the backgrounds of knowledge and experience and the levels of vision and discernment appropriate for making these important decisions."

This challenge to speak out is wise, lest by default of the well-informed we find the children and the teachers of the nation being led by the less-informed. This article and the ones which follow may be viewed as an effort to meet this challenge.

This article spells out in considerable detail a rationale for selecting content in mathematics appropriate for talented children and presents the position that there is an over-abundance of appropriate content, and hence it was not necessary to take material from the secondary school mathematics program in order to have material for grades kindergarten through six. Rather we hold that it is a more educationally sound practice to extend and enrich the program through topics that are *related to* and *continuous with* the basic school mathematics program. It is also pointed out that these topics are suggestive, not prescriptive, and illustrative, not exhaustive, of the reservoir of available material.

REFERENCES

1. AUSUBEL, DAVID P. "Learning by Discovery: Rationale and Mystique." *Bulletin of the National Association of Secondary School Principals* 45: 18–58; December 1961.
2. BROWNELL, WILLIAM A. "Arithmetic . . . in 1970." *The National Elementary Principal* 39: 42–45; October 1959.
3. BRUNER, JEROME S. *The Process of Education.* Cambridge, Mass: Harvard University Press, 1960. 97 pp.
4. CLARK, JOHN R. "Looking Ahead at Instruction in Arithmetic." *Arithmetic Teacher* 8: 388–94; December 1961.
5. D'HEURLE, ADMA; MELLINGER, JEANNE C.; and HAGGARD, ERNEST A. "Personality, Intellectual, and Achievement Patterns in Gifted Children."

Psychological Monographs Vol. 73, No. 13. Washington, D. C.: The American Psychological Association, 1959. 28 pp.

6. FEHR, HOWARD F. "Trends in the Teaching of Arithmetic." *Frontiers of Elementary Education VII* (Vincent J. Glennon, editor). Proceedings of a Conference on Elementary Education. Syracuse: Syracuse University Press, 1960. Chapter III, pp. 29–37.

7. FOSHAY, ARTHUR W. "Foreword." *The Self-Contained Classroom.* Washington, D. C.: Association for Supervision and Curriculum Development, a department of the National Education Association, 1960. 88 pp.

8. FOSHAY, ARTHUR W. "From the Association." *Balance in the Curriculum.* Yearbook 1961. Washington, D. C.: Association for Supervision and Curriculum Development, a department of the National Education Association, 1961. pp. iii-iv.

9. GAGNÉ, ROBERT M. "Implications of Some Doctrines of Mathematics Teaching for Research in Human Learning." *Research Problems in Mathematics Education.* U.S. Department of Health, Education, and Welfare, Office of Education, (OE-12008). Cooperative Research Monograph No. 3. Washington, D.C.: Government Printing Office, 1960. 130 pp.

10. GLENNON, VINCENT J. "Updating the Theory of Supervision." *Frontiers of Elementary Education VII* (Vincent J. Glennon, editor). Proceedings of a Conference on Elementary Education. Syracuse: Syracuse University Press, 1960. Chapter V, pp. 44–55.

11. HENDRIX, GERTRUDE. "A New Clue to Transfer of Training." *Elementary School Journal* 48: 197–208; December 1947.

12. HLAVATY, JULIUS H., editor. *Mathematics for the Academically Talented Student in the Secondary School.* Washington, D. C.: National Council of Teachers of Mathematics, a department of the National Education Association, 1959. 48 pp.

13. JONES, PHILLIP S. "The Mathematics Teacher's Dilemma." *The University of Michigan School of Education Bulletin.* Ann Arbor: University of Michigan, School of Education 30: 65–72; January 1959.

14. JUNGE, CHARLOTTE W. "Depth Learning in Arithmetic—What Is It?" *Arithmetic Teacher* 7: 341–46; November 1960.

15. JUNGE, CHARLOTTE W. "Gifted Children and Arithmetic." *Arithmetic Teacher* 4: 141–46; October 1957.

16. KERSH, BERT A. "The Adequacy of 'Meaning' as an Explanation for the Superiority of Learning by Independent Discovery." *Journal of Educational Psychology* 49: 282–92; October 1958.

17. KOUGH, J., and DEHAAN, R. *Teacher's Guidance Handbook.* Chicago: Science Research Associates, 1955.

18. NATIONAL COUNCIL OF TEACHERS OF MATHEMATICS, a department of the National Education Association. "On the Mathematics Curriculum of the High School." *Mathematics Teacher* 55: 191–95; March 1962.

19. NATIONAL EDUCATION ASSOCIATION, Department of Elementary School Principals. "Resolutions." *National Elementary Principal* 39: 55–58; October 1959.

20. SOBEL, M. A. "Concept Learning in Algebra." *Mathematics Teacher* 49: 425–30; October 1956.

21. STODDARD, GEORGE D. "The Dual Progress Plan—After Two Years." *Frontiers of Elementary Education VII* (Vincent J. Glennon, editor). Proceedings of a Conference on Elementary Education. Syracuse.: Syracuse University Press, 1960. pp. 1–12.
22. SUELTZ, BEN A. "A Time for Decision." *Arithmetic Teacher* 8: 274–80; October 1961.
23. THELEN, HERBERT A. *Education and the Human Quest.* New York: Harper and Brothers, 1962. 131 pp.
24. WEAVER, J. FRED, and BRAWLEY, CLEO FISHER. "Enriching the Elementary School Mathematics Program for More Capable Children." *Journal of Education* 142: 1–40; October 1959.
25. WEAVER, J. FRED. "Improvement Projects Related to Elementary School Mathematics." *Arithmetic Teacher* 7: 311–15; October 1960.
26. WOOLCOCK, CYRIL W. *New Approaches to the Education of the Gifted.* Morristown, N. J.: Silver Burdett Co., 1961. 112 pp.

2

NON-DECIMAL NUMERATION SYSTEMS

LONIE E. RUDD

Tufts University
Medford, Massachusetts

A key to learning the Hindu-Arabic system of notation is the fact that grouping is in powers of ten. However, the learner should gain a much broader insight into the structure of a place value system and the incidental nature of the selection of the size of the group. Therefore, a study of non-decimal systems of numeration is very desirable for both teachers and pupils.

There are several outcomes from a study of non-decimal systems: (a) Teachers may realize for the first time that it is possible for a place value system to use one of several grouping plans. (b) Teachers may experience, through working with non-decimal systems, a struggle similar to that which children encounter in working with the decimal plan. As a result, these teachers gain a better understanding of the problems of learning place value notation, and consequently they have more patience with children. (c) Children extend their knowledge of the principle of place value to the reading and writing of numerals in other bases. (d) Children develop additional insight into the fundamental operations of addition, multiplication, subtraction, and division, through performing these operations in a non-decimal numeration system.

Rudd has provided the reader with three lesson plans showing how the base five numeration system can be developed with talented children in the primary, middle, and upper elementary grades. Suggestions are made for teaching this unit.

POINTS STRESSED

This discussion of non-decimal systems gives attention to two separate place value plans: one, which uses five as the basic group; and another, which uses two as the basic group. In each case emphasis is placed upon the four fundamental operations: addition, subtraction, multiplication, and division. The following four characteristics of a place value system are stressed: (a) Place value eliminates the need for the development of new symbols beyond one less than the basic group. (b) A place value system uses the same number of different symbols as the size of the basic group. For example: In base five, there are five different symbols; and in base two, there are two different symbols. (c) The total value of a number symbol in any one place is determined by multiplying the face value of the symbol by the value assigned to the place. (d) The value represented by a numeral is determined by adding the total value of all the symbols in that numeral.

GROUPING BY FIVES

It is surprising that the group of five was relatively unnoticed by many early civilizations. The reason for the choice of ten as base—ten fingers on both hands—makes it doubly surprising that five was not used more often, since five fingers on one hand provide a more natural group. The notion of one-to-one correspondence, or matching, likely led to the practice of using five as a group. Figure 1 shows the distinct advantage of tallying by groups of five in contrast to matching without grouping.

$$\cancel{||||}\ \cancel{||||}\ \cancel{||||}\ \text{equals}\ ///////////// $$

FIGURE 1

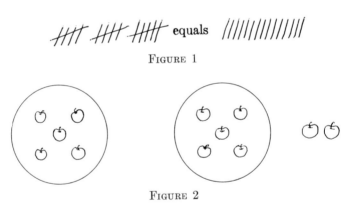

FIGURE 2

Obviously when we use tallies, a grouping plan facilitates the recognition of the size of the group. Another means of grouping by five is that of circling groups as in Figure 2. However, a system of symbolic notation must have more economy than is provided by either tallying or by circling groups, since both plans retain the disadvantage of cumbersomeness in representing large numbers. Clearly there are *two ways* for achieving the economy of representing numbers with fewer symbols than with one-to-one correspondence. One plan provides a separate symbol for a group, and still another symbol for groups of groups, etc. This is precisely the plan used by many early civilizations such as the Egyptian and Roman. However, this plan has the serious disadvantage of an upper limit where it is necessary to invent a new symbol. The second alternative, that of assigning group names to the places in which a symbol may be located, makes possible the use of only a few basic symbols, beginning with zero. Re-use of these basic symbols in places with group names makes possible the easy representation of large numbers.

In any place value plan of notation, a symbol in the first place represents the number of ones indicated by its face value; however, a symbol in the second place represents the number of base groups. In fact, the essence of place value is the naming of places in consecutive powers of the base group, that is in base ten, 10^0, 10^1, 10^2, etc., and in base five, 5^0, 5^1, 5^2, etc.

Since 10 means one base (one five in base five) and zero ones, the face value symbol with the largest value is the symbol for one less than five, or four. The first eleven base five numerals when placed on a number line are as follows:

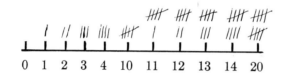

Since the Hindu symbols tend to promote confusion in bases other than ten, a new set of symbols—0, A, B, C, D—will be used here. The number line then becomes

0	A	B	C	D	$A0$	AA	AB	AC	AD	$B0$

The reading and writing of numerals requires both a knowledge of the face value symbols and an understanding of place value. For the base five (quinary) system, the face value of the basic symbols must be established. An effective plan for remembering the value of each symbol is that of associating the symbol with its group, as illustrated by the following example.

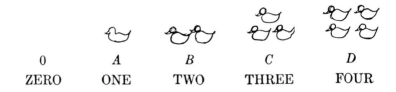

0	A	B	C	D
ZERO	ONE	TWO	THREE	FOUR

In reading quinary numerals it is desirable to read only face value symbols, thus avoiding the confusion of names which overlap with names used for base ten numerals. For example, *DB* is read "four, two" (meaning *four* fives and *two* ones) and not forty-two. Reading place value numerals, then, is a matter of reading the symbols and interpreting these symbols in accordance with values assigned their positions. In *DAB* base five, *D* represents *four* twenty-fives, *A one* five, and *B two* ones. Adding these values, the total is the number 107. That is, $DAB_{\text{base five}} = 107_{\text{base ten}}$.

Extensive use of objects which may be grouped by fives and groups of fives, and pictures which show groups of fives are very valuable as introductory work in the base five system of place value notation. Furthermore, work with a five-bead abacus, followed by work with a paper-and-pencil abacus is highly recommended. Examples of the paper-and-pencil abacus are given in Figure 3. Pennies, nickels, and quarters may be used to develop the notion of the five-to-one relationship which

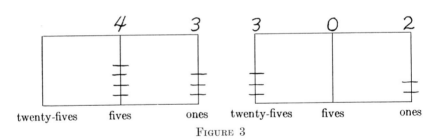

FIGURE 3

exists between places, and also to develop the thought pattern of groups of ones, fives, and twenty-fives. However, a clear distinction must be made between coins, which have "object-value" without regard for their position, and symbols which have an additional value *because* of their position. For example, three nickels and two pennies are equal to seventeen cents, irrespective of their positions. Therefore it is very misleading to claim that place value is being demonstrated with coins. Figure 4 gives a simple illustration of a five-to-one relationship between places.

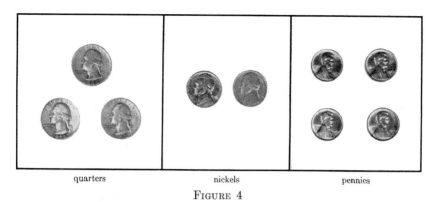

quarters   nickels   pennies

FIGURE 4

Once this notion of "five-to-one" has been developed, the use of coins should be discontinued.

Excess of D

In place value systems of notation (the binary system excepted) the sum of the digits shows the excess of groups of size one less than the base. This is a basic characteristic of such systems. In regrouping to express a number more economically, such as regrouping *ten* ones for *one* ten, a group of nine is removed. This may be dramatically illustrated with objects on an "exchange box." When this characteristic is applied to the base five plan, the sum of the digits indicates the excess of groups of four or D. For example, B plus C may be expressed as $A0$, or one more than D; C plus C is AA, or two more than D, etc. This "excess" characteristic of a place value system may be used advantageously in checking work with each of the four basic operations of addition, subtraction, multiplication, and division.

THE BINARY SYSTEM: BASIC OPERATIONS

The simplest of all place value plans of notation, from the standpoint of number of symbols used, is a place value plan which uses *two* as a group. Its simplicity is inherent in the small number of different symbols, and in the ease with which the operations may be performed. The two symbols 0 and 1 are used. The number line for the first eleven numerals in this system is as follows:

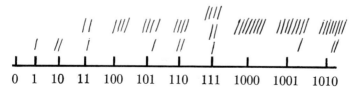

| 0 | 1 | 10 | 11 | 100 | 101 | 110 | 111 | 1000 | 1001 | 1010 |

This use of only two face value symbols dramatically illustrates the effectiveness of place value.

An understanding of the operations with numbers may be developed by performing these operations with objects. Therefore, children should be provided with many opportunities for doing each of the operations with materials which illustrate such basic notions as re-grouping, addition of equal-sized groups, and repeated subtraction of a designated group. A place value exchange box and an abacus may be used effectively for this purpose. Following the work with objects, children should work with symbols with the full realization that operations are performed with numbers and that the algorisms with symbols are simply an indication of the results of these operations. Examples of the algorisms with symbols in the binary system are as follows:

Addition

The four possible combinations of addition facts are as follows:

$$
\begin{array}{cccc}
0 & 0 & 1 & 1 \\
+0 & +1 & +0 & +1 \\
\hline
0 & 1 & 1 & 10 \\
\end{array}
$$

The addition-subtraction chart of these facts is—

$$
\begin{array}{c}
11 \ (\text{three}) \\
+10 \ (\text{two}) \\
\hline
101 \ (\text{five}) \\
\end{array}
\qquad
\begin{array}{c}
101 \ (\text{five}) \\
+111 \ (\text{seven}) \\
\hline
1100 \ (\text{twelve}) \\
\end{array}
$$

±	0	1
0	0	1
1	1	10

Subtraction

$$
\begin{array}{c}
111 \ (\text{seven}) \\
-11 \ (\text{three}) \\
\hline
100 \ (\text{four}) \\
\end{array}
\qquad
\begin{array}{c}
1010 \ (\text{ten}) \\
-111 \ (\text{seven}) \\
\hline
011 \ (\text{three}) \\
\end{array}
$$

Multiplication

The four possible combinations for multiplication facts are as follows:

$$
\begin{array}{cccc}
0 & 0 & 1 & 1 \\
\times 0 & \times 1 & \times 0 & \times 1 \\
\hline
0 & 0 & 0 & 1 \\
\end{array}
$$

The multiplication chart is—

×	0	1
0	0	0
1	0	1

$$
\begin{array}{c}
11 \ (\text{three}) \\
\times 11 \ (\text{three}) \\
\hline
11 \\
110 \\
\hline
1001 \ (\text{nine}) \\
\end{array}
\qquad
\begin{array}{c}
111 \ (\text{seven}) \\
\times 101 \ (\text{five}) \\
\hline
111 \\
11100 \\
\hline
100011 \ (\text{thirty-five}) \\
\end{array}
$$

Division

$$
\begin{array}{r}
1) \\
1000) \quad\quad 1001 \text{ (nine)} \\
\text{(three) } 11 \quad \overline{)11011} \quad \text{(twenty-seven)} \\
11000 \\
\hline
11 \\
11
\end{array}
$$

QUINARY PLAN

Examples of the algorisms in the quinary (base five) plan are as follows:

Addition

$$
\begin{array}{l}
AB \text{ (seven)} \\
+BA \text{ (eleven)} \\
\hline
CC
\end{array}
\qquad
\begin{array}{l}
C \\
+C \\
\hline
AA
\end{array}
\qquad
\begin{array}{l}
BD \text{ (fourteen)} \\
+CB \text{ (seventeen)} \\
\hline
AAA \\
\quad C
\end{array}
\qquad
\begin{array}{l}
B \\
+A \\
\hline
C
\end{array}
$$

$B \longleftrightarrow B$

Check is by excess of D.

The 25 addition-subtraction facts, base five, are easily illustrated in a chart:

±	0	A	B	C	D
0	0	A	B	C	D
A	A	B	C	D	A0
B	B	C	D	A0	AA
C	C	D	A0	AA	AB
D	D	A0	AA	AB	AC

Subtraction

$$
\begin{array}{l}
BA \text{ (eleven)} \\
-AA \text{ (six)} \\
\hline
A0 \text{ (five)}
\end{array}
\qquad
\begin{array}{l}
C \\
-B \\
\hline
A
\end{array}
$$

$$
\begin{array}{l}
DC0 \\
-BBA \\
\hline
B0D
\end{array}
\qquad
\begin{array}{l}
C \\
-A \\
\hline
B
\end{array}
$$

Check is by excess of D.

Multiplication

The 25 possible combinations for multiplication facts are conveniently illustrated with the following chart.

×	0	A	B	C	D
0	0	0	0	0	0
A	0	A	B	C	D
B	0	B	D	AA	AC
C	0	C	AA	AD	BB
D	0	D	AC	BB	CA

$$
\begin{array}{r}
AB \quad \text{(seven)} \\
\times \quad DA \quad \text{(twenty-one)} \\
\hline
AB \\
AC \\
D \\
\hline
A\,0DB \quad \text{(one hundred forty-seven)}
\end{array}
$$

$$
\begin{array}{r}
C \\
A \\
\hline
C
\end{array}
$$

Check by excess of D.

Division

$$
\begin{array}{r}
B\ A \\
AB\overline{)\,BDAB} \\
BD \\
\hline
AB \\
AB \\
\hline
\end{array}
\qquad
\begin{array}{r}
B\,0A \\
\times\quad AB \\
\hline
D\,0B \\
B\,0A \\
\hline
BDAB
\end{array}
$$

Check by multiplication.

Thus the basic characteristics of a place value system have been illustrated with two non-decimal plans of notation, and emphasis has been given to both reading and writing of number symbols and to the basic operations.

SUGGESTIONS FOR TEACHERS

In teaching a unit on non-decimal numeration systems, extensive use should be made of such exploratory materials as loose objects, the place value box, the number frame, and the abacus. With these materials, children should be given many opportunities for grouping objects and for developing a clear understanding of place value and its application to symbols in any grouping plan. Practice with the paper-and-pencil abacus provides an excellent transition from work with objects to work with symbols. Reading and writing number symbols should be done with care, and frequent reference should be made to the number of objects being represented. Each of the operations should be performed with objects on both the place value box and the abacus. The "excess" check is best explained with loose objects on the place value box. Teachers and pupils are likely to find that a unit on non-decimal numeration systems is both interesting and challenging. Certainly it provides a depth of understanding, particularly for talented children.

SAMPLE LESSONS

SAMPLE LESSON No. 1—FOR PRIMARY GRADES

Objective: To develop an understanding of grouping by fives as the basis for the base five place value plan of notation.

Materials:

1. Loose objects such as beads, tongue depressors, small blocks, and small cardboard discs
2. A place value box
3. Five-fives number frame
4. Guide sheets.

Background: It is assumed that the children have a good understanding of the place value characteristic of the Hindu-Arabic system.

Teaching the Lesson: The children will work first with loose objects, then with a five-fives number frame and eventually with the paper-and-pencil abacus.

Each child will be given a quantity of objects to work with. The teacher will guide the children in selecting a number of objects, and in grouping these objects by fives. Several numbers between five and twenty-four should be used first; then, numbers above twenty-five will be used to help the children to see that five-fives may be considered as *one* large group.

The next step is for the children to select a number of cardboard discs and to use them on Guide Sheet No. 1. The top part of the sheet will be used first, since it represents a special placement of the discs. The big step from group value to place value is taken by guiding the children in representing each number, first by grouping, then with fewer discs on the place value section. For example, eight discs are first placed on the sheet as one group of five and three singles, then the same number is represented with only four discs in the place value section by placing one disc in the "five" section and three discs in the "ones" place. The step from one group of five to one in the five place dramatically illustrates the economy of place value, and this point must be strongly emphasized.

Experiences with the five-fives number frame should now be provided to help children to move from work with loose objects to work with objects structured by fives. Children should be guided in showing numbers such as illustrated in Figure 5.

The use of loose objects on the place value box may be very effective in helping children to represent numbers by placing the correct number of objects in the appropriate places. For example, see Figure 6.

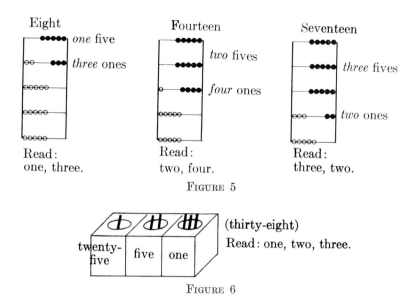

FIGURE 5

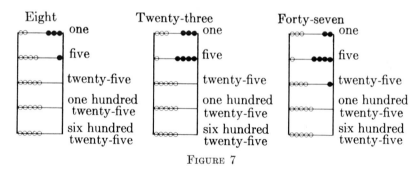

FIGURE 6

Then, the economy of place value is illustrated again by using the number frame as a five-bead abacus and guiding the children in representing numbers on the abacus as in Figure 7.

FIGURE 7

Children should have the opportunity to represent several numbers on the abacus.

Finally, Guide Sheet No. 2 should be used to provide the next step in moving from work with objects to work with symbols. The first example should be discussed, then the children should represent numbers on the remaining drawings, together with the symbol for each number at the top of each drawing.

EVALUATION

Some measure of the effectiveness of the lesson may be made by spending 5 minutes asking the children to show numbers on either the five-bead abacus or the place value box. Also, observation of the children's work with the guide sheets should help in deciding the extent to which they understand grouping by fives.

GUIDE SHEET NO. 1—PRIMARY GRADES

Groups of objects   Single objects

Group Value

Fives   Ones

Place Value

GUIDE SHEET NO. 2—PRIMARY GRADES

SAMPLE

Thirteen

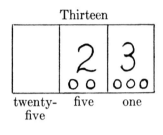

twenty-
five   five   one

Twenty-three

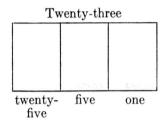

twenty-
five   five   one

Fourteen

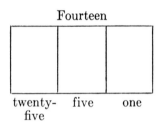

twenty-
five   five   one

Twenty-eight

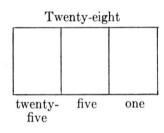

twenty-
five   five   one

Thirty-six

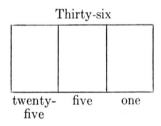

twenty-
five   five   one

Forty-four

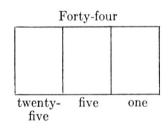

twenty-
five   five   one

Fifty-five

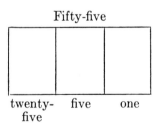

twenty-
five   five   one

Seventy-five

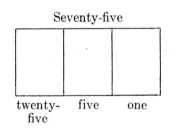

twenty-
five   five   one

SAMPLE LESSON NO. 2—FOR MIDDLE ELEMENTARY GRADES

Objective: To provide children with both an understanding of the operation of addition with numbers in base five and practice with the addition algorism with symbols.

Materials:

1. Loose objects, such as beads, blocks, sticks
2. Place value box
3. Five-fives number frame
4. Five-bead abacus
5. Guide sheets.

Background: It is assumed that the pupils have an understanding of grouping by fives, and are experienced with both reading and writing base five numerals.

Teaching the Lesson: The sequence for experiences in this lesson is first to work with objects, then to work with the paper-and-pencil abacus, and finally to work exclusively with symbols. The notion of addition as the operation of combining groups is reviewed with groups of loose objects. The combined group is then considered in base five numeration. An example is the combining of four and three for a group of seven. The seven is then considered as a group of five and two ones. By working with several different groups, the pupils learn the addition facts in base five. Further practice with addition will be provided with sticks on the place value box. The necessary regrouping to express the final group most economically is given special emphasis, for example see Figure 8.

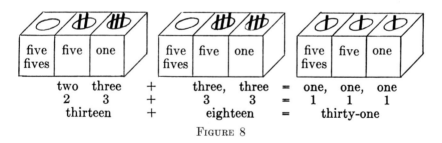

FIGURE 8

Each of the 25 basic addition facts in base five will then be demonstrated on the five-fives number frame. An example is given in Figure 9. Pupils will then be given an opportunity to demonstrate addition facts on the number frame, and to add numbers whose sum does not exceed 25. It should be noted that operations on the number frame are operations with group values which correspond one-to-one with the numbers

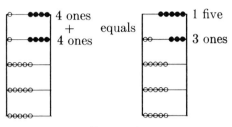

FIGURE 9

represented. The next step is to operate with place value representation of numbers, and thus parallel more closely the algorisms with symbols. This place value representation may be done by using the number frame as an abacus with assigned place values of ones (5^0), fives (5^1), twenty-five (5^2), one hundred twenty-five (5^3), and six hundred twenty-five (5^4).

The operation of addition may be demonstrated on the abacus, first with numbers where regrouping is unnecessary, then with regrouping. Either the term "regrouping" or "exchanging" should be used to describe the more economical representation of numbers, but the word "carrying" should not be used. Pupils may be given an opportunity to add in base five on the abacus. For example see Figure 10. (Note that 5^4, 5^3, etc. are

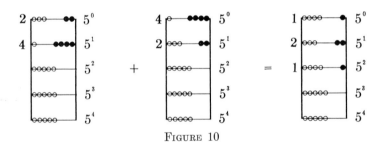

FIGURE 10

not numerals using the base five numeration system, for in that system the only digits are 0, 1, 2, 3, and 4. See the previous paragraph for a *definition* of these symbols.)

Guide Sheet No. 1 will then be used to provide the pupils an opportunity to do addition with the paper-and-pencil abacus.

Guide Sheet No. 2 will be used to give the children practice in working with base five symbols. The addition chart will be explained, and children will be asked to try to do the examples without using the chart, but to use it when absolutely necessary to do so. The sequence estimate, then solve, and then check by excess of fours will be encouraged. Children will also be asked to double check by adding in the opposite direction.

EVALUATION

The children's work on these lessons and guide sheets should provide a good indication of their understanding of addition in base five, and of their ability to handle the addition algorism. Brief oral exercises would furnish additional information.

GUIDE SHEET NO. 1—MIDDLE ELEMENTARY GRADES

SAMPLES

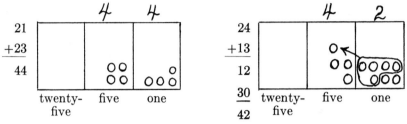

Work each example both with the algorism and on the abacus.

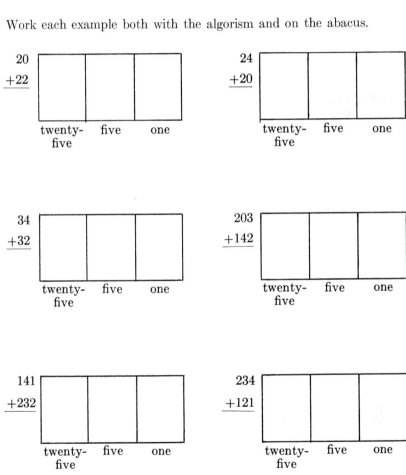

GUIDE SHEET NO. 2—MIDDLE ELEMENTARY GRADES

CHART OF ADDITION FACTS IN BASE FIVE

	0	1	2	3	4
0	0	1	2	3	4
1	1	2	3	4	10
2	2	3	4	10	11
3	3	4	10	11	12
4	4	10	11	12	13

SAMPLES

Estimate: 40   Estimate: 110

```
  22     0 (excess        43     3 (excess
 +21    +3  check)       +20    +2  check)
  43     3               113    10
   3                       1 ←→ 1
```

(a)   20   (b)   31   (c)   24
    + 12   + 13   + 11

(d)   42   (e)   30   (f)   32
    + 31   + 20   + 14

(g)   21   (h)   10   (i)   312
        10   32   + 423
    + 13   + 42

SAMPLE LESSON NO. 3—FOR UPPER ELEMENTARY GRADES

Objective: To provide children with both an understanding of the operation of multiplication with numbers in base five and practice with the multiplication algorism with symbols.

Materials:
1. Five-fives number frame
2. Five-bead abacus
3. Place value box
4. Guide sheets.

Background: It is assumed that the pupils have an understanding of the base five grouping plan, and have developed reasonable proficiency with reading and writing numerals and with addition and subtraction.

Teaching the Lesson: The five-five number frame will be used to develop the notion of multiplication as addition of equal-sized groups. For example:

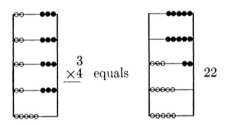

The remaining facts for multiplication will be demonstrated on the number frame. Then the place value box will be used to demonstrate multiplication as group addition. Thus the close relationship between addition and multiplication will be demonstrated. The commutative principle, or rule of interchangeability, will be demonstrated on both the number frame and the place value box. Also, the principle of association will be carefully developed with these two visual aids.

The five-bead abacus will be used to reinforce further the notion of multiplication as repeated addition of equal-sized groups. Then, the abacus will be used to illustrate the distributive law for multiplication. The abacus is an outstanding aid for developing an understanding of the distributive law, since even a multi-place multiplicand may be added repeatedly, or each place may be added independently the appropriate number of times for the cumulative total.

Guide Sheet No. 1 will be used to facilitate the transition from multiplication with objects to work with symbols. The success of the children with this guide sheet will be a strong indication of their understanding of the work with the five-bead abacus.

Guide Sheet No. 2 will provide practice with symbols, and should help the pupils to understand the relationship between multiplication with objects and manipulation of symbols to agree with the results of work with objects. Furthermore, the importance of the basic facts of multiplication is made clear. Since it is assumed that the pupils understand the multiplication algorism in base ten and also the "excess of four" check from their work with addition, a small amount of explanation of the algorism and the excess check should be sufficient.

EVALUATION

The success of the lesson may be rated in terms of pupil response and the extent to which they seemed to understand what they were doing. Also, their work with the guide sheets should provide an excellent measure of the effectiveness of the lesson.

GUIDE SHEET NO. 1—UPPER ELEMENTARY GRADES

SAMPLES

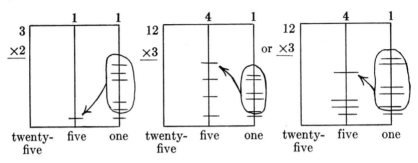

Work each of the following examples on the paper-and-pencil abaci, as was done in the samples above.

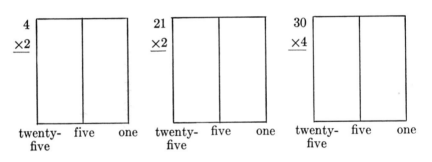

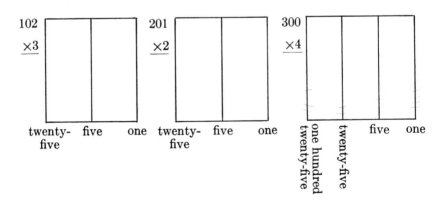

GUIDE SHEET NO. 2—UPPER ELEMENTARY GRADES

CHART OF MULTIPLICATION FACTS IN BASE FIVE

	0	1	2	3	4
0	0	0	0	0	0
1	0	1	2	3	4
2	0	2	4	11	13
3	0	3	11	14	22
4	0	4	13	22	31

SAMPLES

Estimate: 110   Estimate: 2200

```
  21      3 (excess          42      2
 ×3     ×3  check)          ×23     ×1
 113    14                  231      2
        1←──→1              134
                           2121 ╱   (excess
                              2 ╱    check)
```

For each of the following examples: estimate, solve, and check.

(a)   21   (b)   30   (c)   42
     × 2   × 3   × 4

(d)   41   (e)   33   (f)   40
     × 23   × 21   × 24

ARITHMETIC
WITH FRAMES*

UICSM PROJECT STAFF

University of Illinois
Urbana, Illinois

The following article is based on work undertaken by the Project of the University of Illinois Committee on School Mathematics, a project which is primarily concerned with students in grades 9 through 12. Frequently the project staff is asked if its work with high school students has implications for students in earlier grades, that is, if in attempting to work out better ways of presenting material to high school students, ideas have occurred for better ways to present mathematics to elementary school students. It should be pointed out that the authors have not tested these proposals in grade school classes as they have their ideas for high school students. However, they have done some informal experimenting with groups of two or three children ranging in age from 4 to 11 years. The authors offer the following suggestions not as a comprehensive new program nor as answers to any specific problems, but rather as informal notes for those teachers who are interested in trying new ideas.

Although the ideas presented here were developed for use with all children, many teachers will find them particularly useful for enriching the program for the talented.

ARITHMETIC AND ALGEBRA

At each stage in their learning of arithmetic, children should be taught to do some of the corresponding kinds of "algebra." In this way, algebra

---

* The ideas expressed in this article are derived from the experimental textbooks developed by the University of Illinois Committee on School Mathematics, published in 1955. Revised editions for students and for teachers are available from the University of Illinois Press, Urbana, Illinois.

This article originally appeared in the April 1957 issue of the *Arithmetic Teacher* 4:119–24, published by the National Council of Teachers of Mathematics, a department of the National Education Association.

grows naturally out of arithmetic and will not be viewed in high school as an entirely new subject in which, as too often happens, the student thinks he "adds and subtracts letters." If a student knows that

$$5 + 3 = 8,$$

then he is ready to be asked to replace the question mark by a numeral[1] in each of the following three sentences so that in each case the resulting statement is true:

$$? + 3 = 8$$
$$5 + ? = 8$$
$$5 + 3 = ?$$

Also, the student should be asked to make other replacements for each question mark so that the resulting statement is false. In an attempt to teach students correct arithmetic facts, many teachers make students feel that the writing of a statement such as:

$$5 + 6 = 8$$

is an immoral act. Rather than this, a student should examine such a statement and simply declare that it is false (or "not-true," as some children prefer to say). When the student gets to high school mathematics, he will no longer be able to live in a Utopia where false statements are never allowed to occur. Even in grade school work, he should be encouraged to make estimates and enlightened guesses; and when he checks such a guess, he must be prepared to find that he obtains a false statement which tells him that his guess was not good enough.

* * *

An early elementary school student becomes too accustomed to the pattern:

Add:

2

3

5

and should work with, at the same time, the pattern:

$$2 + 3 = 5$$

---

[1] For a discussion of the use of the word 'numeral' as distinguished from the word 'number' and of the use of semi-quotation marks, see *High School Mathematics—First Course* (Urbana, Ill.: UICSM, 1956) or the article "Words, 'Words', "Words"" in the *Mathematics Teacher*, April 1955. p. 213.

and should also work with the pattern:

$$5 = 2 + 3.$$

One of the authors asked several second and third graders if the following statement is true:

$$2 + 2 + 2 = 2 + 8$$

A surprising fraction of them, "knowing that only one numeral can follow an equals sign," transformed it mentally into '2 + 2 + 2 = 8' and said it was true. Students need to learn the symmetry of the use of ' = ' from the very outset of their work with it.

USING FRAMES

Rather than using question marks to give open sentences such as '3 + ? = 5', we propose the use of large frames. Thus, students might be told to first complete the following sentences, so that each of them becomes true, and then so that each of them becomes false.

Add:

After students have carried out these exercises, ask them to tell for a given exercise how many choices for replacement there are which will give true statements. If you want to introduce a bit of philosophical discussion in the arithmetic class, ask them to support the answer that there is only one number such that replacement of the '△' in '4 + △ = 9' by a name for this number makes a true statement. (A clever student may say that either a '5' or a '2 + 3' can go in the box to give a true statement. In this case he is giving two names for one number, just as 'William' and 'Bill' may be two names for one boy.) Also, ask them how many choices they have for replacements which give false statements. Accept answers such as "many," or better yet "as many as you please"; or "so many you couldn't even write them all down"; but do *not* introduce the word 'infinity'.

At a later stage, students can be introduced to sentences which contain more than one frame, for example:

$$\boxed{} + \boxed{} = 12$$

When more than one frame occurs, the student must learn an important rule of the game. Whatever numeral he puts in one of the two boxes above, he must put a copy of it in the other box. Thus, if he writes a '3' in one of the boxes, to follow the rule, he must write a '3' in the other box; he must not write a '3' in one of the boxes and, say, a '9' in the other. (This rule of "like" replacements holds only for frames having the same shape.) Again, ask students to consider how many choices they have for making '$\square + \square = 12$' into a true statement, and how many choices they have for making it false.

When students have had considerable experience in playing the "replacement game," they will find it convenient to use the word *satisfy* in such cases as:

the number 6 satisfies the sentence '$\square + \square = 12$'

and:

7 does not satisfy '$\triangle + 3 = 11$'

Notice that only one number satisfies '$\square + \square = 12$' although any one of that number's names could be used in changing the open sentence into a true statement. For example, if a student writes a '5 + 1' in each box, he has followed the replacement rule correctly, and has found that the number 6 satisfies the sentence. At various stages of his work in arithmetic, it would be appropriate for a student to be confronted with a problem such as the following.

Find all numbers which satisfy each sentence given below.

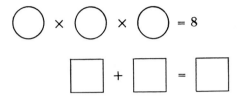

Such a list can be extended indefinitely and as much difficulty as is appropriate can be introduced using fractions, decimals, and "big numbers." Mixed in with such exercises should be some, such as:

which are satisfied by every number. The student has solved the problem when he points out that every number satisfies it (or that no number fails to satisfy it). The fact that every number satisfies such a sentence can be disguised as intricately as you please by increasing the complexity of the sentences. For example:

$$16.2 + \Box + 8.3 + \Box = \Box + 36.9 + \Box - 12.4$$

Also, among such problems should be some like this:

$$\langle\!\rangle + 3 = \langle\!\rangle$$

Here, the student should observe that there is no number which satisfies this sentence and, correspondingly, every number will make it false.

Another interesting exercise is

$$\diamondsuit \times \diamondsuit = \diamondsuit$$

Here, the student should observe that each of the numbers 0 and 1 satisfies the sentence and that no other numbers do. Encourage students to do a considerable amount of thinking in support of such a conclusion. Look for statements such as, "Well, if you multiply any number bigger than 1 by itself, you get an even bigger number. If you multiply a number between 0 and 1 by itself, you get a smaller number." Arrival at such conclusions can be promoted through questions such as:

If you multiply 8 by a number larger than 1, what do you know about the answer?

If you multiply 1,632 by a number less than 1, what do you know about the answer?

In fact, as soon as a child can tell *without computation* which of the numbers $\frac{981}{972}$ and $\frac{981}{972} \times \frac{981}{972}$ is the greater, and whether $\frac{2355}{2356}$ is greater or less than $\frac{2355}{2356} \times \frac{2355}{2356}$, you know he has discovered the important ideas mentioned above, whether he can give precise statements of them or not.

USING FRAMES OF DIFFERENT SHAPES

If you give the student an exercise such as:

$$10 \ = \ \square \ + \ \hexagon$$

then, since you are using differently shaped frames, the agreement is that he may refer to different numbers in making the replacements. Now candidates for replacement are *ordered pairs* of numbers rather than single numbers. You might speak of candidates for the box as *first numbers*, and candidates for the hexagon as *second numbers*; since in left-to-right order the box appears first and the hexagon appears second. Thus, the pair of numbers with first number 7 and second number 3 satisfies the given sentence. Students will soon discover that, for every pair which satisfies the sentence, the pair obtained by interchanging first and second numbers also satisfies the sentence. They have discovered the *commutative principle for addition*, one of several fundamental laws of the number system.

Many first grade children through working with, say, parts of 5 as being 3 and 2, 2 and 3, 4 and 1, 1 and 4 have intuitively discovered the commutative or order property. (Students do not need to learn the word 'commutative' but they ought to be aware of the principle.)

If you give students a sentence such as:

$$10 \ = \ \left[\ 4 \times \square \ \right] \ + \ \hexagon$$

they will quickly find that it is not true that, for every pair that satisfies this sentence, the "reverse" also does. (But you can ask them to find a pair such that it and its reverse both satisfy the sentence. The pair (2, 2) is the only such pair.) Again, ask students to give pairs which "work"—that is, which satisfy the sentence—and pairs which do not work, and discuss the number of choices they have for pairs which do and for pairs which do not work. Problems of this type can be constructed with any desired degree of computational complexity.

USING FRAMES IN MORE PLACES

Once students are familiar with the ideas of working with these differently shaped frames and when they understand the replacement rule, you can use frames in many more places. For example, when introducing students to the ideas of rational numbers, you can make precise the idea of, say, $\frac{1}{3}$ by telling them that this is *the one number* which satisfies the sentence:

$$\Box + \Box + \Box = 1.$$

This is a step toward a mathematically sound formulation of the intuitive idea of cutting a pie into three pieces of the same size. (We are not recommending that the pie idea be dropped but rather that it be used to support the statement above.)

Frame-notation provides a convenient tool for the introduction of *inverse operations*. Students can learn that the number

$$5 - 3$$

is the number which satisfies the sentence:

$$5 = \bigcirc + 3.$$

Thus, students come to realize that two ways of setting the same problem, or, more precisely, two sentences which are satisfied by the same number are

$$5 - 3 = \bigcirc \quad \text{and:} \quad 5 = \bigcirc + 3.$$

Similarly, when students are introduced to division, they can learn that the number $6 \div 3$ is the number which satisfies the sentence:

$$6 = 3 \times \Box.$$

When you present the idea of square root, students can learn that $\sqrt{9}$ is the number which satisfies the sentence:

$$\bigcirc \times \bigcirc = 9.$$

You may move in another direction which is not common in grade school but which we think might be profitably undertaken, particularly in the later grades. You can give students problems which become more difficult, as indicated by the following sequence. Of course, you would give many more problems than those listed, taking several days in going from problems like the first one to problems like the last one.

$$5 + \boxed{} = 13$$

$$10 = \left\langle\right\rangle + \left\langle\right\rangle + 3$$

$$\tfrac{1}{2} = \left[2 \times \diamond \right] + \tfrac{1}{4}$$

$$\left[2 \times \boxed{} \right] + 6 = 12$$

$$\left[\tfrac{1}{3} \times \diamond \right] - 4 = 1$$

$$\frac{\left[2 \times \boxed{} \right] + 3}{5} = 3$$

$$\left[2 \times \boxed{} \right] + 1 = \boxed{} + 4.$$

You will realize, of course, that students are solving equations in one variable. Students enjoy solving these equations as straight puzzle problems, asking themselves, "Can I find a number which satisfies it?" *Do not spoil their fun by giving rules for solving equations.* Let each student work out his own methods, and discourage attempts to give procedures for obtaining answers, such as transposing or dividing both sides, etc. Tell students that a method which they work out for themselves will be more meaningful and longer remembered than one which they are told by teacher or classmate. Again, it should be noticed that here is another method for giving practice in computation as well as preparing a student for a real understanding of algebra.

Frame-notation[2] is suggestive of many game-type activities which should appeal to elementary school children. Here is one which includes a self-checking feature. The student is given a list of "equations" such as:

(1) $4 + \triangle = 7$

(2) $(9 \times \triangle) + (2 \times \square) = 37$

(3) $4 \times \square = (6 \times \square) - \diamond$

(4) $(3 \times \diamond) + 12 = (4 \times \diamond) + 2$.

His first task is to find a number which satisfies (1). He discovers that such a number is 3. So, he writes a '3' in the '$\triangle$' of (2) and obtains:

(2′) $(9 \times \boxed{3}) + (2 \times \square) = 37$.

Next, he must discover a number which satisfies (2′); 5 satisfies (2′). So, he writes a '5' in each '$\square$' in (3) and obtains:

(3′) $4 \times \boxed{5} = (6 \times \boxed{5}) - \diamond$.

10 satisfies (3′). Sentence (4) is so constructed that it is satisfied by 10.

CONCLUSION

As indicated in our opening remarks, the suggestions made in this article do not prescribe a new mathematics curriculum for the elementary school. But they do reflect the belief that children are willing to spend time on mathematics for the sheer intellectual challenge it can be made to offer. Such challenge is present whenever a child is encouraged to use his imagination and intuition. Games which are rich in mathematical (and not necessarily complex computational) content almost always contain challenges to which children respond.

---

[2] In preparing lists of exercises in duplicated form (by spirit or stencil duplicating processes) you will find it convenient to use a template containing various sizes of frames (circular, triangular, square, and hexagonal). You can make such a template out of stiff cardboard or obtain one from a drafting supply house.

4

MODULAR ARITHMETIC

FRANCIS J. MUELLER

University of Rhode Island
Kingston, Rhode Island

The topic of a simple mathematical system, here called modular arithmetic (sometimes referred to as clock arithmetic), may be explored profitably at various levels. In the elementary school, a study of modular arithmetic can provide insight into a type of activity that is basic to the creative advance of mathematics and science: the search for pattern. Conscious awareness, often intuitive, of some kind of pattern or orderliness in an area of study is the usual springboard for fruitful hypotheses; and hypotheses-to-be-tested are at the growing edge of every living science.

The teacher who has taught or is going to teach a base five numeration system as well as a modular 5 system needs to take extra care to avoid confusion.

WHY TEACH MODULAR ARITHMETIC?

School mathematics has traditionally played down the role of imagination and intuition on the part of the learner. This, we now suspect, operates at cross purposes to greater mathematical understanding. Properly slanted, work with modular arithmetic with an emphasis upon pattern can arouse the elementary student's mathematical imagination and intuition and, in the process, provide a possible foretaste of the thrill of scholarly discovery.

A second and perhaps more frequently developed aspect of modular arithmetic has to do with the underlying abstract structure of our number system. At an elementary level, modular arithmetic provides finite, miniature number systems which may be explored rather thoroughly. These systems afford pointed illustrations of such basic structural properties as closure, commutativity, associativity, distributivity, and the like. At more advanced levels, in modern abstract algebra and number theory, for instance, modular systems serve again as key illustrations. In

73

fact, at such levels, modular systems are themselves objects of considerable study.

In this discussion of modular arithmetic, earlier attention centers on the aspect of pattern, while later emphasis is on the axiomatic or structural aspects of the subject. Between the two there is a twilight zone of considerable breadth. How these parts should be blended for a given group of students is necessarily a matter of local option. Thus, no attempt is made to peg any of the material to a specific grade level. Moreover, this presentation is intended for the teacher. Its mode has been chosen to make it easier for that teacher to project his understanding of the topic into his own classroom situation. He can then best decide whether or how much modular arithmetic should be used with his students.

AT THE OUTSET

Since a great deal of the work in modular arithmetic at an elementary level can be handled by the use of tables, it may be well to begin with a comparable table of familiar facts—that for addition of whole numbers as given in Table I.

After it has been demonstrated on the chalkboard how the sequence of numerals along the left and top margins represent addends (Table I), and how, for classical arithmetic, the sums of pairs of addends may be recorded in the box at the intersection of the horizontal row and vertical column containing the addends, the students can be directed to complete their individual addition tables. The students' tables need not be square

TABLE I
Decimal Addition Facts

+	0	1	2	3	4	5	6	7	8	9	10	11	12	13	14	15	16	17
0	0	1	2	3	4	5	6	7	8	9	10	11	12	13				
1	1	2	3	4	5	6	7	8	9	10	11	12						
2	2	3	4	5	6	7	8	9	10	11	12	13	14	15	16	17		
3	3	4	5	6	7	8	9	10	11	12	13							
4	4	5	6	7	8	9	10	11	12	13	14	15						
5	5	6	7	8	9	10	11	12	13	14	15							
⋮	⋮	⋮	⋮	⋮	⋮	⋮	⋮	⋮	⋮	⋮	⋮	⋮	⋮					
16	16	17	18	19	20	21	22	23	24	25								
17	17	18	19	20	21	22	23	24	25									
18	18	19	20	21	22	23	24	25	26	27	28	29	30					

arrays, though it is advisable to have the addends go well into the 'teens. Also, there may be some value in leaving empty the last few rows and columns to suggest the fact that the table could be extended at will, granted a large enough sheet of paper. This is not the case, of course, with modular systems.

As the class proceeds with the table-completing assignment, almost certainly some students will begin to race ahead and finish more quickly than others. Along the way they will discover some recurring pattern in their work and elect to follow it, rather than tediously to compute each sum individually. This circumstance offers a natural point of departure for subsequent discussion, because here the student's discovery of pattern has served him advantageously.

Attention can be drawn to another practical aspect of pattern by turning the discussion to the correctness of the sums entered in the table . . . *How might we check our answers?* . . . Some students will suggest comparing tables with a neighbor, or recomputing each sum, or checking the ordered sequence in each row or column (but that is the way many got their sums in the first place). The observation we are waiting for, however, involves the diagonal pattern of like sums that appears to parade across the body of the table, from lower left to upper right. Any numeral which violates that pattern had better be checked for accuracy.

Ordinarily, this should be enough in the way of preliminaries before setting the students off to prospect for new and different patterns. It is probably best to have them work independently, and write down discovered patterns on another sheet of paper so as not to dim the glow of personal discovery for the other pupils. Discovered patterns should certainly be discussed later on by the class as a whole. Among the patterns or consistent characteristics likely to be noted (and one may hope for more) are the following:

1. No sum is repeated in any given row or column.
2. The numbers represented by numerals are in counting sequence across each row and down each column.
3. The values represented by the numerals down each diagonal, from upper left to lower right, "jump" by two.
4. Sums in the top row and left-most column are the same as the addends across the top and down the left margin, respectively.
5. Corresponding pairs of addends have the same sums, regardless of order (e.g., the same numeral occurs at the intersection of column 6 and row 3, and at intersection of column 3 and row 6).
6. Counting sequence is maintained when one moves across any row

and then turns down any column; or moves down any column and then right across any row. (Vice-versa for a reversed counting sequence.)

7. Triangular, rectangular, and other geometric arrays of numerals—and their sums along the sides—reveal certain interesting and persistent patterns.

CLOCK ARITHMETIC

If the students have had experience with addition tables along the lines suggested above, transition into clock arithmetic can be made easily by calling attention to the potential endlessness of the addition table . . . *Can any one think of some use of numbers where the supply of numbers is limited?* . . . The key answer we are looking for, of course, is "the clock" (which might have to be prompted by a leading question: "What about the clock on the wall?").

The usual clock face carries twelve numerals: 1, 2, 3, $\cdots$ 12, that serve to mark the passing hours. Questions such as the following lead into the topic rather well. If it is 2 o'clock now and I work 3 hours, what time will I finish? How did you arrive at that answer—by counting the hours, or by adding 2 + 3? . . . What time is it 4 hours after 6 o'clock? Did you use the addition fact 6 + 4 = 10? . . . If it is 8 o'clock and I work 4 hours, what time should I finish? Did you use the addition fact, 8 + 4 = 12? . . . How about if I started at 8 o'clock and worked 6 hours, what time would I finish? 2 o'clock? . . . Would you say 8 + 6 = 2 is a correct addition fact when dealing with clocks? . . . and so on.

A possible follow-up to such a discussion would be to have each student prepare a special "clock addition" table, which when completed will look like the one in Table II.

Again, the students should be encouraged to seek out patterns . . . . Does the diagonal pattern we used to check our sums in the regular addition table (Table I) appear here? . . . Can it be used to check our "clock" work? . . . How does this clock addition table differ from the regular one? . . . How are they alike?

Likely to be detected will be the fact that no zero appears in the clock table; numerals in rows and columns are in sequence, but "circular" rather than ever-increasing; observations (1) through (7) listed above are seen to hold, so long as "counting sequence" is interpreted to mean "circular counting sequence."

One of the more striking differences between Table I and Table II, of course, is the limited or finite extent of the clock addition table. Addends beyond 12 are not needed. Of particular mathematical significance is the

TABLE II
Clock Addition Facts

+	1	2	3	4	5	6	7	8	9	10	11	12
1	2	3	4	5	6	7	8	9	10	11	12	1
2	3	4	5	6	7	8	9	10	11	12	1	2
3	4	5	6	7	8	9	10	11	12	1	2	3
4	5	6	7	8	9	10	11	12	1	2	3	4
5	6	7	8	9	10	11	12	1	2	3	4	5
6	7	8	9	10	11	12	1	2	3	4	5	6
7	8	9	10	11	12	1	2	3	4	5	6	7
8	9	10	11	12	1	2	3	4	5	6	7	8
9	10	11	12	1	2	3	4	5	6	7	8	9
10	11	12	1	2	3	4	5	6	7	8	9	10
11	12	1	2	3	4	5	6	7	8	9	10	11
12	1	2	3	4	5	6	7	8	9	10	11	12

fact that no number beyond 12 is needed, either as an addend or to express a sum. This reflects a mathematical characteristic known as "closure." *A set of numbers is closed for an operation (here addition) when that operation, performed with any of the numbers of the set, invariably results in one, and only one, of the numbers of that set.* Since every addition of two numbers in our clock arithmetic yields a unique number in the original set, 1, 2, 3, $\cdots$ 11, 12, (see Table II), we say this set of numbers is *closed* for the operation of addition. Except for the trivial case of a set containing only the single number, zero, a similar statement cannot be made for any limited or finite set of whole numbers in our usual arithmetic where some additions always result in numbers not in the set. (It should be pointed out, however, that the *infinite* set of whole numbers that we use in ordinary arithmetic *is* closed for addition; for multiplication, too.) Whether or not the term "closure" is used with the students in the classroom depends upon local circumstances; but certainly the characteristic should be given some degree of conscious attention.

Before the class leaves the 12-numeral clock addition table it may be well to anticipate what is to follow by discussing the use of 0 in the place of 12. Actually this amounts to nothing more than a mere substitution of symbols, 0 for 12. But to put the students' minds at ease on the matter, they might be encouraged to replace the 12's in their table with 0's, then check the validity of the substitution by testing some or all of the cases on the clock.

OTHER MODULAR SYSTEMS

The system of arithmetic we have been discussing here, in terms of the clock, is a *modular* system. Since the system contains but 12 basic elements or numbers, which repeat, we say the system has a *modulus* of 12. The system may also be referred to as *modular 12* or *modulo 12*. Accordingly, a modulo 4 system would contain just the four numbers: 0, 1, 2, 3; a modulo 3: 0, 1, 2; a modulo 2: 0, 1; a modulo 5: 0, 1, 2, 3, 4; etc.

Modulo 4 and 5 are very useful systems for introductory purposes— both here and in the classroom. In the classroom, the teacher would do well to construct a circular "clock face" with a "free-to-rotate" single hand. Blackboard paint on the face of the clock makes it easy to display around its edge different sets of chalked-on numerals. Figure 1 shows a clock face appropriate for a discussion of a modulo 4 system: four points or tick marks divide the circumference into quarters and each is labeled, in clockwise succession, by a numeral: 0, 1, 2, 3.

As with the usual clock, let movement of the hand "clockwise" imply addition. (Subtraction by reverse motion is discussed later.) To determine the sum of 1 + 3 in a modulo 4 system, say, the hand starts at the first addend: 1. Then it is rotated clockwise through a number of points equal to the second addend: 3. This brings the hand to rest at the point labeled 0 Hence, in this modulo 4 system, 1 + 3 = 0.

After another illustration or two, the students should find it an interesting project to compute the remaining parts of the modulo 4 addition table (Table III). A similar approach can be used to produce the addition table for a modulo 5 system (Table IV).

As a practical matter of technique, a small deck of cards may be

Modulo 4 Clock Face
FIGURE 1

TABLE III
Modulo 4 Addition Facts
ADDEND

+	0	1	2	3
0	0	1	2	3
1	1	2	3	0
2	2	3	0	1
3	3	0	1	2

ADDEND

TABLE IV
Modulo 5 Addition Facts
ADDEND

+	0	1	2	3	4
0	0	1	2	3	4
1	1	2	3	4	0
2	2	3	4	0	1
3	3	4	0	1	2
4	4	0	1	2	3

(Row label at left, running vertically: ADDEND)

substituted for the more cumbersome clock face—either playing cards
from a regular deck, or a specially constructed deck of 3 x 5 index cards.
In either case, each card should represent one numeral on the clock face,
whatever the modular system under consideration. For modulo 4, say,
there should be four cards, each bearing one of the numerals, 0, 1, 2 and
3, initially in that order from top to bottom of the deck. The cards can
be made to effect the rotation of the clock hand by moving the top card
of the deck to the bottom of the deck. (Caution: At no time is the deck
"shuffled"; the circular order of the cards must always be maintained.)
Thus, to find the sum of 1 + 3 with a modulo 4 deck, we start with the
deck showing the 1-card on top—under it will be the 2-card; below that,
the 3-card; on the bottom will be the 0-card. To add 3, three top cards
are rotated from top to bottom, one by one, as follows: "bury" the top
card (1); "bury" the next top card (2); "bury" the new top card (3). On
top now should be the card marked 0, representing the sum of 1 + 3,
modulo 4.

In the classroom, the students might use the clock face first, to construct the modulo 4 and 5 addition tables, then be introduced to the
cards. An initial student experience with the card technique might be to
verify the addition tables arrived at via the clock. By simply adding
more cards to the deck, or removing certain of the higher ones, a ready
means is at hand for the construction of new modular tables.

Leading Questions in a Search for Patterns

As before, constructing the tables is only part of the activity. Looking
for patterns among the numeral entries in the table should be an item of
major interest. . . . What consistencies or patterns can be seen among the
entries in the table? . . . How do the tables of even-numbered moduli

differ from odd-numbered moduli? . . . How might the modulo 7 table
be used if the addend numerals across the top are replaced with the names:
Sunday (for 0), Monday (for 1), Tuesday (for 2) . . . Saturday (for 6)?
. . . If 0, wherever it is written in the modulo-2 table, is interpreted to
mean "even number," and 1 to mean "odd number"—does the table
tell the truth? . . . How could the modulo 7 deck of cards be used to tell
what day of the week the 19th day after Tuesday will be? . . . How could
the modulo 12 deck be used to tell what month is the 27th month after
March; or how could it be used to tell what the 8th month before
December was? . . .

MULTIPLICATION

There are various paths that lead into modular multiplication. One is
a sequence paralleling that discussed earlier with respect to the addition
table: Start with a multiplication table based on the usual whole numbers
of arithmetic, then go to the clock or cards. Or, one might start directly
with the clock or the cards. In any case, multiplication is defined in
terms of repeated addition, i.e., $3 \times 5 = 5 + 5 + 5$, or $6 \times 8 = 8 + 8 + 8 + 8 + 8 + 8$. Thus, to show 3×2 on a modulo 4 clock, one
starts at 0 and moves the hand two spaces at a time: $0 \frown 2 \frown 0 \frown 2$,
which means for modulo 4: $3 \times 2 = 2$. With cards, one starts the pack
with 0 on top, buries two cards (in order) three times. Tables V and VI
show the multiplication tables that should result for modulo 4 and 5,
respectively.

Again, certain identifiable patterns within the table can be readily
discovered, and the students should be encouraged to search for them.

If the students have been introduced to the concepts of commuta-

TABLE V
Modulo 4 Multiplication Facts
FACTOR

×	0	1	2	3
0	0	0	0	0
1	0	1	2	3
2	0	2	0	2
3	0	3	2	1

TABLE VI
Modulo 5 Multiplication Facts
FACTOR

×	0	1	2	3	4
0	0	0	0	0	0
1	0	1	2	3	4
2	0	2	4	1	3
3	0	3	1	4	2
4	0	4	3	2	1

tivity, associativity, and distributivity, then modular systems offer additional illustrations.

An operation on a set of members is said to be _commutative_ when it makes no difference in which order the numbers of the set undergo that operation. We know this to be so in classical arithmetic (e.g., $42 + 36 = 78$ and $36 + 42 = 78$, so $42 + 36 = 36 + 42$; $6 \times 25 = 25 \times 6$). Hence, our ordinary system of arithmetic numbers is said to be commutative with respect to each of the operations, addition and multiplication. Are the modular systems we have been discussing here commutative with respect to addition? With respect to multiplication? The answers are "yes," but the limited nature of these modular systems makes it possible for the student to discover this for himself by testing all possible combinations—something utterly impossible for our usual system of arithmetic numbers.

An operation on a set of numbers is said to be _associative_ when the grouping of the numbers of the set, as they undergo the operation, has no effect upon the result. We express our confidence in the fact that our set of numbers used in ordinary arithmetic is associative with respect to addition when we know that $3 + 2 + 7$ will lead us to the same sum of 12, whether we group and add the first two numbers separately, as $3 + 2 = 5$, and then add the 7, or add the 2 and 7 separately, and then add the 3. Similarly for multiplication: $3 \times 2 \times 4$ will lead us inevitably to 24, whether we group and multiply the factors 3 and 2 for 6, and then multiply 6 by the factor 4; or multiply the factors 2 and 4 together for 8, then multiply the 8 by 3. All the modular systems of the type we are discussing are associative with respect to both addition and multiplication; but, again, the students should discover the fact of associativity themselves by checking their various tables. For example, in the system, modulo 5:

$$
\begin{array}{ccc}
& ? & \\
3 \times (4 \times 2) & = & (3 \times 4) \times 2 \\
\downarrow \quad \downarrow & ? & \downarrow \quad \downarrow \\
\underbrace{3 \times 3}_{4} & = & \underbrace{2 \times 2}_{4} \quad (\text{mod } 5)
\end{array}
$$

In classical arithmetic we have another important property, _the distributive property for multiplication with respect to addition_. In symbols this is fairly easy to express:

$$a \times (b + c) = a \times b + a \times c.$$

But in words it is more awkward. When the distributive property for multiplication with respect to addition holds for a set of numbers, it means the sum of two (or more) of these numbers in the set, when multiplied by another number, must yield the same result as would be had by multiplying each addend in the sum by the multiplying number, then adding the resulting products. For instance,

$$3 \times (7) = 3 \times (2 + 5)$$
$$(3 \times 2) + (3 \times 5) = 6 + 15 = 21;$$

or for that matter,

$$\tfrac{1}{2} \times (1) = \tfrac{1}{2} \times (\tfrac{1}{3} + \tfrac{2}{3})$$
$$(\tfrac{1}{2} \times \tfrac{1}{3}) + (\tfrac{1}{2} \times \tfrac{2}{3}) = \tfrac{1}{6} + \tfrac{2}{6} = \tfrac{3}{6} = \tfrac{1}{2}.$$

The modular systems of the type we have been discussing are also distributive for multiplication with respect to addition. In modulo 5, for example,

$$3 \times (4 + 2) = 3 \times 1 = 3 \qquad\qquad \text{(mod 5)}$$
$$3 \times (4 + 2) = (3 \times 4) + (3 \times 2) \qquad\qquad \text{(mod 5)}$$
$$= 2 + 1 = 3; \qquad\qquad \text{(mod 5)}$$

or in modulo 4,

$$2 \times (3 + 1) = 2 \times 0 = 0 \qquad\qquad \text{(mod 4)}$$
$$2 \times (3 + 1) = (2 \times 3) + (2 \times 1) \qquad\qquad \text{(mod 4)}$$
$$= 2 + 2 = 0. \qquad\qquad \text{(mod 4)}$$

Rather obviously, the modular tables developed by the students can be used to test the presence of the distributive, commutative, and associative properties in situations that are relatively foreign to the students' general experience. This is likely to induce a greater appreciation of the properties than the sole use of comparable exercises within the usual arithmetic structure.

SUBTRACTION

Subtraction in modular systems provides many interesting and illuminating mathematical opportunities. For instance, with our usual set of arithmetic (non-negative) numbers, certain subtractions have no results, e.g., $4 - 5$, $18 - 24$, etc. But there is no such problem in our modular systems, as we shall see.

Mechanically, modular subtraction means running the hand "backwards" (counter-clockwise) on the clock face, or "dealing off the bottom" with the modular card deck. Figure 2 shows a modulo 5 clock face, Table VII a modulo 5 subtraction table, and Table VIII a modulo 4 subtraction table.

By way of illustration, the remainder for 2 − 3 would be found by starting with the hand at 2 (the minuend) and ticking off, counterclockwise, 3 (the subtrahend) points, 2, 1, 0, 4, to indicate 4 as the remainder for 2 − 3.

With the modulo 5 deck of cards, this computation can be performed by starting with the 2-card on top of the deck in its usual top-to-bottom sequence. Then by "dealing" one card at a time, *from the bottom to top*, three times (1; 0; 4), the final top card will show the remainder 4. (An alternative, of course, is to reverse the sequence of cards in the deck and "rotate" them as in addition—i.e., continuously burying each top card, one by one. Reversing the deck, however, is less likely to be a meaningful means of showing subtraction as the inverse operation of addition than is reversing the deal.)

In constructing the subtraction tables, greater care needs to be exer-

TABLE VII
Modulo 5 Subtraction Facts
MINUEND

−	0	1	2	3	4
0	0	1	2	3	4
1	4	0	1	2	3
2	3	4	0	1	2
3	2	3	4	0	1
4	1	2	3	4	0

SUBTRAHEND

TABLE VIII
Modulo 4 Substraction Facts
MINUEND

−	0	1	2	3
0	0	1	2	3
1	3	0	1	2
2	2	3	0	1
3	1	2	3	0

SUBTRAHEND

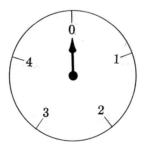

Modulo 5 Clock Face
FIGURE 2

cised than in the cases for addition or multiplication, where numerals at the left and across the top are both addends or factors. Some stock phrase or "patter" can be helpful in keeping things uniform in interpreting the table. For instance, if the subtrahends are arrayed down the left side of the table (as shown in Tables VII and VIII) the stock phrase for the modulo 5 example, $2 - 3$, might be: *2* (looking at the 2 in the top margin) *take away 3* (looking at the 3 in the left margin) *is 4* (looking at the numeral at the intersection of column 2 and row 3). A uniform thought-and-word sequence of this sort is also helpful when working to complete the subtraction table via clock face or card deck.

Again, interesting number patterns can be detected. Compare tables for even and odd moduli. . . . Compare diagonals of subtraction tables with those of addition for that modulus. . . .

As for the structural properties, there are other interesting aspects. The modular systems we have been discussing are, of course, closed for subtraction. (An allied question: What must be done to introduce closure for subtraction into our system of arithmetic numbers? Answer: Introduce negative numbers.) The properties of commutativity and associativity do not hold for subtraction in these modular systems, as indeed is also the case in our usual arithmetic. The distributive property for multiplication with respect to subtraction does hold, however. Students should be encouraged to show these things.

Modular arithmetic can also be very useful in bringing home the inverse relationship between addition and subtraction. By looking at subtraction as a means of finding a missing addend, the corresponding modular subtraction table can be constructed directly from that of addition. Thus, in ordinary arithmetic, the subtraction,

$$5 - 2 = n$$

may be paraphrased as

$$n + 2 = 5$$

or, "What number (the remainder) added to 2 (the subtrahend) will yield 5 (the minuend)?"

Correspondingly, for the now-frequently-cited modulo 5 fact,

$$2 - 3 = n \qquad (\mathrm{mod}\ 5)$$

we can find the correct value for n by paraphrasing the number sentence to read:

$$n + 3 = 2 \qquad (\mathrm{mod}\ 5).$$

Then, by consulting the modulo 5 addition table, note in that system,

$$4 + 3 = 2 \qquad (\text{mod } 5).$$

Another way to get at this concept is the well-known formula for checking subtraction: *remainder + subtrahend = minuend*. The same relationship also holds for modular subtractions, as the students should be encouraged to verify.

DIVISION

In constructing a table of division facts for the modular systems under discussion, there are two basic approaches. One involves application of the inverse relationship between division and multiplication to the table of multiplication facts for the system. As subtraction may be thought of inversely as an operation for finding the value of a missing addend in an addition, division may be thought of as an operation for finding the value of a missing factor in a multiplication. Thus, in ordinary arithmetic, the division, $8 \div 4 = n$, may be paraphrased in multiplication form as $4 \times n = 8$, which asks "What number (the quotient) multiplied by 4 (the divisor) yields 8 (the dividend) as the product?"

Similarly, to find the modulo 5 quotient, n, for $2 \div 3 = n$, say, we hunt in our multiplication table for the corresponding product, $n \times 3 = 2$, and find that $4 \times 3 = 2$. Moreover, 4 is the only possible value for n. Hence $2 \div 3 = 4$, in modulo 5.

Whether or not there is one and only one result for a division is important, since our definition of closure for an operation calls for a unique result. Not only is division by zero excluded in modular systems, for the very same reasons that exclude it in our usual arithmetic (quotient either non-unique, e.g., $0 \div 0$, or non-definable in terms of multiplication, e.g., $7 \div 0$), but in certain of these modular systems there are other instances involving nonzero divisors.

For systems having prime-number moduli, there is a unique quotient for every number in the system for any nonzero divisor. Tables IX and X illustrate this. But for non-prime or composite moduli, some numbers divided by nonzero divisors fail to have unique quotients. For example in modulo 4 (Table V), we see that $2 \times 1 = 2$ and $2 \times 3 = 2$. It would follow, then, that $2 \div 2 = 1$, and $2 \div 2 = 3$. Hence, there is no unique quotient for $2 \div 2$ in a modulo 4 system.

Because of the absence of a corresponding multiplication fact, still other divisions in non-prime modular systems are impossible. In modulo 4, for example, there is no corresponding multiplication fact $(2 \times ? = 1)$ for $1 \div 2$. (Of course, in our usual arithmetic, not all divisions involving

TABLE IX
Modulo 4 Division Facts
DIVIDEND

÷	0	1	2	3
0	–	–	–	–
1	0	1	2	3
2	–	–	–	–
3	0	3	2	1

DIVISOR

TABLE X
Modulo 5 Division Facts
DIVIDEND

÷	0	1	2	3	4
0	–	–	–	–	–
1	0	1	2	3	4
2	0	3	1	4	2
3	0	2	4	1	3
4	0	4	3	2	1

DIVISOR

whole-number terms are possible within the set of whole numbers, and so our set of whole numbers is not closed for division either. We may attain closure there, however, by introducing fractions.)

Tables of division facts can also be constructed (or perhaps verified) by clock face and modular card deck. Here, the relationship between division and subtraction is basic: division may be conceived as repeated subtraction. For example, in our usual arithmetic,

$$12 \div 3 = n$$

may be solved by subtracting the divisor (3), repeatedly, as shown below,

$$
\begin{array}{r}
12 \\
-3 \ \checkmark \\
\hline
9 \\
-3 \ \checkmark \\
\hline
6 \\
-3 \ \checkmark \\
\hline
3 \\
-3 \ \checkmark \\
\hline
0
\end{array}
$$

until the dividend is exhausted or reduced to some number less than the divisor. The number of times the divisor is subtracted corresponds to the quotient. Since, in our example, four 3's by subtraction exhaust the original dividend of 12, the quotient is 4. So,

$$12 \div 3 = 4.$$

But we know that in our usual arithmetic of whole numbers not all divisions are even; some involve remainders. Such an example appears below.

$$19 \div 5 = 3 \; R4$$

$$\begin{array}{r} 19 \\ -5 \; \sqrt{} \\ \hline 14 \\ -5 \; \sqrt{} \\ \hline 9 \\ -5 \; \sqrt{} \\ \hline 4 \end{array}$$

In modular arithmetic the repeated subtraction concept applies to division, but there are never any remainders: either the division is even, or there is no quotient!

For instance, consider first this simple modulo 5 example:

$$4 \div 2 = 2 \qquad (\bmod 5).$$

To check this division on the clock face (see Fig. 2) we would start with the hand at 4 (the dividend), then subtract sets of two (the divisor) points at a time by a counter-clockwise movement of the hand. The number of such hand movements it takes to bring the hand to rest at 0 corresponds to the quotient. Thus, since two of these two-point-counter-clockwide moves are needed to bring the hand from 4 to 0 $(4 \smile 2 \smile 0)$, the modulo 5 quotient of $4 \div 2$ is 2.

Now consider the modulo 5 division:

$$2 \div 4 = 3 \qquad (\bmod 5).$$

Here we start at 2 (the dividend) and subtract sets of four (divisor) points until the hand *comes to rest* at 0. You will note that the first counter-clockwise move of four points from 2 carries the hand *past* 0 and to rest at 3. The next counter-clockwise sweep of four points carries the hand again past 0 and to rest at 4. The third counter-clockwise move of four points brings the hand to rest at 0. Those three sweeps indicate that in modulo 5 arithmetic,

$$2 \div 4 = 3 \qquad (\bmod 5).$$

This fact can be verified by reference to Table X.

The card-deck parallel should be fairly obvious by now. With the deck in the usual circular sequence, we start with the dividend as the top card. Then, by rotating the deck as in subtraction, in sets of cards (though card-by-card) equal in number to the divisor, we count the number of sets we must pass through in order to produce the 0-card on top at the end of a pass. The number of sets we must go through corresponds to the quotient.

For example, to find the modulo 5 quotient for $1 \div 3$, we would start with the 1-card on top (the dividend); then rotate the cards, one by one, in subtraction fashion, in sets of three (because the divisor is 3). At the end of the first pass, 3 is on top; at the end of the second pass, 0 is on top. Thus, for modulo 5, $1 \div 3 = 2$.

We have seen that with non-prime moduli, such as 4, certain divisions have no quotients. On the clock face, this is exemplified by the fact that for those divisions the hand will never come to rest at 0. Or in the deck, the 0-card will never come up at the end of a pass. To illustrate, consider the modulo 4 division:

$$3 \div 2 = n \qquad (\text{mod } 4).$$

On the clock, this means start at 3, then move the hand two points at a time in counter-clockwise fashion. At the end of the first sweep of two points, the hand rests at 1; at the end of the second sweep of two points, the hand is back again at 3; the next sweep brings it again to 1; then to 3; and so on, *ad infinitum*. Never, under these circumstances, will the hand come to rest at 0.

A similar effect is obtained with a modular 4-card deck.

Whether modular division is commutative, associative, or distributive (it isn't the last) should be left to the students to discover. There are many more interesting properties related to modular division, but for the most part they are found at higher mathematical levels.

RESIDUES AND CONGRUENCES

Related to the modular systems we have been discussing here is the important number-theory concept of residue and congruence. Although its most fundamental expressions are not beyond the ken of elementary youngsters, extensions do lead well into the mathematical stratosphere.

Residue, for our purposes, may be considered as synonymous with *remainder* in division. The number whose residue we are interested in, in effect, is the *dividend*; and the *divisor* we call the *modulus*. Granted these loose definitions, it is possible to classify or group all of our whole numbers into the same number of categories as the modulus. For example, all whole numbers when divided by 3 will have for a remainder either 0, 1, or 2. Thus, we have three distinct yet comprehensive categories into which all whole numbers can be placed according to the scheme:

Remainder 0:   0, 3, 6, 9, 12, $\cdots$
Remainder 1:   1, 4, 7, 10, 13, $\cdots$
Remainder 2:   2, 5, 8, 11, 14, $\cdots$

On the basis of this, we say

$0, 3, 6, 9, 12, \cdots, 99, \cdots, 162, \cdots$ are equal to 0 (mod 3)
$1, 4, 7, 10, \quad \cdots, 100, \cdots, 163, \cdots$ are equal to 1 (mod 3)
$2, 5, 8, 11, \quad \cdots, 101, \cdots, 164, \cdots$ are equal to 2 (mod 3).

Since this type of modular equality is different from our usual arithmetic equality, we use the symbol, $\equiv$, for modular equality. Thus, we can state

$$10 \equiv 1 \text{ (mod 3)}; 14 \equiv 2 \text{ (mod 3)}; 300 \equiv 0 \text{ (mod 3)};$$
$$8 \equiv 2 \text{ (mod 6)}; 44 \equiv 0 \text{ (mod 11)}; 100 \equiv 10 \text{ (mod 15)}.$$

In short, the modular equivalent of any number can be found by dividing the number by the modulus, ignoring the quotient, but using the remainder as its indicator.

Two numbers that are the modular equivalents of each other (for a given modulus) are said to be *congruent*. Thus 9, 17, 41, and 77 are congruent to each other, *mod 4*, because upon division of each by 4, the remainder is uniformly 1. However, the first three are all different, *mod 5*, since $9 \equiv 4$ (mod 5), $17 \equiv 2$ (mod 5), and $41 \equiv 1$ (mod 5). Because $77 \equiv 2$ (mod 5), it follows that 17 (mod 5) $\equiv$ 77 (mod 5).

One consequence of this viewpoint is that each of the modular systems discussed earlier becomes a potential model of finite dimension for our infinite system of whole numbers. For instance, if we start with our modulo-5 deck of cards with the 3-card on top, and count forward 11 cards in the usual rotating fashion, the final top-card will reveal 4 as the sum. This is not too surprising, for $11 \equiv 1$ (mod 5) and $3 \equiv 3$ (mod 5); and 1 (mod 5) + 3 (mod 5) $\equiv$ 4 (mod 5).

Students who have had experience checking computation by "casting out 9's" will find work with congruences fascinating. Basically, the 9's check works because of a modular characteristic that is by no means peculiar to the number 9. The checks would work for any other number as well. What makes 9 of special interest is the fact that we don't have to divide a number by 9 to find its residue or remainder—we can get it quickly by adding the digits of the number's decimal numeral in one of the usual ways. This short cut is possible for 9 because of a special relationship that exists between 9 and the base of our decimal system of number notation, 10. But computation checks can also be performed in essentially the same way with any modulus, say 7, as shown below.

SOME COMPUTATIONS INVOLVING CONGRUENCES MODULO 7

Addition

$$32 \equiv 4 \pmod 7$$
$$24 \equiv 3 \pmod 7$$
$$+\ 16 \equiv 2 \pmod 7$$
$$72 \equiv 9 \pmod 7 \equiv 2 \pmod 7$$

Check: $72 \div 7 = 10\ R\ 2$
or $72 \equiv 2 \pmod 7$

Multiplication

$$25 \equiv 4 \pmod 7$$
$$\times\ 13 \equiv 6 \pmod 7$$
$$325 \equiv 24 \pmod 7 \equiv 3 \pmod 7$$

Check: $325 \div 7 = 46\ R\ 3$
or $325 \equiv 3 \pmod 7$

Subtraction

$$310 \equiv 2 \pmod 7 \qquad \text{since } 310 \div 7 = 44\ R\ 2$$
$$-\ 126 \equiv 0 \pmod 7 \qquad \text{since } 126 \div 7 = 18\ R\ 0$$
$$184 \equiv 2 \pmod 7$$

Check: $184 \div 7 = 26\ R\ 2$
or $184 \equiv 2 \pmod 7$

Division

$$\begin{array}{r} 21\ R\ 2 \\ 6\overline{)128} \end{array}$$

The dividend $128 \equiv 2 \pmod 7$
the divisor $6 \equiv 6 \pmod 7$
the quotient $21 \equiv 0 \pmod 7$
the remainder $2 \equiv 2 \pmod 7.$

Now we have the formula

$$\text{dividend} = [\text{divisor} \times \text{quotient}] + \text{remainder}.$$

Then in decimal notation we have

$$128 = [6 \times 21] + 2$$
$$= \quad 126 \quad + 2.$$

In modulo 7 congruences we have

$$2 \ (\text{mod } 7) \equiv [6 \ (\text{mod } 7) \times 0 \ (\text{mod } 7)] + 2 \ (\text{mod } 7)$$
$$\equiv \ 0 \ (\text{mod } 7) \qquad\qquad + 2 \ (\text{mod } 7).$$

Actually, modular computation as suggested in the checks could be performed *entirely* within the addition, multiplication, and subtraction tables for modulo 7 (e.g. in the addition check: 4 (mod 7) + 3 (mod 7) $\equiv$ 0 (mod 7); 0 (mod 7) + 2 (mod 7) $\equiv$ 2 (mod 7).

Over and above these checking activities, students who explore the basic notions of congruences might also be interested in testing for the existence or non-existence of associativity, commutativity, and distributivity.

Congruence concepts play an important role in the more sophisticated study of divisibility, prime numbers, and other areas of number theory and advanced algebra. Reference to any of a large number of standard works in these fields may be helpful to teachers wishing to expand their knowledge of the subject.

SHORT CUTS AND BYPATHS

FOSTER E. GROSSNICKLE

Nutley, New Jersey

JOHN RECKZEH

Jersey City State College
Jersey City, New Jersey

Short cuts in arithmetic when properly taught (that is, when taught in such a manner as to capitalize upon and illuminate the fundamental properties of our number system) provide the teacher with a most useful source of enrichment material; and, as Grossnickle and Reckzeh state, "constitute a vital part of the arithmetic program for the talented pupil."

The approximate grade level suitable for learning each type of short cut is indicated by a Roman numeral after each example, thus facilitating use of the specific types of short cuts presented in this article.

PREPARATION FOR LEARNING SHORT CUTS

The term *short cut* is descriptive of the concept. In arithmetic a short cut usually refers to a computational procedure which is either a quicker or a shorter way to solve an example than by applying the conventional algorism. To illustrate, divide 237 by 25. First, divide 237 by 100 by pointing off two places from the right in 237. Then multiply 2.37 by 4. The quotient of 237 divided by 25 is 4×2.37 or 9.48. This sequence of steps may be shortened as follows: multiply 237 by 4 and point off two decimal places in the product. A procedure of this kind requires less time for computation than the conventional way of dividing.

Before a short cut is used, the conventional way should have been learned. Several decades ago Ballard of the University of London described the conventional procedures or operations as the Kings Highway. This highway is so well marked that the traveler can easily find his way.

After the traveler becomes familiar with the route, he may discover detours and bypasses which will reduce travel time in numberland.

A pupil should understand and use the conventional algorism before he uses a short cut. This plan does not imply that a short cut is not to be attempted until a pupil understands all mathematical relationships involved in the operation. This understanding is desirable, but sometimes complete comprehension is not possible at the time a pupil could learn and profit from the use of certain computational short cuts. Learning to square a two-place number ending in 5 illustrates the point at issue. From observing a few examples, the pupil should be able to discover the pattern for squaring such numbers as 25, even though he does not have the needed background in algebra to understand the steps in the procedure. Therefore, there may be types of short cuts which the pupil could use effectively without complete mathematical understanding of these operations. However, the number of these applications should be limited.

In dealing with the program for the gifted pupil, most writers in the field of arithmetic encourage the use of computational short cuts. Many of the widely used short cuts are applications of the basic laws governing the structure of our number system. The pupil who discovers the application of a mathematical law by using a short cut shows deep insight into number patterns and relationships. Many of the short cuts should be applied mentally. The pupil who is able to perform mental computations with unseen numbers develops a valuable trait for success in mathematics. The gifted pupil should be encouraged to operate with unseen numbers.

CLASSIFICATION OF SHORT CUTS

There are so many different short cuts in arithmetic that a classification of various kinds seems advisable. It is not possible to discuss all the applications of each classification, but a representative sampling of different types will be given. One of the great difficulties encountered in selecting short cuts is to differentiate between a short cut and another method of solving an example. To illustrate consider the example: 16 $\times$ 48. The pupil may multiply the numbers in the conventional way. He may express the example as 16 $\times$ 16 $\times$ 3, which equals $16^2 \times 3$ or 256 $\times$ 3. The question arises whether either of the two factor methods constitutes a short-cut procedure. The factor methods of finding the product of two numbers demand a much higher level of mathematical understanding than finding the product of the two numbers in the conventional manner. However, it is debatable whether either factor method is as short a procedure as the regular algorism for multiplication. Therefore, at times it is difficult to make a distinction between computational procedures that are short cuts and those that demand greater insight into number than conventional algorisms.

Computational short cuts in arithmetic may be classified as representative of the following:

1. The Law of Grouping
2. The Law of One
3. The Distributive Law
4. Number patterns
5. Operations requiring a high degree of insight into number relationships.

The Roman numeral in parentheses after each illustrative example represents the approximate grade level at which the short cut may be presented. Due to great variation among grades, the grade level given should be interpreted to include at least one grade above or below the designated grade.

LAW OF GROUPING

According to the *law of grouping*, numbers may be rearranged or regrouped in addition or multiplication without affecting either the sum or the product. The law of grouping is a consequence of the commutative law and the associative law. The commutative law states that *the order of adding or multiplying two numbers does not affect the answer*. The associative law states that *the manner of grouping three numbers in either addition or multiplication does not affect the answer*. The law of grouping is derived from these two basic laws.

Examples

Some of the applications of the law of grouping and grade levels for presentation are as follows:

1. $2 + 3 + 6 = (2 + 3) + 6 = 2 + (3 + 6)$ (II, III)

2. $23 + 18 + 27 = (23 + 27) + 18$ (IV)

3. $36 + 52 + 64 + 28 = (36 + 64) + (52 + 28)$ (V)

4. $4 \times 17 \times 25 = (4 \times 25) \times 17$ (V)

5. $25 \times 3 \times 4 \times 7 = (25 \times 4) \times (3 \times 7)$ (V)

6. $5\frac{1}{3} \times 2\frac{1}{8} \times 6 = (5\frac{1}{3} \times 6) \times 2\frac{1}{8}.$ (VI)

The list can be extended to include decimals as well as common fractions and mixed numbers. The teacher should have the pupil look for groupings that will simplify the computation. It is important that the pupil should discover the law governing the operation or procedure.

LAW OF ONE

The law of one sometimes is known as the *identity property of one*. According to this law: the *product of a number multiplied by 1 is that number*. A familiar application of this law consists in changing a fraction to higher or lower terms, as $\frac{3}{4} = \frac{6}{8}$. Both terms of the fraction $\frac{3}{4}$ are multiplied by 2 to form the equal fraction $\frac{6}{8}$. Multiplying by $\frac{2}{2}$ is the same as multiplying by 1. The numerals, $\frac{2}{2}$, $2 \times \frac{1}{2}$, and $\frac{1}{2} \times 2$ are different expressions for the number 1.

Examples

Some of the short cuts based on the law of one are as follows:

1. Factors of 100 or 1000:

(a) $25 \times 36 = 4 \times 25 \times 36 \times \frac{1}{4}$ (V, VI)

(b) $33\frac{1}{3} \times 36 = 3 \times 33\frac{1}{3} \times 36 \times \frac{1}{3}$ (V, VI)

(c) $125 \times 72 = 8 \times 125 \times 72 \times \frac{1}{8}$ (VI)

(d) $333\frac{1}{3} \times 72 = 3 \times 333\frac{1}{3} \times 72 \times \frac{1}{3}$ (VI)

2. $5\frac{3}{8} \times 32 = (8 \times 5\frac{3}{8}) \times (32 \times \frac{1}{8})$ (VI)

The quotient of a number and one of its factors is a whole number.

Certain factors of 100 or 1000 are useful in computation. To multiply a number by 25, multiply that number by 100 and divide the product by 4. This statement is a short-cut application of the law of one. In the example, 25×36, first multiply 25 by 4 to get the product 100. Annex two zeros to 36 to multiply 36 by 100. To keep from changing the product of the example, divide 3600 by 4. Dividing by 4 is the same as multiplying by $\frac{1}{4}$. The symbolization of the application of the law of one in the example, 25×36, may be expressed as follows:

$$25 \times 36 = (4 \times 25 \times 36) \times \frac{1}{4}.$$

Division is the inverse of multiplication. Therefore, the sequence of steps to divide by 25 will be the inverse of the steps to multiply by that number. To divide 347 by 25, proceed as follows: divide 347 by 100, and multiply that quotient by 4. The representation of the steps in division in the example, $25/\overline{347}$, is as follows:

$$347 \div 25 = 347 \times \tfrac{1}{25} = 347 \times (\tfrac{1}{25} \times \tfrac{1}{4}) \times 4.$$

The plan for multiplying or dividing by 25 applies to all the applications of factors of numbers that are powers of 10.

The law of one may apply to multiplying a mixed number and a whole number, as $5\frac{3}{8} \times 32$. In this example, the 32 is a multiple of the denominator of the fraction $\frac{3}{8}$. An easy way to solve the example is to multiply $5\frac{3}{8}$ by 8. The product of a mixed number and the denominator of the fraction of the mixed number always will be a whole number. If one of the factors of the given example is multiplied by 8, the other factor must be divided by 8 to keep the product unchanged. Therefore,

$$5\tfrac{3}{8} \times 32 = (8 \times 5\tfrac{3}{8}) \times (32 \times \tfrac{1}{8}).$$

Multiplying one of two factors by n and dividing the other factor by n illustrates the law of one.

THE DISTRIBUTIVE LAW

The *distributive law* applies to the operations of multiplication and addition. To solve the example 3×42 or $3 \times (40 + 2)$, multiply both 40 and 2 by 3, then add the products. Some of the applications of the distributive law as applied to short cuts are as follows:

1. $19 \times 34 = (20 - 1) \times 34$ (V)

2. Special application of 15 as a multiplier   (V)

3. $4\frac{2}{7} \times 21 = (4 + \frac{2}{7}) \times 21$ (VI)

4. $\frac{1}{8} \times 19 + \frac{1}{8} \times 61 = \frac{1}{8}(19 + 61)$ (VI)

5. $158 \div 7 = \frac{1}{7} \times (140 + 18)$.   (V)

It is easy for the pupil to discover how to multiply by 9 or a number ending in 9. A very convenient short cut for the multiplier 15 is to multiply by $10 + 5$. The method to follow may be illustrated by the example, 15×86. Multiply 86 by 10 by annexing a zero to 86. Since 5 is half of 10, divide 860 by 2 and add that quotient to 860.

The solution of the example, $4\frac{2}{7} \times 21$, may illustrate either the law of one or the distributive law. The two solutions may be represented as follows:

$$4\tfrac{2}{7} \times 21 = (7 \times 4\tfrac{2}{7}) \times (21 \times \tfrac{1}{7}) \quad \text{(Law of One)}$$

$$4\tfrac{2}{7} \times 21 = (4 + \tfrac{2}{7}) \times 21 \quad \text{(Distributive Law)}.$$

The multiplication of a two-place number by a one-place number, as 2×34, is a familiar illustration of the distributive law. Similarly, division should be a familiar representation of this law. To divide a number which is not a multiple of the divisor, express the divisor as the sum of two or more addends. All but one of these addends should be a multiple of the divisor, as in the example $6/\overline{178}$, which may be written as $\frac{1}{6} \times (120 + 30 + 28)$. The quotient of $6/\overline{178}$ is equal to the sum of the indicated products, or $20 + 5 + 4\frac{2}{3}$. The same example could be expressed as $\frac{1}{6} \times (180 - 2)$ which is equal to $30 - \frac{1}{3}$.

NUMBER PATTERNS

A variety of short cuts in computation in arithmetic are based on number patterns. Some number patterns adapted for short cuts are given below.

Examples:

1. Finding the product of two unequal factors if each is equally removed from a multiple of a decade number, as in the example, 21×19 (VI)
2. Squaring a number ending in 5   (VI)
3. Finding the square of a number if the square of the preceding number is given   (VI)
4. Finding the product of two two-place factors in the same decade when the sum of the digits in ones' places of the factors is 10   (VII)
5. Same as in 4 above, but the sum of the digits in ones' place is not 10   (VIII)
6. Finding the product of two factors near 100, as 98×97.   (VII)

PROCEDURES FOR ABOVE EXAMPLES:

(1) The number pattern for finding the product of two factors of the type, 19×21, can be discovered from the table shown below.

$$10 \times 10 = 100$$

$$11 \times 9 = 99$$

$$12 \times 8 = 96$$

$$13 \times 7 = 91$$

The first product of unequal factors is 1 less than the product of the equal factors. The differences between the product of two equal factors and the product of two unequal factors are squares. Hence, if $n = 10$, then $n^2 = 100$ and $(n + 1)(n - 1) = n^2 - 1$. Therefore, $21 \times 19 = 20^2 - 1^2$. Similarly, $22 \times 18 = 20^2 - 2^2$.

(2) The pupil should experience little difficulty in discovering how to

square a two-place number ending in 5. The squares given below show that each square ends in 25.

$$15^2 = 225 = 200 + 25$$
$$25^2 = 625 = 600 + 25$$

The number of hundreds is equal to the product of the numerical value of the digit in tens' place and the numerical value of the tens' digit increased by 1. Thus, $200 = 10 \times 20$, and $600 = 20 \times 30$. The algebraic pattern for writing the sequence can be derived by squaring any two-place number ending in 5, as $(10t + 5)^2$.

$$(10t + 5)^2 = 100t^2 + 100t + 25 = 100t \, (t + 1) + 25$$

(3) The difference between the squares of consecutive whole numbers is equal to the sum of these whole numbers as shown by the table below.

$6^2 = 36$	11	$10 + 1$
$5^2 = 25$	9	$8 + 1$
$4^2 = 16$	7	$6 + 1$
$3^2 = 9$		

This difference also is 1 more than twice the smaller of the two numbers. Therefore, if the square of a number is given, as $13^2 = 169$, the square of 14 will be 27, $(13 + 14)$, more than 169. Also $14^2 = 13^2 + (2 \times 13 + 1)$. If n represents any number, the next whole number is $n + 1$. The difference between $(n + 1)^2 - n^2 = n^2 + 2n + 1 - n^2$ or $2n + 1$ which is the same as $n + n + 1$.

(4) Finding the product of two two-place factors in the same decade, when the sum of the digits in ones' place is 10, is similar to finding the square of a two-place number ending in 5. Thus, the product of $23 \times 27 = 621$. This product is equal to $(20 \times 30) + (3 \times 7)$. The algebraic representation of the steps in the solution can be shown in the example, $(10t + x) (10t + y)$ in which $x + y = 10$,

$$(10t + x) (10t + y) = 100t^2 + 10t \, (x + y) + xy$$

$$= 100t \, (t + 1) + xy.$$

(5) Two two-place factors in the same decade may have the sum of the digits in ones' place equal to, less than, or greater than 10. The method just described deals with these factors when the sum of the ones' digits is 10. A similar pattern is followed when the sum is not equal to 10. If the sum is not equal to 10, multiply the difference between

10 and this sum by the decade number. Then add or subtract this product to the product found by using the pattern when the sum of the digits in ones' place is 10. Thus, the product of $22 \times 27 = (20 \times 30) + (2 \times 7) - (1 \times 20)$. The product of $24 \times 28 = (20 \times 30) + (4 \times 8) + 2 \times 20$.

(6) If each of two factors is near 100, a similar pattern to that described above is used to find the product. Thus, $97 \times 98 = (100 - 3)(100 - 2) = 100^2 + (2 \times 3) - 100 \times 5$. In the same way the product of $104 \times 103 = 100^2 + (4 \times 3) + (100 \times 7)$.

OPERATIONS REQUIRING INSIGHT

Some operations demand either a high level of ability to work with unseen numbers or keen insight into the structure of our number system. A limited sampling of computational procedures of this kind is given below.

Examples:

1. Multiplying by 11   (V)
2. Using a two-place multiplier as a one-place multiplier   (VII)
3. Multiplying by doubling and halving.   (VII)

PROCEDURES FOR ABOVE EXAMPLES:

(1) A quick way to multiply by 11 is to write in ones' place in the product the digit in ones' place of the number multiplied. To find each succeeding digit in the product, add in order the digit on the right of the multiplicand to the next digit on the left. In the example shown, $11 \times 476 = 5236$, proceed as follows: Write 6. Add $6 + 7 = 13$; write 3 and carry 1. Add $1 + 7 + 4 = 12$; write 2 and carry 1. Add $1 + 4 = 5$; write 5. Adding numbers in the sequence described is the same as adding 10×476 to 476. Therefore, this method of multiplying by 11 also is an application of the distributive law.

(2) A pupil must be able to work with unseen numbers in order to treat most two-place multipliers as one-place numbers. In the example at the right, the thought pattern to follow to find the product is as follows: Find $4 \times 6 = 24$; write 4 and carry 2. Find $4 \times 3 = 12$; add $12 + 6 + 2 = 20$; write 0 and carry 2. Find $4 \times 4 = 16$; add $16 + 3 + 2 = 21$; write 1 and carry 2. Finally, add $4 + 2 = 6$; write 6. The product is 6104.

$$\begin{array}{r} 436 \\ \times 14 \\ \hline 6104 \end{array}$$

(3) The final illustration of an application of a short cut is shown at the right. In the example 21×34, the factors in A are divided by 2 but the remainders are disregarded. The factors in B are multiplied by 2. The product of 21 and 34 is equal to the sum of the products in B that are opposite odd numbers in A. Hence, $21 \times 34 = 34 + 136 + 544$, or 714.

A	B
21	34 $\checkmark$
10	68
5	136 $\checkmark$
2	272
1	544 $\checkmark$

The ancient method of finding the product of two factors as illustrated can be explained in either of two ways. First, the factor halved, or 21, is equal to the sum of certain powers of 2, as $2^4 + 2^2 + 2^1$, or $16 + 4 + 1 = 21$. The products in B that are checked represent 34 multiplied by the given powers of 2. Second, dividing 21 by 2 as shown in A is a means of expressing 21 in base two. Thus, 21 in base two is equal to 10101. The value of each 1, from left to right, in the numeral 10101 (base two) is $2^4 + 2^2 + 1$, or 21.

CONCLUSION

As previously indicated, the list of short cuts given is not intended to be an exhaustive list. The sampling given is adequate to help the teacher discover the kinds of short cuts that illustrate basic laws governing the structure of our number system. Again the teacher is directed to see that the discovery of the principle involved in a procedure is the important part of the learning program and not the short cut *per se*. Then short cuts should constitute a vital part of the arithmetic program for the talented pupil.

6

A METHOD OF
FRONT-END ARITHMETIC

ANDRÉ J. DE BÉTHUNE

Boston College
Chestnut Hill, Massachusetts

Front-end arithmetic, like the short cuts which appear in the previous article, requires the use of one or more of the fundamental laws of commutation, association and distribution, and the concepts of place value and regrouping. As such, front-end arithmetic is a worthwhile addition to the teacher's inventory of enrichment materials.

Following a careful reading of this article, the teacher will readily find application of the ideas contained in it to the topics usually taught at elementary grade levels, from grade 3 up.

Whereas the original article by de Béthune[1] contained applications to all four fundamental processes, the editorial committee limited this article to addition and multiplication. Subtraction, by front-end arithmetic, involves negative numbers; division is a front-end process.

A particularly challenging exercise for the talented child would be figuring out how to work subtraction by front-end arithmetic, and division by successive subtractions. The teacher can find help on these topics by reading the original article.[2]

ADVANTAGES OF FRONT-END ARITHMETIC

Front-end arithmetic is not anything really new. Many businessmen have resorted to front-end techniques, in one form or another, to get significant answers more quickly. The author's oilman regularly computes the bill for a fuel-oil delivery by a technique of front-end multiplication of his own devising.

Front-end arithmetic transfers computational fatigue from the most significant to the least significant columns. Furthermore, it can be made

[1] Originally appeared in *Arithmetic Teacher* 6: 23–29; February 1959.
[2] *Ibid.*

to eliminate carry-overs and their errors. And by laying out in plain sight the structure of the operation in its component parts, it makes errors and mistakes easier to ferret out and to correct.

The front-end methods described below represent techniques that the writer and his students have found convenient and practical. Other and better techniques may yet be devised. Those given here are offered as suggestions for an interesting variation to the conventional rear-end approach.

FRONT-END ADDITION

The numbers are lined up as usual in their proper vertical columns (Fig. 1). The first column is totalled and the subtotal, $3 + 6 + 9 = 18$, written in its proper place under the line. At this stage we already know that the final sum will exceed $180. The second column is summed, its subtotal, $7 + 3 + 7 + 8 = 25$, is written in its proper place under the line. The first two subtotals add up to $205, a second approximation to the final answer. The third column subtotal, 24, is again written in its proper place under the line. The first three subtotals give the partial sum $207.4, a third approximation to the final answer. The fourth column subtotal, 22, is again written in its proper place under the line. Added to the previous partial sum, it yields the final answer $207.62.

The four column subtotals as they appear under the line can also be added by rear-end addition at this stage to yield the sum $207.62. Or a second round of front-end subtotals can be obtained (1, 10, 7, 6, and 2) and written in their proper places under a second line (Fig. 1). This second round of column subtotals can then be added, front-end or rear-end, to yield once again the final answer $207.62.

This same addition, worked out from the rear-end, yields the following partial sums: after one column, 22 cents; after two columns, $2.62; after

$$
\begin{array}{r}
\$37.55 \\
63.86 \\
97.23 \\
8.98 \\
\hline
\end{array}
$$

$$
\left.\begin{array}{r}
18 \\
25. \\
2.4 \\
.22
\end{array}\right/
$$

$$
\left.\begin{array}{r}
1{-}7.62 \\
10
\end{array}\right)
$$

$$
\begin{array}{r}
\hline
207.62
\end{array}
$$

FIGURE 1

$275.33,\quad 62.18,\qquad 59.95,\ 177.33$

$$\text{xxx.xx}$$
$$\begin{pmatrix} 3 \\ 25 \\ 23. \\ 1.6 \\ .19 \end{pmatrix}$$
$$\overline{574.79}$$

FIGURE 2

$$\text{xxx.xx}$$
$$\begin{pmatrix} 3-1.6 \\ 25-.19 \\ 23. \end{pmatrix}$$
$$\overline{574.79}$$

FIGURE 3

three columns, $27.62, and then the final answer $207.62. In contrast, at comparable stages of progress, front-end arithmetic yields the partial sums: $180, $205, $207.40, and finally $207.62, a much more "convergent" approach to the desired result.

Front-end arithmetic can also be used to advantage in summing numbers that are not lined up vertically (Fig. 2). The numbers in the hundreds place are summed first, and their total, 3, written in the hundreds place. The tens add up to 25 and this number is written on the next line in its proper place. The partial sum at this stage is $550. The units add up to 23, this is written on the next line in its proper place, and the partial total is now $573. The tenths add up to 16, this is written in its proper place, and the partial total now is $574.6. Finally the hundredths add up to 19, this is written in its proper place, and the final total is $574.79. The column subtotals can also be added, front-end or rear-end, to yield once again $574.79.

Figure 3 illustrates a repetition of this sum by front-end arithmetic but with the column subtotals entered in such a way as to save space.

FRONT-END MULTIPLICATION WITH A GUIDE LINE

To find $3\frac{1}{4}\%$ of $96.95, i.e., to multiply it by 0.0325: Line up the two factors by their front digits (Fig. 4), and then draw a vertical guide line after the front digits. The decimal point is located by counting places right and left of the guide line, i.e., *one place to the right for the multiplicand plus two places to the left for the multiplier* locates the decimal point *one place to the left of the guide line for the product*. Start multiplying from the front end, writing the subproducts in their proper places: $3 \times 9 = 27$, $3 \times 6 = 18$, $3 \times 9 = 27$, $3 \times 5 = 15$, then $2 \times 9 = 18$, $2 \times 6 = 12$, etc., until every digit of the multiplicand and of the multiplier has been combined into a subproduct. To locate the proper place of a subproduct, count places *to the right of the guide line*, e.g. the subproduct $5 \times 6 = 30$ occurs with its last digit in the *third place to the*

```
        9 │ 6.95                              9 │ 6.95
×   0.0 3 │ 2 5                      ×   0.0 3 │ 2 5
   ───────┼────────                    ───────┼──────
    x.x   │ x xxxxx                     x.x   │ x x
    2.7   │                            2.7   │ 2 7
     .1   │ 8                           .1   │ 8 2
          │ 2 7                         .1   │ 8 2
          │   15                              │ 1 2
     .1   │ 8                                 │ 4 5
          │ 1 2                               │   3
          │   1 8                    ───────┼──────
          │     10                 ⎛ 2.9   │ 2 1 ⎞
          │   4 5                  ⎝  .2   │ 3   ⎠
          │     3 0                   ───────┼──────
          │       45                ⎛ 2.   │ 5 1 ⎞
          │       25                ⎝ 1.1   │     ⎠
   ───────┼────────                    ───────┼──────
 ⎛ 2.9   │ 1 9       ⎞                  3.1   │ 5 1
 ⎝  .2   │ 3 1 875   ⎠
   ───────┼────────
 ⎛ 2.    │ 4   875   ⎞
 ⎝ 1.1   │ 1 0       ⎠
   ───────┼────────
    3.1   │ 5 0 875
```

| FIGURE 4 | FIGURE 5 |

right of the guide line since *three* is the sum of *two places to the right* for the 5 *plus one place to the right* for the 6. The subproducts can be summed by either front-end or rear-end methods. The front-end method yields the column subtotals 2, 9, 23, 19, 18, 7, and 5. These have been written in Figure 4 on only two lines to save space. The next round of column subtotals, from the front end, yields 2, 11, 4, 10, 8, 7, and 5, written also on only two lines. The next set of subtotals finally yields the complete answer as $3.150875 without any discarded digits.

Figure 5 gives a repetition of the front-end multiplication of Figure 4, carried out so as to yield an answer to the nearest cent, i.e., the multiplication is cut off after the third (or tenths of a cent) decimal place, i.e., after two places beyond the guide line. Note that the initial subproduct, $3 \times 9 = 27$, is written immediately to the left of the guide line. The next subproduct, $3 \times 6 = 18$, is written one place to the right of the guide line. The next subproduct, $3 \times 9 = 27$, occurs two places to the right of the guide line, and may be written next to the first 27 to save space. The next subproduct, $3 \times 5 = 15$, is rounded off to a 2 in the third decimal place, and written next to the 18 to save space. Then the $2 \times 9 =$ a second 18 is written one place to the right of the guide line; the $2 \times 6 = 12$ occurs two places to the right of the guide line, then the second $2 \times 9 =$ a third 18 is rounded off to a 2 in the third decimal place and written next to the second 18 to save space. The next

```
      6 | 0.3
 ×    3 | 0.5
    --------------
   XX | x x x x
   18 |   9.
      | 3 0.15
    --------------
   18 | 3 9.1 5
```

FIGURE 6

```
      6 0.3
 ×    3 0.5
   -----------------
   x x x x.x x
   1 8    9.
        3 0.1 5
   -----------------
   1 8 3 9.1 5
```

FIGURE 7

subproduct, $2 \times 5 = 10$, falls entirely beyond the third decimal place and may be ignored. Next $5 \times 9 = 45$ is written two places to the right of the guide line, then $5 \times 6 = 30$ contributes a 3 to the third decimal place and all further subproducts may be ignored. The column subtotals, from the front end, are 2, 9, 23, and 21, and are written on two lines only, to save space. A second round of subtotals gives, from the front end, 2, 11, 5 and 1. The third round of subtotals gives answer 3.151, or to the nearest cent, $3.15.

The occurrence of *zeros* in a factor must be carefully watched. Zeros contribute no subproducts, but they do affect the location of subproducts in their proper places. By counting the proper number of places *to the right* of the guide line, subproducts can be easily located. In Figure 6 the decimal point is located *one plus one* or *two places* to the right of the guide line. The subproducts 18, 9, 30 and, 15 occur, respectively, to the left of, and two, two, and four places to the right of the guide line. With a guide line, the initial subproduct (e.g., $3 \times 6 = 18$ in Fig. 6) occurs always immediately to the left of the guide line. Further subproducts occur in the proper number of places to the right of the guide line, as indicated in Figures 6 and 7.

FRONT-END MULTIPLICATION WITHOUT A GUIDE LINE

A guide line is not strictly necessary, and subproducts can still be placed properly by reference to the decimal point only. Places to the right of the decimal point can be counted off in the usual way. But the calculator should be on his guard for the places of numbers to the left of the decimal point. Numbers immediately to the left of a decimal point should be always counted as being in the *zero-th place*. Numbers in the tens place should then be counted as *zero-one places* to the left of the decimal point; numbers in the hundreds place, as *zero-one-two places* to the left and so on.

Figure 7 illustrates this rule by repeating the front-end multiplication of Figure 6 without a guide line. The 6 and the 3 are both *zero-one places* to the left of the decimal point. *One plus one* make *two* and the subprod-

```
      2 | 0.3                        2 0.3
  ×   3 | 6.2                    ×   3 6.2
    x x | x x.x x                  x x x x.x x
      6 | 9.                       6   9.
      1 | 2 1.8                    1 2 1.8
        | 4.   6                       4.   6
    ────┼──────                   ──────────
      7 | 2   .8 6                 7 2   .8 6
      1 | 4.                       1 4.
    ────┼──────                   ──────────
      7 | 3 4.8 6                  7 3 4.8 6
     FIGURE 8                       FIGURE 9
```

uct 18 is written *zero-one-two places* to the left of the decimal point.
Next the subproduct 3 × 3 = 9 should be placed *zero-one to the left
plus one to the right* equals *zero places to the left* or immediately to the
left of the decimal point. Similarly the product 5 × 6 = 30 occurs *one
to the right* plus *zero-one to the left* equals *zero places to the left*. The final
subproduct 5 × 3 = 15 occurs *one to the right* plus *one to the right* equals
two places to the right of the decimal point.

Figures 8 and 9 illustrate another front-end multiplication carried
out both with, and without, a guide line. Subproducts are located by
reference to the guide line in Figure 8, and by reference to the decimal
point in Figure 9, by use of the *zero-one-two* . . . rule for places to the
left of the decimal point, and the *one-two* . . . rule for places to the right
of the decimal point.

The reason why two separate rules are needed in dealing with places
to the left and to the right of the decimal point has been admirably
discussed by Mueller, in his paper "The Neglected Role of the Decimal
Point" (*Arithmetic Teacher* 5: 87–89; March 1958). Mueller points out
that place numberings are symmetrical, not with respect to the decimal
point, but with respect to the units' place. If the *units' place* is given
place number zero, the *tens'* and *tenths' places* are then given the *place
number one*, to the left, and to the right, respectively. Similarly, the
hundreds' and *hundredths'* places are given the *place number two*, to the
left, and to the right, respectively; the *thousands'* and *thousandths' places*
are given the *place number three*, to the left and to the right, respectively.
The reader will readily correlate these place numbers with the powers
of ten in the sequence: . . . $1000 = 10^3$, $100 = 10^2$, $10 = 10^1$, $1 = 10^0$,
$0.1 = 10^{-1}$, $0.01 = 10^{-2}$, $0.001 = 10^{-3}$, ... The counting off of places
in multiplication is equivalent to the algebraic rule that $10^a \times 10^b = 10^{(a+b)}$.

With a guide line, however, the leading subproduct occurs immediately
to the left of the guide line. All other subproducts are located by use of

the *one-two* . . . rule for places to the right of the guide line. The decimal point is located by use of the *one-two* . . . rule for places, both left and right of the guide line.

SUMMARY

The methods of front-end arithmetic suggested here provide an interesting variation for, and supplement to, the classical rear-end approach. The rear-end methods are still those which are well established, universally understood, and most economical of space. To someone who is already a skilled calculator with rear-end arithmetic, front-end arithmetic offers a stimulating challenge in speed, in versatility, and in accuracy also. By eliminating carry-overs, and by laying out in plain sight the skeletal structure of the operation, front-end arithmetic makes it easier to ferret out mistakes—mistakes that occur frequently with multiple figures, with long sums, etc. These same mistakes often remain concealed in the rear-end method, and are much harder to dig out. Front-end arithmetic provides a swiftly convergent approach to the answer, it gives the operator a "feel" for his answer long before the answer is completely worked out. It puts off computational fatigue to the less important columns. On the whole, it provides the calculator with a challenging alternative path to the same final result and with an opportunity for reviving interest in what, to many, has become the deadly and dull routine of arithmetic. Finally, since all numerical work should always be checked for accuracy, front-end and rear-end methods provide alternative paths for the checking of results. Working with numbers has been a particular activity of man since earliest times. Front-end arithmetic provides us with new methods for carrying out a perennial human task, a task which the Lord already gave to Moses: "Take yee the summe. . ." (Numbers: 1; 2).

7

CONCEPTS OF MEASUREMENT*

J. HOUSTON BANKS

George Peabody College
Nashville, Tennessee

> In addition to his familiarity with the traditional kinds and uses of measures, the child is increasingly being made aware of the new kinds of measures used in the adult world. On television he watches the lift-off of a missile and hears references to measures and units of measure: "250,000 pounds of thrust," "Mach 3," and others that were not part of his world a few years ago.
>
> Increasingly, therefore, the elementary school teacher needs to direct effort toward developing both the principles of measurement and the applications of these principles to the physical world. The topic of measurement is one for which the teacher can find a great many supplementary books which can be used for individual and small-group work with talented children.

WHY TEACH MEASUREMENT?

In the strict sense of the word, the science of arithmetic is in no way dependent upon measurement or upon such allied topics as graphical and tabular representation of data. However, a surprisingly large proportion of both adult and child applications of arithmetic is related to measurement. Fifteen of the twenty-nine items on the check list of mathematical competence contained in the final report of the Commission on Post-War Plans[1] are definitely related to some aspect of measurement.

---

* By special permission of Allyn and Bacon, Boston: reprinted from *Learning and Teaching Arithmetic* by J. Houston Banks, 1959. 405 pp.

[1] Commission on Post-War Plans of the National Council of Teachers of Mathematics, a department of the National Education Association. "Guidance Report of the Commission on Post-War Plans." *Mathematics Teacher* 40: 315–39; July 1947.

Competence in the abstract arithmetical processes is necessary but not sufficient for gaining the arithmetic goals of social utility. Furthermore, situations from the child's experiences which involve aspects of measurement provide some of the best motivation for introduction of new arithmetical processes. We may properly think of measurement as representing a significant part of *applied* arithmetic. The two, application and theory, should proceed together. We cannot assume that the skills and concepts of measurement will be mastered incidentally.

NATURE OF MEASUREMENT

The most basic concept concerning measurement is the idea that a measurement is a comparison, a comparison of magnitudes. The rudiments of measurement are present in many of the child's earliest experiences. In such phrases as "this is too big," "a short time," "it fits just right," "real fast," "slow," "a long way," "too late," there is an implicit comparison with some reference unit. The referent to a small child is often inappropriate, at least from the adult standpoint. To a 6-year old, a 150-pound 40-year-old man is indeed "real big" and "real old." Parenthetically, even to most adults there are times when $5 looks like a lot of money. The appropriateness of such indefinite comparisons, wherein the reference unit is merely implied, is a relative thing. Its appropriateness depends upon the past experience of the observer as well as the particular situation. The speed of a propeller-driven airplane is slow if the implied referent is a supersonic jet. But compared to a man on a bicycle it is quite fast. The referent which the young child uses is of necessity limited by his experience. An ordinary apple would appear quite large to one who had never seen anything but crabapples.

Opportunities should be provided beginning in the earliest grades for development of more appropriate and more precise referents, including the standard units of measure. Definitions alone are of little value in developing the child's conception of the standard units such as pound, foot, minute, and mile. Even though one has a fairly precise conception of a foot, it is not too much help in visualizing a mile to know it contains 5280 feet.

Development of such concepts must be an outgrowth of experience. Concepts of *inch*, *foot*, and *yard* can be developed from use of the foot rule and yardstick. But something more than the measuring instrument is needed to help the child form concepts of *ounce*, *mile*, *acre*, and the like. An elementary school child will gain a better notion of an acre if he knows it is about the size of a football field, than if he knows it is $\frac{1}{640}$ of a square mile.

Nonstandard units such as: a pace, the surface covered by the hand, the time it takes to play a phonograph record, and a glassful can be used to advantage in developing such concepts as: distance, area, time, and capacity. They may also be used to advantage in demonstrating such properties of measurement as: (a) all measurements are comparisons, (b) no measurement is exact, (c) objects may be compared with each other indirectly by comparing each with a measuring unit, (d) the preciseness of a measurement, (e) the desirability of standard reference units.

UNITS OF MEASURE

In the past a great amount of time and effort was devoted in arithmetic instruction to memorizing innumerable tables of measures.[2] Modern arithmetic texts contain such tables in varying numbers. They are carried in an appendix and are presumed to be used for reference.

The question of memorizing tables of measures is somewhat similar to that of memorizing the basic addition and multiplication facts. We may never require the latter to be memorized in serial order. However, there are arguments in favor of it. In any event, the ultimate goal is automatic response. Ideally, this goal is reached through meaningful development of the facts coupled with sufficient familiarity through use.

A similar point of view may be adopted with regard to tables of measure. But there are important differences. We should not take the position that it is unnecessary to know any measure facts simply because they can be found in reference sources. One should not have to use a reference source to find that 12 inches equal 1 foot, 4 quarts equal 1 gallon, or 60 minutes equal 1 hour. On the other hand, it is hardly worth the effort for the child to commit to memory such facts as 4 inches equal 1 hand; 40 rods equal 1 furlong; or 1 barrel, standard, equals 105 dry quarts. By contrast, the multiplication fact 7×8 should be recognized on sight even though it may not be needed as frequently as 6×5.

A sensible goal is automatic recall of the most commonly used facts plus competence in the use of reference sources to find the less familiar ones. This poses the difficult question as to which facts belong in which category. In this respect it is impossible for the text to be an adequate guide. Some facts are so universally used as to require automatic recall regardless of geographical locality. Others may be in this category in some localities but not in others. Rural children have more need to know

[2] Spitzer, Herbert F. *The Teaching of Arithmetic*, 2d ed. Boston: Houghton Mifflin Co., 1954. p. 139.

that 4 pecks equal 1 bushel, or that 1 cord of wood equals 128 cubic feet, than do city children. The teacher's responsibility does not end with those scales of measure which are extensively used in all parts of the country. Grades of cotton staple are extremely important measures in cotton-producing areas, but they are seldom encountered in midwestern metropolitan areas. On the other hand, it is much more important for the midwesterner to know how many pounds there are in a bushel of wheat. The duration of a spring tide is more significant information to the sea coast dweller than to the mountaineer. The teacher must assume responsibility for inclusion of those facts which are significant for the particular area but are not included in the textbook.

The metric system of measures has many ardent advocates. There are many individuals who feel that our refusal in the United States to abandon our present system of weights and measures in favor of the metric system is not far short of a calamity. An entire professional yearbook has been devoted to descriptions of the metric system and discussions of its superiority.[3]

The advantages of the system come, in the main, from two sources. First of these is its decimal nature. We can convert from one unit to a different-size unit of the same kind by merely shifting the decimal point. The fact that ten is the base of the system is not important. The fact that it employs the same base as our system of notation is of extreme importance. The second characteristic is the manner in which linear measure, capacity, and weight are interrelated. There is no denying the fact that use of the metric system would greatly simplify computation. Common fractions would become obsolete fractions.

Other claims for the system are somewhat exaggerated. For example, as is well illustrated by American industry, it is possible to obtain just as high a degree of precision under our present system. The system also has disadvantages. Our anatomical predisposition to *ten* as a frame of reference for counting does not carry over to measurement. The human race has shown a decided preference for both *two* and *twelve*. Subdivisions into a half, half of a half, half of a half of a half ($\frac{1}{2}$, $\frac{1}{4}$, $\frac{1}{8}$), and so on, are easily visualized. *Twelve* is superior to *ten* in that we are as apt to be concerned with any one of the fractional parts, $\frac{1}{3}$, $\frac{1}{4}$, $\frac{1}{6}$, as we are with the part $\frac{1}{5}$.

Regardless of the merits of the argument, the teacher must take into account the situation as it is and as it is apt to become within a genera-

---

[3] The National Council of Teachers of Mathematics, *The Metric System of Weights and Measures*. Washington, D. C.: The Council, a department of the National Education Association, 1948.

tion. In the United States we operate under a dual system. The metric system is legal; in fact the units of the older system are legally defined in terms of metric units. We use both systems side by side. We use both centigrade and Fahrenheit temperature scales. We use both cubic centimeters and fluid ounces. Calories are probably more widely used and understood than B.T.U. With increasing frequency one hears reference to such units of measure as kilograms, liters, kilometers, and megacycles. Incidentally, we are not alone with regard to dual systems. Many countries which have adopted the metric system officially use it in foreign commerce but continue to use their native system in everyday affairs.

The situation is not apt to change materially in the foreseeable future. The proponents of the metric system have about the same chance of success in obtaining its adoption to the exclusion of other systems as the advocates of the duodecimal number system have in replacing our present base-ten notation. The arithmetic program certainly should not ignore the metric system, although an exhaustive study of it is hardly justified. A minimum program should include something of its decimal nature and familiarity with systematic use of prefixes to indicate multiples and submultiples of the basic units. It should also include knowledge of the approximate English equivalents of the most common metric units. For example, a yard is slightly less than a meter, a liter is between a liquid quart and a dry quart, a kilometer is a little over $\frac{1}{2}$ mile, and a kilogram is a little over 2 pounds.

There are other widely used units of measure which most children may never have occasion to put to direct use. But they are so ingrained in our culture the child should know what kind of unit they are and have some conception of their magnitude. Included in this category are a number of compound units such as foot-pound, light-year, passenger-mile, knot, acre-foot, and kilowatt-hour. Typical of other units which might be included are carat (weight of precious stones), radian and mil (angle), wire gauge, nail size, square (100 square feet of roofing), vision ratio, horsepower, volt, watt, ampere, and nautical mile. Although direct experience with most of these units is highly unlikely for most elementary school children, these units can be used to advantage in problem material.

MEASURING INSTRUMENTS

Children will be aided in learning many measurement concepts through carefully directed use of measuring instruments. For this reason, as well as the fact that it is a legitimate educational aim in its own right, children should be provided with the opportunity to learn their proper use. The simple task of reading a clock or watch must be learned. The great majority of first grade children have not learned it. There is much for the child to learn about the correct use of a yardstick. Apparently there are many adults who cannot read a simple device such as an electric meter. Some rural electric cooperatives do not read the customer's meter every month. The customer is asked to send the reading in. A facsimile of the face of the meter is printed on a card, and the customer is asked to draw in the hands in the correct position.

Any measuring instrument contains a smallest unit, a smallest graduation. This fact should be emphasized in developing the approximate nature of measurement. A child can make linear measurements to the nearest inch with a yardstick as soon as he can read the numbers. After the concept of fractional parts has been developed, he can use the same instrument with greater precision; he can measure to the nearest eighth of an inch. But in either case he is confronted with the necessity of estimating the nearest whole unit, be it an inch or an eighth of an inch. If he desires still closer results the instrument is inadequate. The approximate nature of measurement may also be emphasized by measuring the same object, using two different scales of measure. Suppose a distance is measured with both a yardstick and a meter stick. If we then convert the results to the same scale we will not get exactly the same number of yards or the same number of meters.

The existence of a limit of precision inherent in the instrument brings to focus the appropriateness of the unit of measure, as well as the instrument, we select. One would not attempt to weigh either a carload of coal or a precious stone on ordinary bathroom scales. The width of this page is most appropriately measured in inches, the dimensions of a room in feet, the distance between cities in miles, and the distance to a fixed star in light-years.

There is of course a limit to the measuring instruments which can be used in the classroom. But the child should be provided with practice in making measurements of length, time, temperature, capacity, angles, and weight. Improvised instruments and units of measure should be used in conjunction with the standard ones. An unmarked stick or a length of string can be used to make linear measures. This will help in bringing to focus both the nature of measurement and the arbitrary character of standard units of measure. When we measure, we compare the unit with the thing measured. We must count the whole number of times the unit, unmarked stick or length of string, is used and we must estimate the fraction of a unit needed to complete the measurement. We use a *standard* unit so that different measurements may be compared.

We measure the room at home for a new rug. We use a standard unit in order that the measurements may be duplicated in the store where the rug is bought. But how long is a yard? Why is it the same in Maine and in California? A yard is 3 feet, and a foot is 12 inches. But what is an inch? Such questions may be used to show that any scale of measure must have a basic unit. All other units in the scale may be defined as multiples or fractional parts of the basic unit. But how is the size of the basic unit determined? To the extent that time permits, the story of the development of measurement is excellent enrichment material. But we certainly should not fail to bring out the fact that the size of the basic unit in any scale of measurement is arbitrarily selected and established by law.

Weight and *mass* are not identical. The weight of an object is the measure of its gravitational pull, a quantity which varies with altitude and atmospheric conditions. Mass is a measure of the quantity of matter contained by the object. It is constant. The two measures are proportional and often are mentioned interchangeably. In fact, when we commonly refer to the weight of an object we actually mean its mass. The distinction between weight and mass need not concern the children.

The two types of scales in common use are the spring balance and the platform balance. The former measures weight, the latter mass. The platform balance should be used in developing the concepts of weight measurement and the standard units of weight. A workable balance can be constructed easily. All that is needed is an upright stand, a balance arm pivoted at its center with pans (small ash trays will do) suspended at its ends.

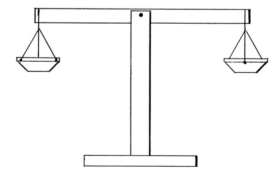

Children are familiar with the principle involved from their experience with the seesaw. When the seesaw is pivoted in the middle, the heavy child goes down. The children must be of approximately the same weight if the seesaw is to work (unless the distances from the pivot are altered).

When using the balance, checkers, buttons, loose counting beads, and bottle caps make good reference units. The object to be weighed is placed on one pan, and sufficient reference units placed on the other to counterbalance it. Here again we may emphasize the comparison aspect of measurement, we compare the pull of the object weighed with that of the reference units. We must use enough reference units to match the pull of the object. Merely to clarify the distinction between weight and mass which has been stated, we observe that the scale is in balance when each pan contains the same weight. But we measure mass because the constant mass of the reference units is known and equal weights have equal masses (at the same time and place).

Experiments with the balance also emphasize the measurement concepts of precision, error (inexactness), and indirect comparison. If the balance is sufficiently sensitive, that is free to move easily about its pivot, we find the need for reference units of different sizes. A tablet may not weigh 4 checkers, or 3 checkers and 1 button, but it may weigh 3 checkers, 1 button, and 2 bottle caps—thus, the notion of a least unit of measure or precision. In fact, we may not be able to bring the scale into perfect balance with the reference unit at hand—showing the presence of error. We can illustrate indirect comparison or substitution by weighing a quantity of modeling clay which balances the reference units, then showing that the clay and the tablet balance each other.

Experiences involving standard units of weight should also be provided. We may know the relationships between units—16 ounces = 1 pound, 2000 pounds = 1 ton—without having any conception of just how heavy a pound is. Children enjoy helping find heights and weights for their health records. Practice in estimating lengths, weights, and areas should be provided. The habit of estimating before measuring gives meaning to the results of measurement.

Time measurement does not end with ability to read a clock or watch. Young children's concept of time is hazy. Participation in planning the daily schedule—when we rest, when we play, how long we have for lunch, when it is time to go home—is helpful in sharpening the concept of an interval of time. The order of the days of the week and months of the year must be learned. In the lower grades, a valuable class project consists of making a large calendar for the bulletin board each month. Special school events and holidays for the month may be indicated, and the class may hold a discussion concerning the days and weeks before an event and between events.

In the middle grades, determination of age in years, months, and days should be included in the work with denominate numbers.

Standard time and time zones are topics which should be included in the study of the measurement of time. Although the reason for time zones is properly a part of geography or natural science, unless these topics are understood, arithmetic work in this area can only lead to confusion. Discussion of television programs, network and local, and study of a map showing time zones are helpful aids in fixing the ideas. Since the sun travels from east to west (or appears to) sunrise occurs in New York before it does in California. Then at any given time the clock indicates an earlier hour farther west. As we go west we advance our watches one hour each time we enter a new time zone.

For the upper grades a study of the history of the development of timepieces and the calendar make excellent enrichment materials. Such devices as a burning candle, three-minute timers of the hour glass type, and a second-pendulum are easily obtained. They may be used effectively as illustrations of methods which have been employed to measure the passage of time. In studying the history of the calendar, the interrelationships of days, weeks, months, and years may be emphasized. This type of activity might well include the need for calendar reform and a study of the currently proposed World Calendar.

INDIRECT MEASUREMENT

A surprisingly large number of our measurements are made indirectly. An indirect measurement is one which is obtained from some other direct measurement. The direct measurement may and may not be of the same kind. Linear measure is direct if the standard unit is applied directly to the object to be measured. Many linear measures are made indirectly from other direct linear measures. The solution of a triangle by numerical trigonometry is an example of indirect measurement. So also is the application of proportion to similar triangles as in shadow reckoning. The height of the house (Fig. 1) is obtained indirectly from the direct measurements of the height of the post and the length of the two shadows.

Measurement of areas and volumes are almost invariably indirect.

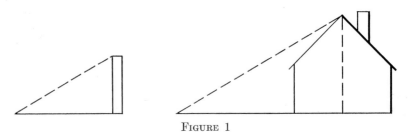

FIGURE 1

We make direct linear measures from which we obtain measures of area and volume. On the other hand, capacity is more often determined directly. If we wish to find how much a jar holds we can fill it with water, then measure the water in a measuring cup. This is direct measurement. If the jar is cylindrical we can measure its inner diameter and height, from which we can compute its capacity. This is indirect measurement.

The ease with which we can find area as an indirect measure constitutes a threat to meaningful learning. We measure two consecutive sides of a rectangle and multiply their lengths to obtain the area. "Feet times feet equal square feet." There is nothing in this procedure to suggest what a square foot is. The measurements were linear, why should not their product be linear also? As a matter of fact, if we multiply two line segments geometrically the product is a line segment, not an area.

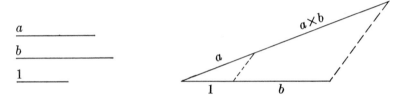

Before any area formulas are taught, the child should (a) have an opportunity to learn what "area" means, and (b) see *why* the formula gives the correct result.

How many pairs of hands are needed to cover the desk? How many tiles (if the floor is tiled) does it take to cover the floor? Experiences suggested by such questions will develop the idea of area as surface covered. We find area when we find how many units of area—hands, dominoes, tiles—are needed to cover the surface.

Preliminary to the development of the formula for the area of a rectangle, the selection of the unit of area should be considered. We cannot completely cover the desk top with hands unless there is overlapping. How many checkers does it take to cover the page? A circle cannot be our unit because circles will not fill the surface about a point. The class is asked to find how many different shapes the unit may take so as to fill the surface about a point. Triangles, rectangles, and squares probably will be suggested. These and the hexagon are the only possibilities. If we select from these the ones that will completely fill a rectangle we have only rectangles and squares. Further experimentation can show that a rectangular unit of area which will fill one rectangle may not fill a dissimilar one (Fig. 2).

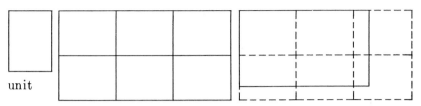

FIGURE 2

Of course, the unit square may not fill the rectangle either. But we can succeed in filling it with smaller squares. The incommensurable case

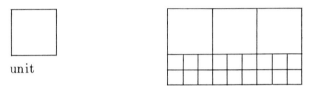

should be ignored in elementary school instruction. For that matter, it is not pertinent in the context of measurement. Since measurement is practically approximate we may with justification assert that any two lines are commensurable, for we can always find a length that will measure both lines to any desired degree of accuracy.

Having selected the square as our unit of measure, we find the area of a rectangle by finding the number of unit squares required to cover it. When we multiply the length by width we are multiplying the number

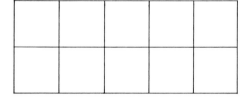

of squares per row (column) by the number of rows (columns). It is well to require drawing solutions early in the study of area. A drawing solution should continue to be required occasionally as a check.

There is much less need for the formula for the area of a parallelogram than for a triangle or circle. It should, however, be included as an aid in the meaningful development of other formulas.

A nonrectangular parallelogram cannot be filled with squares, no matter how small. In spite of this we can find its area by finding that of a rectangle which has the same amount of surface. If the formula for the area of a parallelogram is to have a meaningful basis, the concept

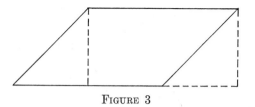

FIGURE 3

of congruent figures, as well as that of a unit of area, is essential. In Figure 3 the right-hand triangle has the same area as the left, even though neither can be filled with unit squares. This is so because if one is superimposed on the other they fit, they cover the same surface. It is advisable when deriving the area for the parallelogram to challenge the class to cut up a cardboard parallelogram and rearrange the parts in such a way that the area can be found.

Once the area of the parallelogram is accepted, that of the triangle follows from the fact that any triangle may be considered half a parallelogram with the same base and altitude. Here again we cannot demonstrate

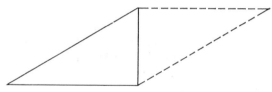

the formula by covering the triangle with unit squares. Some writers advocate the inductive approach. With this approach the class estimates the number of unit squares necessary to cover the triangle. This approach has serious limitations, though a reasonably good estimate can be made. There is no particular reason why this should be related to half the product of base times height. The procedure provides no clue to the desired generalization. On the other hand, if the child can be led to discover the relationship between the triangle and some other figure whose area can be found, then the generalization is easy.

The relationship between the radius and the circumference of a circle should be established before its area is considered. This will have to be done empirically. By actual measurement of circular objects we can obtain the ratio of circumference to diameter. A number of such observations should be made. The results will not be identical, but they will show that $\frac{22}{7}$ (or 3.14) is a good representative value. Care should be exercised to emphasize that the ratio is obtained from measurement, and hence our result is only approximately correct. This applies not only to the individual results but to the chosen representative value as well.

We may now show the area of the circle as the equivalent of the sum of the areas of many triangles, all of which have the radius of the circle as altitude and the sum of whose bases equals the circumference. Then

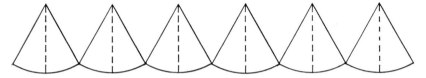

the area of the many triangles is found by the formula $\frac{1}{2} \times c \times r$, where c is the circumference and r is the radius of the circle. But since $c = \frac{22}{7} \times 2r$, we have for area $\frac{1}{2} \times \frac{22}{7} \times 2r \times r = \frac{22}{7}r^2$.

As the exceptionally observant pupil will note: no matter how many parts the circle is cut into, they never actually become triangles. Any kind of formal discussion of the limit process even in grade eight would be inappropriate for all but the most gifted pupils. The following is one way out of the dilemma presented. It is quite true that the circular sectors never become triangles. But neither is the ratio of circumference to diameter exactly $\frac{22}{7}$. After relatively few divisions the sectors so nearly approach triangles the error is well within the limits of precision imposed by our measurements. This approach is an evasion of the point at issue—the limit process. But it is in keeping with the fact that $\frac{22}{7}r^2$ is not an exact expression of the area. We have already evaded the issue when we accept $\frac{22}{7}$ as the value of π.

COMPUTATION WITH MEASUREMENT NUMBERS

Formal rules for computation with approximate numbers are seldom included in the elementary arithmetic curriculum. This does not relieve the elementary teacher of all responsibility relative to the matter.

The ability to round off numbers is an invaluable aid when one's work is checked by approximating the answer as well as when finding trial divisors. Children should learn how to round off whole numbers early in the arithmetic program. It is recommended that the rule for dropping a terminal 5 be that the last digit held is increased by 1 if the digit immediately to its right is 5, 6, 7, 8, or 9, including those cases when it is a terminal 5. The concept can easily be extended to fractions and mixed numbers when decimal fractions are studied. This does not imply that elementary school children should or must learn the meaning of significant digits. Rounding off will consist of rounding to a named digit. For example, 17,863 is to be rounded *to the nearest hundred* rather than to three significant digits.

If the concept of precision in measurement has been emphasized consistently when measurement is studied, rounding off in the manner indicated above should cause no great difficulty when extended to measurement. Measurements will be rounded to a named unit of measure, a named degree of precision. For example, we round 17,863 miles to the nearest hundred miles rather than to three significant digits. Similarly, 12 years 8 months 17 days can be rounded to the nearest month, as 12 years 9 months, or to the nearest year as 13 years. It is not unreasonable to expect seventh or eighth grade children to understand both the meaning of and the justification for the rule which requires the sum to have the same precision as the least precise addend. The safest and simplest procedure is to require each addend to be rounded to the precision of the least precise measurement, then add. This procedure will avoid "ragged decimals"; it is in keeping with the notion that only like things can be added, inches to inches, feet to feet, and so on; the sum automatically has its justifiable precision and it requires no rounding off.

Multiplication of approximate numbers poses a more difficult situation. Here we are concerned with accuracy rather than precision. The concept of significant digits is unavoidable. It is recommended that the teacher not attempt to develop these concepts. But the child should be given some kind of guidance and direction as to the proper way the result should be left. For example in posing the problem, the precision required of the product can be specified: Find the area of a rectangle 2.3 feet by 3.5 feet, *to the nearest square foot*, or *to the nearest tenth*. In dealing with measurement computation common sense should be the guiding rule. The child should be helped to determine the appropriateness of his answer. We do not know the area of the 2.3-by-3.5 foot rectangle is 8.05 square feet. The child unaided is not apt to see the inappropriateness of this answer.

RELATED TOPICS

Topics rather closely related to measurement include scale drawings, charts, graphs, and tables. A casual perusal of the daily newspaper is sufficient evidence of their importance from the standpoint of social utility.

Major instructional emphasis should be placed on reading and interpreting these items rather than on construction. This does not imply that no attention should be given to construction of scale drawings, graphs, and tables. If for no other reason, children should have experience in their construction because such experience brings out quite clearly the basic ideas necessary for intelligent use.

Since a scale drawing is a miniature, the basic ideas are the ratio of lengths and the preservation of shape. Children will have had experience with scale drawings from their study of maps in social science, where the "scale of miles" is given as a line equal a given number of miles. The ratio factor is sometimes given as, for example, "scale $\frac{1}{15}$." Still another means of indicating ratio is the substitution of units, such as "1 inch equals 1 mile." This is probably the best method for developing the ratio concept. We employ the idea that the small unit in a drawing "stands for" the large unit in an object.

Instruction in the use of scale drawings should begin with a situation in which the class has a real interest, in which a need for the drawing is apparent. For example, the class is to go to some member's home for a picnic. Discussion of how to find the place suggests the drawing of a map to scale. A picture of the route is drawn, one inch in the drawing takes the place of one mile (or block) on the route.

Practice should be provided in reading and interpreting floor plans, road maps, and diagrams such as drawings of a basketball court or baseball diamond. There are three essentials to interpretation of a scale drawing: (a) orientation as to direction, (b) awareness of the preservation of shape, and (c) knowledge of the ratio employed.

Substitution of units is also basic to the construction and reading of graphs. The substitution is usually of a more general kind. In a scale drawing a small *linear* unit is substituted for a larger *linear* unit. When a bar graph is used, the linear unit employed in determining the height of each bar may be used to represent either a count or any kind of unit of measure. When a pictograph is used, each picture represents a specified number of the pictured objects. Temperature is indicated on a continuous-line temperature chart by the height of the curve above the base line. Thus, a linear unit represents a given number of units of temperature.

The arithmetic program should provide competence in the construction and interpretation of bar graphs, circle graphs, pictographs, and line graphs. Graphs are used primarily to show comparisons, trends and fluctuations, and relationships. The best type of graph to use will depend upon the kind of information it is to convey. The bar graph and pictograph are used primarily to show comparisons. The circle graph is used to advantage in comparing the various parts with the whole. The line graph best shows trends, rise and fall, and rate of change of a changing quantity.

Two related sets of things are pictured in a graph. If the graph shows the wheat production of the nations of the world, the nations constitute

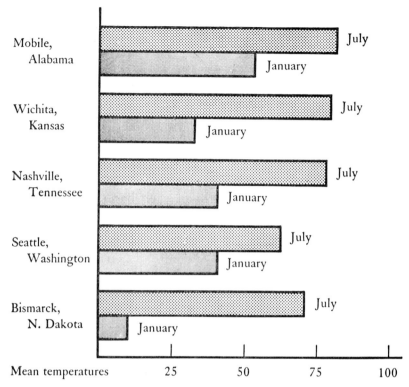

Comparative mean temperatures in selected cities. Unit of bar length is substituted for degrees of temperature.

FIGURE 4

one set and the quantity of wheat produced in each nation the other. On the other hand, we may have a graph showing the wheat production for the state of Kansas over a period of years. In this case one set of things is the years, and the other the quantities of wheat for each year. Wheat production by nations is best shown with a bar graph. But wheat production in Kansas should be shown with a bar graph if we wish to read from the graph a comparison of specific years. A line graph would be more meaningful if we wish to picture the manner in which production has changed with the passage of time.

One of the sets of things shown in the graph may be geographic locations, makes of automobiles, seasons of the year, or types of house construction—sets of things which may not be comparable numerically. Associated with each member of the first set, there is a member of another

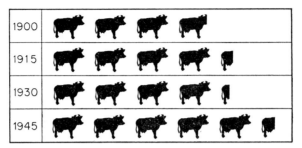

*Cattle production in the United States. Picture of cow is the
substitution unit. One cow represents 15,000,000 head.*

FIGURE 5

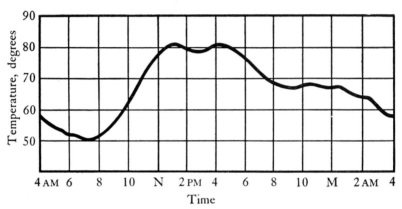

*Fluctuations in temperature. One horizontal unit is substituted for two hours of
time. One vertical unit is substituted for ten degrees of temperature.*

FIGURE 6

set of things which are numerically comparable. The graph gives a pic-
torial representation of the set of pairs of things. In a graph showing
wheat produced by nations, the pairs of things shown are the nation
and its wheat production. Each bar represents one nation-production
pair. The temperature chart (Fig. 6) pictures time-temperature pairs.
Each point on the line represents one such pair. Intelligent reading of a
graph requires that we correctly identify the nature of the pairs of things
represented and that we correctly identify the substitution unit or units
employed. When constructing a graph, care should be exercised to assure
that this information is provided.

A table gives essentially the same information as a graph. In a table the pairs are simply written down in some systematic fashion, frequently in two columns, the members of each set of things in a column, and paired by rows. Thus we may think of a graph as a picture of a table.

EXERCISES

1. List units of measure which are very important in some localities but unimportant over the country as a whole.
2. List strengths and weaknesses of a pictograph.
3. The article "Toward Better Graphs" by Edwin Eagle (*Mathematics Teacher* 35: 127–31; March 1942) lists a number of attributes a graph should possess. List those that should be emphasized in the study of graphs in the upper grades.
4. If a small picture of a house is used in a pictograph to represent 100,000 homes, why should we not represent 200,000 homes by a picture whose dimensions are twice those of the other?
5. In the article "When Size 9 Is Size 10" in *The Metric System of Weights and Measures* (see 7: 181–83)[4] reference is made to the fact that there is great lack of uniformity in clothing sizes. Is this a valid argument for the adoption of the metric system by the clothing industry?
6. List arguments both for and against the exclusive use of the metric system.
7. Contrast (a) scale drawings and graphs, and (b) graphs and tables.
8. Under what conditions will the arithmetic mean be misleading as a measure of central tendency of a set of measurements?
9. List the major measurement concepts which should be developed in the elementary school.
10. Describe the skills and concepts which a second grade child must use in making a linear measurement with a yardstick.
11. Describe the probable extent to which a beginning first grade child will have already developed the concepts basic to measurement.
12. Is a table of measures a table in the sense used in the last paragraph of this article? Explain.
13. Is the measurement of time by a watch direct or indirect measurement?
14. List instances of indirect measurement aside from those in the text.

---

[4] The first number gives the reference in the list at the end of this article, the second number gives the page.

15. Should a child be taught that feet times feet equal square feet? Is there any difference between this and the teaching of compound units, such as "feet times pounds equals foot-pounds?" How would you explain this difference to a seventh grade class?

16. Buckingham (see 2:457) states: "Accordingly, every measurement number is incommensurable or irrational." Explain what he means. Do you agree?

SUGGESTED SUPPLEMENTARY READINGS

1. BANKS, J. HOUSTON. *Elements of Mathematics.* Boston: Allyn and Bacon, 1956. pp. 189, 200, 249, 284.
2. BUCKINGHAM, B. R. *Elementary Arithmetic: Its Meaning and Practice.* Boston: Ginn and Co., 1947. pp. 456–735.
3. EAGLE, EDWIN. "Towards Better Graphs." *Mathematics Teacher* 35: 127–31; March 1942.
4. LARSEN, HAROLD D. *Arithmetic for Colleges.* New York: The Macmillan Co., 1950. pp. 162–85.
5. MORTON, R. L. *Teaching Arithmetic in the Elementary School.* New York: Silver Burdett Co., 1938. 2: 411–53.
6. NATIONAL COUNCIL OF TEACHERS OF MATHEMATICS. *Selected Topics in the Teaching of Mathematics.* Third Yearbook. Washington, D. C.: the Council, a department of the National Education Association, 1928. pp. 149–222. (Out of print)
7. NATIONAL COUNCIL OF TEACHERS OF MATHEMATICS. *The Metric System of Weights and Measures.* Twentieth Yearbook. Washington, D. C.: the Council, a department of the National Education Association, 1948. 303 pp.
8. SANFORD, VERA. *A Short History of Mathematics.* Boston: Houghton Mifflin Co., 1930. pp. 351–77.
9. SPENCER, PETER I., and BRYDEGAARD, MARGUERITE. *Building Mathematical Concepts in the Elementary School.* New York: Henry Holt and Co., 1951. pp. 256–96.
10. WHEAT, HARRY G. *How to Teach Arithmetic.* Evanston: Row, Peterson and Co., 1951. pp. 365–96.

8

PROBABILITY IN
THE ELEMENTARY SCHOOL

ROLLAND R. SMITH

Springfield, Massachusetts

Among the several branches of mathematics that have become increasingly important in this century, probability and statistics hold prominent places. Out of a wealth of experience in working with teachers and with children of all age levels, the author has selected that part of the field of probability that could be easily learned by the elementary school teacher and readily used in the classroom. The teacher who wishes to gain further knowledge about the mathematics of probability would do well to study "Probability" by David A. Page in the Twenty-Fourth Yearbook of the National Council of Teachers of Mathematics, *The Growth of Mathematical Ideas: Grades K-12*, 1959. pp. 229–71.

WHY TEACH PROBABILITY IN THE ELEMENTARY GRADES?

Probability in the elementary school! "Too abstract!" you say. Not if we understand what probability means at this stage. We have fractions in the first grade, but the work is nothing like what is done with fractions in the fifth and sixth grades. With this proviso, why should we not have probability in the elementary school? In this connection, remember the words of Brownell:[1]

As a matter of fact their influence (meaning the influence of the groups of mathematicians who have been doing so much in the secondary schools) could be too great if their activities should produce a sterile arithmetic. By this term, I mean arithmetic as the pure science of number, taught with little regard for the ability of children to learn it or for the social purposes to be served. This danger, however, is

---

[1] Brownell, William. "Arithmetic in 1970." *National Elementary Principal* 39: 42–45; October 1959.

unlikely to become a reality, for forces are available within the school to combat it.

We hope that this article may serve as the kind of influence Brownell is talking about.

What the Teacher Should Know

Of course, the teacher should know more about the mathematics of probability than he expects to teach. But even so the amount of knowledge needed is relatively slight. The elementary school teacher does not have to know what the mathematician needs to know about probability. One or two illustrations will help to show what is needed. If one of two dice (the singular of dice, by the way is die) is thrown at random, what are the chances (what is the probability) that a face with a 5 will turn up? There are six possible numbers that can show. Only *one* of these *six* possibilities is a 5. Since each face is equally likely to turn up, the probability is therefore 1 out of 6, or $\frac{1}{6}$. The fundamental idea is as simple as that.

Another illustration will be helpful. Suppose a bag contains four white and five black balls. If one ball is drawn at random, we have the question, "What is the probability that the ball drawn will be black?" Since there are nine balls in the bag, and it is just as likely that one ball may be drawn as another, any one of the nine balls may be drawn. Five of these nine balls could be black. Hence, the probability that the ball drawn will be black is 5 out of 9 or $\frac{5}{9}$.

In general terms we have this statement: If an event can happen in b ways, each way being equally likely, and a of these ways fulfill the conditions of success, then the probability of the event happening successfully is $\frac{a}{b}$.

Answer these questions.

1. From a pack of 52 cards, what is the probability that a particular card will be drawn? (The answer is $\frac{1}{52}$.)
2. Integers 1–10 are written on cards of the same size and shape, put into a bag, and shaken. If one card is drawn, what is the probability that it will be either a 2 or a 5? (The answer is $\frac{2}{10}$ or $\frac{1}{5}$.)

There are a few more things that the elementary school teacher needs to know about probability. Note the following.

If the probability of one event occurring is a, of another is b, then the probability of the two events together is ab not $a + b$. (One event must be independent of the other.) This concept, of course, can be expanded to three or more events. Then the probability of the combination is $abc. \ldots$

Example: What is the probability that I will get two fives in one throw of two dice?

SOLUTION: The probability of getting one five with one die is $\frac{1}{6}$. The probability of getting one five with the other die is $\frac{1}{6}$. Hence, the probability of getting two fives with two dice is $\frac{1}{6} \times \frac{1}{6}$ or $\frac{1}{36}$. It may help to look at it this way. If you throw one die, there are six possible events. For every one of these events there are six possibilities for the other, or thirty-six in all. In only one of these thirty-six possibilities will you get two fives. The answer is therefore 1 out of 36 or $\frac{1}{36}$.

Answer this question.
One bag contains 4 white balls and 5 black balls. Another bag contains 5 white balls and 6 black balls. One ball is taken at random from each bag. What is the probability that both balls will be white? (Answer: $\frac{4}{9} \times \frac{5}{11} = \frac{20}{99}$.)

There are many excellent treatments of probability. Some of them are intended for high school students, and should be simpler to understand than those written for mature mathematicians. The elementary school teacher who wishes to know more about the subject should study one or several of these treatments. She should know, expecially, how Pascal's triangle and the binomial theorem come into the picture, because she may be able to use this information in the elementary school.

Following are suggestions for exercises that' can be given in grades 4, 5, and 6. No attempt has been made to state which grade has been chosen for a given exercise. Any one of them can be adapted to any one of these grades. Much depends upon the background of the teacher and of the pupils.

TO INTRODUCE THE WORD *PROBABILITY*

1. The weatherman says, "The probability of rain is about 1 out of 10." He could have said $\frac{1}{10}$. If you had heard this report would you be likely to carry an umbrella or perhaps wear a raincoat? Can you explain what the report means?
2. Ann and Mary are both good spellers. In a spelling contest with Mary, Ann has a 50-50 chance of winning. Would you understand what we mean if we said, "The probability of Ann's winning is 1 out of 2 or $\frac{1}{2}$?"
3. Dan has three colored pencils on his desk: a red one, a blue one, and a brown one. If he picks up one pencil without looking, what is the probability (chance) that it will be either a red one or a blue one? (Answer: 2 out of 3 or $\frac{2}{3}$.)

TOSSING A COIN

Each pupil should be provided with a coin. He can place it in one hand, cup the other over it, and shake. If now he places the coin on the back of one hand, what is the probability that it will be a *head?*

Theoretically, it is just as likely to be a head as a tail. Out of two possibilities, a head or a tail, one toss *may* give you a head. We say therefore, that the probability is 1 out of 2 or $\frac{1}{2}$.

It would be interesting to see how nearly the theory works in an actual case. If the probability is $\frac{1}{2}$, you may think that you should have one head and one tail after two throws, five heads and five tails after ten throws, twenty-five heads and twenty-five tails after fifty throws, etc. In actual practice, this is not likely to be true, when the number of throws is small; but as the number of throws increases, you should get nearer to a 50-50 chance; that is, a probability of $\frac{1}{2}$.

Let each pupil carry out the experiment as many times as he wishes. A minimum should be 100 times. Record the results and see how nearly he comes to the theoretical probability of $\frac{1}{2}$.

One boy had these results:

After 10 throws	2H	8T
After 20 throws	9H	11T
After 30 throws	18H	12T
After 40 throws	23H	17T
After 50 throws	30H	20T
After 60 throws	32H	28T
After 70 throws	38H	32T
After 80 throws	41H	39T
After 90 throws	45H	45T
After 100 throws	49H	51T

You see that after 10 throws, the number of heads is not anywhere near $\frac{1}{2}$ of 10. After 90 throws the number of heads is exactly $\frac{1}{2}$ of 90. After 100 throws, the number of heads is close to $\frac{1}{2}$ of 100.

It will be interesting to see what the results are as the number of throws increases.

PROBABILITY 2/9

Have nine cards of the same size and shape. Write the integers 1–9 on them. Shuffle them, and place them on your desk with the number sides down. Pick up two cards at random. What is the probability that you will pick a 3 or a 5?

Think this way about it. In how many ways can you pick a card? (The answer is nine.) In how many ways can you be successful? (The

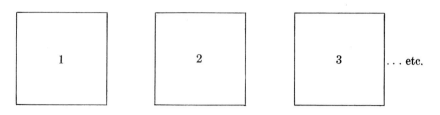

answer is two. If you pick up a 3 or a 5, you are successful.) The probability is then *2* out of 9.

It is interesting to see how near you come to this ratio when you carry out the procedure at least 20 times. Do it and record your results in a table as shown in Table I.

TABLE I

3	5	other
///	//	///// ///// /////

1. The ratio of successes to the total number of tries should be – – –.
2. How will you find the total number of successes?
3. How will you find the total number of tries?
4. Is the ratio near $\frac{2}{9}$?

You would expect to get a probability of $\frac{2}{9}$ if you had an infinite (unlimited) number of tries. The actual ratio will get nearer and nearer to $\frac{2}{9}$ as the number of tries grows larger and larger.

PASCAL'S TRIANGLE AND PROBABILITY

Pascal, a French mathematician who lived about 300 years ago, is well known for his study of probability. Figure 1 shows a triangular arrangement of numbers which he used in his study and which you will use in a very elementary way.

1. You see five rows of numbers. It can be carried on endlessly. The first row has 1 number. The second row has 2 numbers. The third row has 3 numbers. And so on. How many numbers would there be in the tenth row?
2. The first and last number in each row is _1_?
3. Do you see how the other numbers in each row are obtained? If you do not see, discuss the question with your classmates and your teacher.

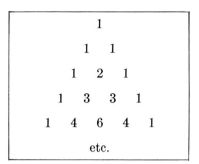

FIGURE 1

Now suppose that two coins are tossed instead of one as previously suggested. In how many ways can they fall? There may be two heads, two tails, a head and a tail, or a tail and a head. If we do not consider the order of a head and a tail, the possibilities look as follows:

1 HH,   2 HT,   1 TT

Note that these numbers 1, 2, 1 are shown in the third row of Pascal's triangle.

You can see that the probability of two heads or two tails is $\frac{1}{4}$ (one success out of four possibilities). The probability of a head and a tail is $\frac{2}{4}$ or $\frac{1}{2}$. The numerators 1, 2, 1 can be taken directly from the triangle. The denominator, 4, may be found by adding 1, 2, and 1.

Try tossing two coins and recording your results. See how many times you need to toss the coins before you begin to get near the ratios shown here. The HH and the TT should be about the same and the HT (not considering the order) should be twice as much.

Now we shall toss three coins. In how many ways can they fall? There can be three heads; three tails; two heads and one tail; a head, a tail, and a head; a head and two tails; a tail and two heads; a tail, a head, and a tail; or two tails and a head. There are eight possibilities in all. If we do not consider the order of a head and a tail, the possibilities look like this:

1   HHH
3   HHT
3   TTH
1   TTT

Once again, we find these numbers in Pascal's triangle. The fourth row has the numbers 1, 3, 3, 1. If you tossed four coins you would find the possibilities in the next row. They would be 1, 4, 6, 4, and 1. And so on.

The probability of three heads or three tails is $\frac{1}{8}$ (one success out of eight possibilities). The probability of two heads and a tail or two tails and a head is $\frac{3}{8}$. The numerators can be taken directly from the triangle. The denominator 8 can be found by adding 1, 3, 3, and 1.

Note that the sum of the four fractions $\frac{1}{8}$, $\frac{3}{8}$, $\frac{3}{8}$, and $\frac{1}{8}$ is 1.

Try tossing three coins to see how many times you need to toss them to get ratios near the ones shown here.

The *probability* suggested here is easy enough for good students in the fourth, fifth, and sixth grades to understand. The pupils will enjoy carrying out the experiments. It is hoped that these suggestions may bring out many more ideas.

9

GEOMETRY IN THE GRADES

IRVIN H. BRUNE

State College of Iowa
Cedar Falls, Iowa

If Arithmetic is, indeed, the "Queen of Mathematics," she receives a great deal of support from a number of "ladies-in-waiting." Most important among these for the school curriculum are algebra and geometry, and of these two, geometry will in all likelihood be the Queen's favorite. Algebra — *qua* algebra, employing an axiomatic approach — has no place in the elementary school program, but geometry is receiving increased recognition.

This article provides the elementary school teacher with a wide variety of activities which he will find useful with talented children, if not with all children. This article originally appeared in the May 1961 issue of the *Arithmetic Teacher* 8: 210–19.

PRACTICAL APPLICATIONS

The proper study of mankind is geometry. Tiny Timothy chose a nickel rather than a dime, but it was worldly wisdom, rather than geometric sense, that he lacked. Timothy's grandfather, however, who was shrewd in money matters, believed that three-foot cubes rather than three-foot spheres should adorn the new courthouse. He surmised that the cubes, covered only on five faces, would require less gold leaf to gild. His geometric guess was almost as naive as Timothy's money muddle.

Myra in grade 5 wanted to grow flowers. Her parents gave her 50 feet of aluminum edging and told her that she could have all the ground that the edging would enclose. What shape of flower bed would give her the most space?

At a sale Mr. Handyman bought a piece of linoleum that was 9 feet wide and 16 feet long. He knew that with only one cut he could fit it on a floor 12 feet x 12 feet. This, as well as the other cases cited, requires geometry.

Or consider the circles in Figure 1. Obviously the black circle on the right is larger than the black circle on the left. But is it?

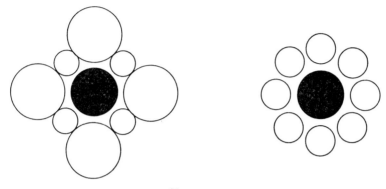

FIGURE 1

Similarly for Figure 2. Which has the greater area, the outer black annulus (or ring) or the inner black circle?

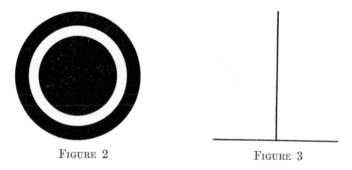

FIGURE 2   FIGURE 3

Another instance is the well-known Figure 3. Which segment, the horizontal or the vertical, is the longer?

Perhaps you think such examples are trivial. At least they are homely, ordinary. But there are more important applications too; you see them daily. When mathematicians use higher geometry to solve complicated problems in the production and distribution of goods, no one concerned thinks the matter trifling. Consider, for example, the problem of selecting the most economical combinations of some 20 ingredients that fluctuate in price daily. This problem arises in sausage making. Which quantities of which meats will satisfy fixed standards of high quality and at the same time cost the least according to today's prices? Here the mathematician employs geometry in an untrivial way.

AGELESS GEOMETRY

To enumerate and to describe man's uses of geometry would take a trip in time from prehistory to the present moment. The subject began in earth-measuring, it grew in planet-observing, it led the way in pure mathematics, and it pioneered in modern mathematics.

Man has always needed geometric principles, however dimly he may at first have perceived them. Similarly, children's lives cannot be devoid of geometry, however unaware they may be of its formal aspects. For, irrespective of its many applications and regardless of its value as a system of reasoning (and both of these phases merit attention), geometry embodies numerous ideas interesting in themselves.

Geometry for All

We suggest, therefore, that geometry deserves a lifetime of interest. To study it in only the tenth grade hardly suffices. At that level pupils presumably study one or more kinds of geometry as deduction. There and in subsequent courses they also learn about applications. But the computing with geometric formulas that frequently represents the only planned experience that pupils have in geometry prior to grade 10 seldom prepares them for grade 10.

Grade School Geometry

Informal geometry in the elementary grades can, therefore, counteract a serious deficiency. In these grades geometry is the study of form. Shapes, sizes, patterns, designs—these are the stuff from which children form concepts. From studying forms children discover numerous geometric relations; from making constructions pupils learn about geometric facts; from measuring figures learners acquire a background of geometric information. The work teems with both classical concepts and contemporary concepts.

Fun *and* Future

We believe, therefore, that children of all ages should get ample opportunities to find out things about geometry. The goal is satisfaction, here and now, with things mathematical, and geometry abounds in such ideas. An accompanying benefit will doubtless be a preparation for a more formalized geometry of proofs. Just as the ancient Egyptians' surveying and the early Chaldeans' star-studying opened a path for the deductions of the Greeks, so the block-arranging of the curious kindergartners and the design-drawing of the enthusiastic upper graders provide understandings for the problems of the older pupils. The pleasures of the moment outweigh the preparations for the future.

Therein lies the heart of the matter. Teachers cherish in their pupils such traits as alertness, preparedness, and willingness. And possibly the greatest of these is willingness. Seldom, though, do these traits develop overnight; rather, they seem to stem from many things that pupils do. Through the situations that teachers encourage them to explore, pupils discover relations, achieve insight, and gain satisfaction for the moment as well as for later studies. Mathematics, you know, is a cumulative subject. For example, clusters of dots, such as those in Figure 4, provide numerous helpful experiences. For infants the dots in Figure 4 are many, whereas those in Figure 5 are few. Later the dots represent nine. Then

FIGURE 4   FIGURE 5

they help with the ancient idea that some numbers are squares: that nine corresponds to three threes. Furthermore, in Figures 6a–e, the square arrays, one, four, nine, sixteen, twenty-five, etc., when ordered and compared via differences, encourage pupils to think about the odd numbers.

Square numbers:

———   1   4   9   16   25   36   49   . . .

Differences between successive square numbers:

———   3   5   7   9   11   13   . . .

Does zero belong in the blank before 1 in the top row?

For upper graders, finding a continuous path containing four line segments and connecting all the nine points will probably be a fascinating challenge. It might also be preparation for later work in simple topology, where, among other things, the study of paths—closed paths and not-closed paths—receives attention.

GEOMETRIC READINESS

Examples such as the foregoing abound. Suppose that we look briefly at the pupils' readiness before we consider further instances.

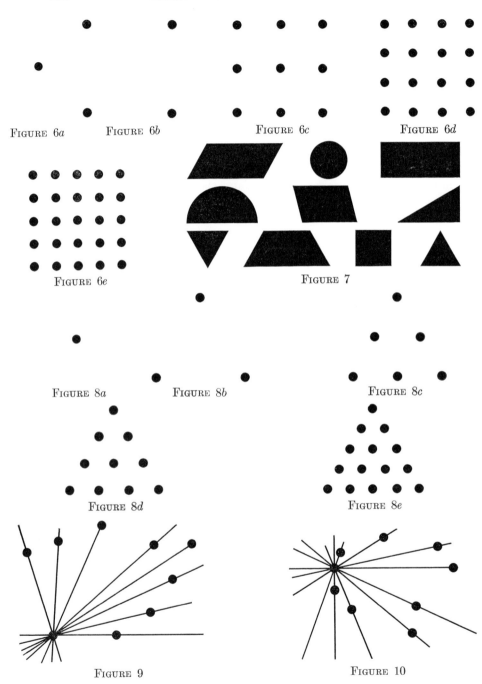

Figure 6a   Figure 6b   Figure 6c   Figure 6d

Figure 6e   Figure 7

Figure 8a   Figure 8b   Figure 8c

Figure 8d   Figure 8e

Figure 9   Figure 10

We have hinted that readiness in geometry implies at least two other "nesses":

1. Preparedness, or adequate mathematical maturity to go on.
2. Willingness, or enough emotional security to go on.

Geometric preparedness begins at an early age. Tots in kindergarten enjoy plopping the cutout figures into their proper places (Fig. 7). Children quickly discriminate between right triangles and equilateral triangles, between squares and oblongs, and between trapezoids and parallelograms. Already these youngsters are shape conscious.

Success in this sort of activity leads young children on. Their handling of squares, cubes, disks, triangles, spheres, and so on, prepares them for further work with forms.

All too frequently, however, such activities terminate abruptly. This occurs because courses of study encourage the pupils to put away "childish things" and settle down to the stern business of memorizing facts and practicing operations with numbers. Since *perfection* in these worthy matters eludes most learners, the study of facts and operations flourishes while the study of forms languishes. Of course, lessons in the upper grades deal with areas and volumes; but computing with numbers, and distinguishing between area and perimeter, and between volume and surface have been known to monopolize the act.

Fortunately, the trend today points to geometry for the sake of geometry, rather than to geometry as further practice in calculating. In the elementary grades informal, or intuitive, studies get the emphasis. Drawing, counting, and measuring lead pupils to observing, inferring, and generalizing. Consciousness of forms continues to grow, and readiness for proofs in geometry also continues to grow.

Let us return momentarily to the tots. By handling wooden, paper, or plastic representations of geometric figures, children appreciate numerous ideas; among these are notions of square corners, straight edges, round edges like faces, roundness of disks, and roundness of balls. These children gain a degree of understanding to go on; they grow in geometric readiness.

But children gain in willingness too. The shapes, the fitting of objects into patterns, the matching, the comparing, and the counting all make children receptive to further activities. One quite ordinary first grader happened onto triangular-number arrangements, as in Figures 8a–e. Pupils do things, learn, and crave to learn some more.

So, as they progress in mathematical maturity (preparedness), pupils tend to seek new mathematical worlds to conquer (willingness). Thus, willingness and also preparedness stem from activities—things done successfully. Junior is likely to wonder what the next step will lead to. If, for example, thirty-six lines can be drawn through nine points such that no three points lie on a line (Fig. 9), then how many lines can be drawn if exactly three of the nine are on one line? (Fig. 10.) In the figures, the joins of one point with each of the others constitute a hint; this is one way to begin the problems.

In sum, teachers seek to challenge pupils, not to frustrate. Sometimes a team approach (or working as a class on a perplexing problem) will prevent defeat. Whether pupils suffer more from frustration than from boredom, however, is moot.

SOME SIMPLE ARRANGEMENTS

Besides patterns of points previously mentioned, we might look at a few other configurations. In Figure 11 a side of the square $ABCD$ measures 3, AE measures 2, and angle DEF measures 90°. In square $KLMN$, a side measures 2, and angle LKP has the same measure as angle BEF. The sections thus cut form two separate squares or one combined square—a kind of readiness for the Theorem of Pythagoras.

In Figure 12 $ABCD$ is a square with a side that measures 6, and $BEFG$ is a square with side 3. $AL = BH = CJ = DK = 1\frac{1}{2}$, and BN measures 3. M is the intersection of HK and JL. Square $MNOP$ has the same measure in square units as squares $ABCD$ and $BEFG$ combined.

Now suppose that we begin with other segments, half-squares and half-oblongs, for example. These, plus a few rectangles, semicircles, and quadrants, form a variety of designs. Figures 13–17 illustrate some of them.

Abundant Triangles

If the pupils start with three equal segments and the question, "What can we do with the segments?", they can soon come up with an equilateral triangle (Fig. 18). By joining the mid-points of the sides of the triangle, they can produce four equilateral triangles (Fig. 19). By repeating the process successively for unshaded triangles, the pupils obtain Figures 20 and 21.

Or, the pupils may choose not to shade any of the component triangles and proceed to successive quartering of all the new triangles. One triangle yields four triangles, four yield sixteen, and sixteen yield sixty-four. Some pupils will move ahead and forge a fifth stage or even further proliferations of triangles. Theoretically, the fast workers need not grow weary of waiting for their slower classmates to finish a step. Endless steps await those who wish them, and the steps get harder.

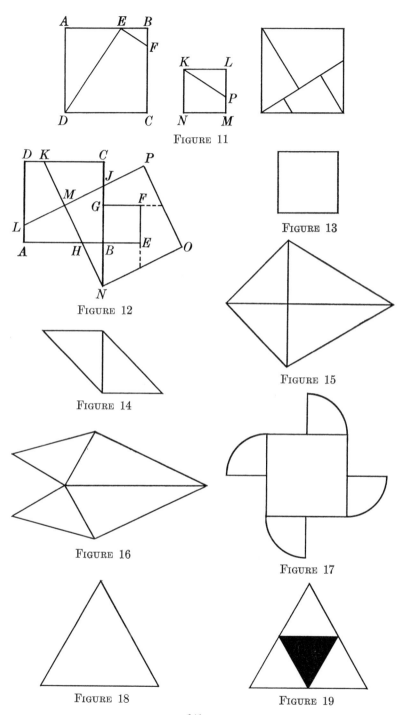

FIGURE 11

FIGURE 12

FIGURE 13

FIGURE 14

FIGURE 15

FIGURE 16

FIGURE 17

FIGURE 18

FIGURE 19

141

Some Pathological Curves

If the pupils begin again with three equal segments, they can form another equilateral triangle (Fig. 22). By trisecting the sides, erecting equilateral triangles on the middle sections, and erasing the intersections of these four triangles, the pupils get Figure 23. Then in Figure 24 further trisections and outward points appear. Some pupils may wish to carry this procedure still further. Although the area of this snowflake-like curve clearly cannot exceed the surface of the page, its perimeter becomes infinitely large.

Similarly for other pathologic curves (Figs. 25–27), the pupils proceed from an equilateral triangle again. Here, however, the open mid-sections are spanned by equal segments that intersect inside instead of outside the original triangle. This gives an inverted snowflake pattern. Here too, the perimeter can be made infinitely large, even though the area will not exceed that of the drawing paper.

Figures 28, 29, and 30 show what results when pupils begin with a circle, divide it into six equal parts, and invert alternate arcs. This procedure, repeated, leads to another figure, the aesthetics of which may be doubtful. It is known as an inside-outside curve. It troubles almost everyone who seeks to determine its curvature.

Tiling Patterns

As pupils soon learn when they begin to work with measures of angles, one full turn measures $360°$. A further interesting investigation results when pupils face the question, "What flat figures will fit around a point and fill in the flat surface?" Regular polygons seem to be needed, although all rectangles will suffice, but not all regular polygons will meet the requirement.

Considerable acquaintance with these polygons can result from such experimenting. How can we draw them? Later the pupils will study the straightedge-compass constructions for regular polygons, and still later they will study criteria of constructibility.

But strictly informal experimenting will reveal some combinations of polygons that, so to speak, cover the floor. Indeed, among sophisticates the whole subject of floor coverings, tilings, and mosaics bears the impressive name of "tessellation."

Six equilateral triangles (Fig. 31), four rectangles (Fig. 32), and three hexagons (Fig. 33) exhaust the possibilities of how many flat figures will fit around a point and fill in a flat surface. If however, the pupils do not limit the problem to polygons of one single sort, then the following serve: two hexagons and two triangles (Fig. 34); two octagons and one

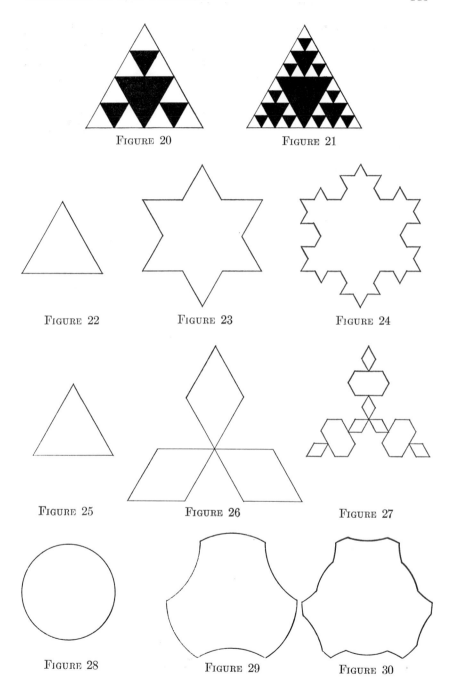

FIGURE 20

FIGURE 21

FIGURE 22

FIGURE 23

FIGURE 24

FIGURE 25

FIGURE 26

FIGURE 27

FIGURE 28

FIGURE 29

FIGURE 30

square (Fig. 35); three triangles and two squares (Fig. 36); one hexagon, two squares, and one triangle (Fig. 37). Still other possibilities, not illustrated here, exist: one hexagon and four triangles; one dodecagon, one hexagon, and one square; two dodecagons and one triangle. Your pupils may want to try them.

It will occur to the pupils that these are possibilities when they experiment and construct the following table, referring to regular polygons:

Number of sides:

| 3 | 4 | 5 | 6 | 7 | 8 | 9 | 10 | 11 | 12 |

Measure of angles:

| 60 | 90 | 108 | 120 | $128\frac{4}{7}$ | 135 | 140 | 144 | $147\frac{3}{11}$ | 150 |

From the increasing sizes of the angles, it appears that regular polygons having a still greater number of sides are not likely candidates.

Centroids

Locating a centroid, or a center of mass of a body, can become a thorny problem. Quite young children, on the other hand, can cut geometric forms from cardboard and readily locate lines and points of balance.

Regular figures balance rather easily on a knife-edge. The intersection of two such lines of balance determines a center of balance, or a center of mass. Equilateral triangles, squares, regular hexagons, regular octagons, and circles illustrate these ideas.

When the figures depart from regularity, the knife-edge procedure may become more difficult, as in the case of a non-convex polygon. There the centroid lies outside the figure (Fig. 38). This might become the germ of the idea of a centroid for a *system* of bodies, which may interest the future physicists and astronomers in your classes.

SIMPLE AND NONSIMPLE

Finding the centroid of an involved, yet technically simple, curve could provide a difficult problem. A cutout again provides an intuitive approach (Fig. 39). Incidentally, when is a curve simple? If the left circle were removed from Figure 39, the curve would remain simple. But if the right circle were removed, the figure would become nonsimple.

Figure 40 is another puzzler. It has two outsides, one of which might seem at first to be inside. When is an outside inside?

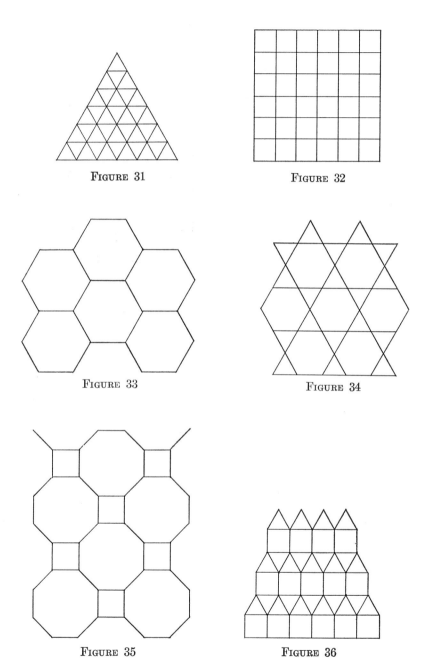

FIGURE 31

FIGURE 32

FIGURE 33

FIGURE 34

FIGURE 35

FIGURE 36

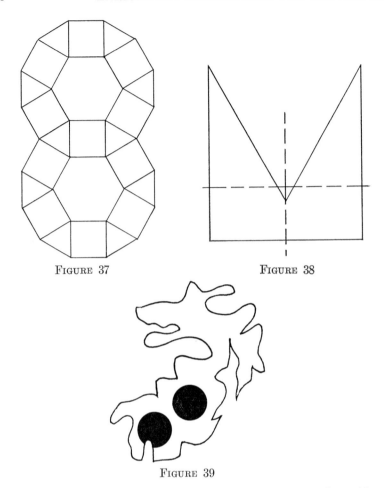

FIGURE 37   FIGURE 38

FIGURE 39

In case you wondered, here is one solution to the path problem, referred to earlier, for nine points arranged as three threes (Fig. 41).

YANG-YIN

Rooted in antiquity, especially in venerable Chinese philosophy, is the symbol represented in Figure 42. From *yang*, literally the south or sunny side of a hill, the unshaded portion of the design represents the bright, good, positive, male principle in Chinese dualism. *Yin*, on the contrary, symbolizes the dark, evil, negative, and female. The shaded part stands for *yin*. Each carries within itself the essence of the other.

Regardless of how pupils and their teachers look on such mystical matters, the emblem appears to be easy to construct and to describe geometrically.

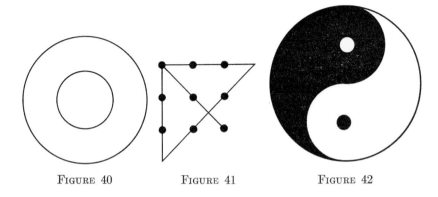

FIGURE 40   FIGURE 41   FIGURE 42

SUMMARY

Suppose, now, that we summarize. From the kindergarten on the concepts, rules, and operations of arithmetic and algebra dominate pupils' experiences in mathematics. Few deny it, and if the teaching has been good, still fewer bemoan it.

New occasions, however, teach new duties. Mathematics grows apace; mathematics education accelerates its search. With reason we urge teachers to learn new mathematics and teach new courses. We see merit in helping young children to gain acquaintance with good mathematics early. We note with pleasure the improvements in textbooks of elementary school mathematics.

In our zeal, however, we run the risk of letting abstractions get out of hand. This we disapprove. The motto "be abstract" should be left to the habitants of Greenwich Village. The race, including those of us who urge mathematics reforms, learned mathematics through practical needs and real problems. The abstractions, the generalizations, and the deductions followed the investigations, the approximations, and the corrections.

Surely, therefore, we should not deny young children the opportunity to explore and learn.

A proper study for all children is geometry—the geometry of form. Here pupils perceive, compare, measure, and generalize. Here they sharpen intuition without plunging too far into abstractions. Above all, children see values in what they do. If we can encourage pupils to discover for themselves some principles in the science of space, then they will bring into their geometry classes a usable store of information about the Euclidean plane. They might also have a good start on three-space.

10

TOPOLOGY

DONOVAN A. JOHNSON

University of Minnesota
Minneapolis, Minnesota

This article on topology provides the elementary school teacher with a background of subject matter and with a set of lesson plans for guiding the teaching and learning of the content. The interested elementary school teacher who studies this article carefully can provide his talented children with a delightful series of experiences. Within this chapter are listed several references for the teacher who wishes to go further in his study of topology. Also, the interested reader should refer to Albert A. Blank's article, "Nets," in the Twenty-Eighth Yearbook, Chapter 7.

WHAT IS TOPOLOGY?

Topology is a branch of mathematics that deals with lines, points, figures, and space. This sounds like geometry but topology is more than geometry. For example, topology deals with properties of position that are not affected by changes in size and shape. Suppose you draw a circle on a rubber sheet, with a dot inside the circle. No matter how you stretch the rubber sheet, the dot will always be on the inside of the "circle." In topology all of the drawings in Figure 1 below are the same, namely a *simple closed curve*.

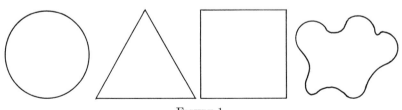

FIGURE 1

Measurement has no meaning in topology. A line 3-inches long can be made 5-inches long by stretching the sheet. Likewise a straight line not only changes in length, but may become a curved line by stretching the sheet. No matter how we stretch or bend the sheet, the path from A to B in Figure 2 remains a path from A to B which does not cross itself. And no matter how we twist or stretch the line from A to B, there are no holes in the line. This idea of no holes in a line sounds simple, but it is a very important idea in mathematics; and it seems sensible to assume that there are no holes in a line. In topology a line or path such as AB is called *arc AB*.

FIGURE 2

Differentiation of Figures

Since shape or size have no meaning in topology, we need a new way of describing how figures are different. The drawings in Figure 1 are examples of *simple closed curves*. A simple closed curve divides the sheet into two regions—an inside and an outside. No matter how you stretch the sheet, to go from the inside region to the outside region we must cross the arc forming the figure.

Not all closed curves are simple curves like those of Figure 1. Some curves have more than one inside. In topology, then, we classify a closed curve in terms of the number of inside regions each has.

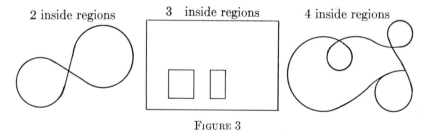

FIGURE 3

The exercises in Sample Lesson 1 at the conclusion of this chapter will introduce you to some of the beginning ideas and tricks of topology.

NETWORKS

Topology is supposed to have been originated by the mathematician Euler, who analyzed networks. A network is made up of arcs and points called vertices. The famous Koenigsberg bridge network is shown in Figure 4.

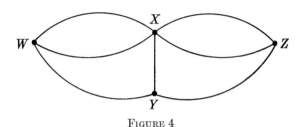

FIGURE 4

The bridge network is made up of the four vertices, W, X, Y, and Z. The number of arcs to vertex W is 3, so the vertex at W is called an *odd vertex*. A network is traveled or traced by passing over all the arcs exactly once. You may pass through the vertices any number of times. Euler discovered that there must be certain types of vertices in any closed network if it is to be traced or traveled in one journey without lifting your pencil from the paper or retracing any arc.

Sample Lesson 2 gives experiments to be tried for discovering network relations.

Network Formulas

Euler discovered the laws for tracing networks which you discovered by experimenting with the networks in Sample Lesson 2. Here are Euler's conclusions:

1. A closed network always has an even number of odd vertices.
2. A closed network of all even vertices can always be traced without traveling any arc twice.
3. If a closed network contains two and only two odd vertices, it can be traced without traveling any arc twice by starting at an odd vertex.
4. If a closed network contains more than two odd vertices, it cannot be traveled in one journey without retracing an arc.

Given a closed network, Euler found an important formula relating to the number of arcs (A), vertices (V), and regions (R) into which the network divides the plane. By analyzing the networks in Sample Lesson 3 you will find that $V + R - A = 2$.

An unsolved problem connected with networks relates to maps. A map is a closed network if the boundaries are considered to be arcs. Suppose we wish to color the countries of a map so that countries with a common border have different colors. How many different colors are needed? So far it has been possible to color in just four colors all maps that have been thought of. However, it has never been proven that four colors are enough for all maps. Try to draw a map that requires more than four colors under the conditions given above.

TOPOLOGY IN THREE DIMENSIONS

In our age of space travel we should, of course, consider how topology applies to three-dimensional objects such as spheres, cubes, cylinders, and doughnut-shaped objects.

In topology a sphere is something like a circle. You recall that a circle divided the surface of a sheet into an inside and an outside. To go from the inside to the outside you had to cut the boundary line. Similarly, a sphere divides space into two regions, an outside region and an inside region. To get from a point inside the sphere to a point outside the sphere, the path or arc must cut the boundary of the sphere. Again, we assume that there is no hole in the surface which is the boundary of the sphere, just as we said there is no hole in a line. Any closed surface which can be obtained by distorting a sphere without tearing or folding is a *simple closed surface*. A cube, football, or shoe box are examples of simple closed surfaces. In topology a box and a football are alike, namely, simple closed surfaces!

Euler's network formula also applies to three-dimensional objects called *polyhedrons*. A polyhedron, such as a box, is a solid made up of faces, edges, and corners. If the faces are considered regions (R); the edges, arcs (A); and the corners, vertices (V); we can test Euler's formula in the exercises on solids (Sample Lesson 4).

MOEBIUS STRIPS

One of the most famous surfaces in topology is called a *Moebius* strip. This is a surface which has only one side. All of the surfaces we know about, such as a sheet of paper, have two surfaces, a front and a back, or a top and a bottom. The Moebius strip has only one side.

You can make a Moebius strip with any strip of paper. Any size or type of paper will do, but gummed tape an inch or two wide and one or two feet long is easy to handle. We use the strip to make a ring or band. But before we glue the ends together, we give one end a half-twist. If you use gummed tape, twist one end so that you stick the gummed side of one end to the gummed side of the other end. Attach the band as illustrated in Figure 15, Sample Lesson 4.

If you draw a line on the surface of your Moebius strip, you will find that you will go all around the entire surface without crossing an edge. Paint or color one surface without going over an edge. Is there another surface that remains to be colored?

For another unusual result, cut the band lengthwise along a line in the middle of the strip. If you make another band, and cut it lengthwise one-third of the way in from an edge, such as at point B in Figure 15, you will get still a different result.

REFERENCES

For further information check these references.

JOHNSON, DONOVAN A., and GLENN, WILLIAM H. *Topology, The Rubber Sheet Geometry*. St. Louis: Webster Publishing Co., 1960. 40 pp.

KASNER, EDWARD, and NEWMAN, JAMES. *Mathematics and the Imagination*. New York: Simon and Schuster, 1940. 380 pp.

MESERVE, BRUCE E. "Topology for Secondary Schools." *Mathematics Teacher* 46: 465–74; November 1953.

SAMPLE LESSONS

Topology, Grade 5 or 6

SAMPLE LESSON 1

Aim

To build an understanding of lines and figures of topology.

Background and Materials

For the Teacher: A knowledge of lines and closed curves as defined in topology.

For the Pupils: A vocabulary of geometric terms such as point, line, circle, square, triangle.

Learning Activities

1. Which of the following figures are paths or arcs (topological lines) from point X to point Y?

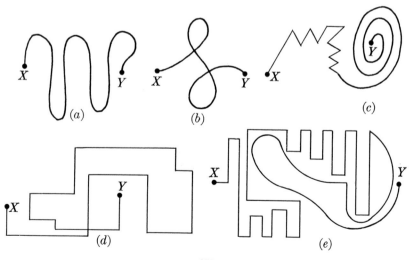

2. Which of the following figures are simple closed curves?

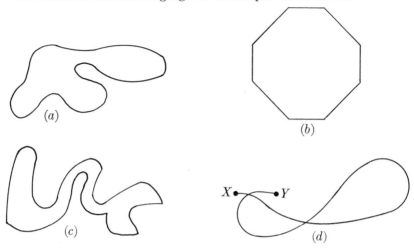

3. How many inside regions does each of these figures have?

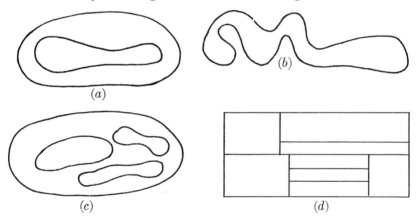

4. Figure 5 shows a simple closed curve: X is inside and Y is outside. Draw lines connecting X and Y to outside points A, B, C, and D. How does the number of crossings differ as you draw lines between points?

(a) From an inside point to an outside point the number of crossings of the closed curve is always an _____ number.

(b) From an outside point to an outside point the number of crossings of the closed curve is always an _____ number.

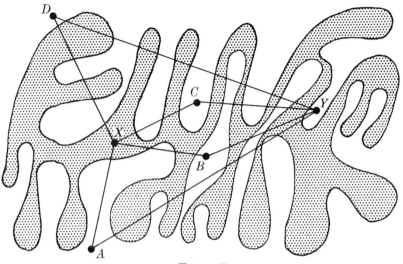

FIGURE 5

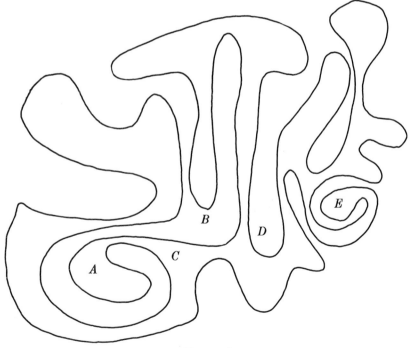

FIGURE 6

(c) From an inside point to an inside point the number of crossings of the closed curve is always an _____ number.

5. Use the facts you discovered in Exercise 4 to decide which of the points (Fig. 6) A, B, C, D, and E are inside this simple closed curve.

SAMPLE LESSON 2

Aim

To develop an understanding of the networks of topology.

Background and Materials

For the Teacher: A knowledge of the networks of topology and Euler's analysis of these networks.

For the Pupils: Readiness should consist of knowing what a closed network is and the parts of a network, namely: vertex, arc, even vertex, and odd vertex.

Development of the Lesson

1. Figure 7 shows a drawing of the islands and bridges of the German town of Koenigsberg where topology got started.
   Can you take a walk which will take you over each of the seven bridges and cross each bridge only once?

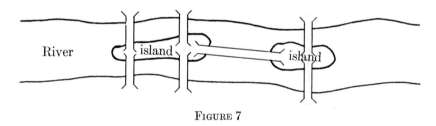

FIGURE 7

2. Figure 8 shows the floor plan of a house. Can you take a trip through every door of this house without passing through any door more than once?

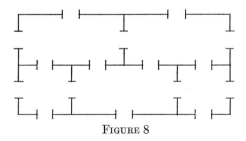

FIGURE 8

3. Discuss the tracing or traveling of the networks in Figure 9. Identify the arcs and vertices of each. Decide which vertices are odd or even vertices.

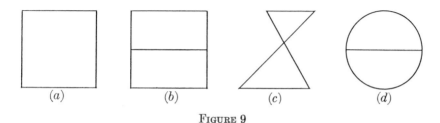

FIGURE 9

Can these networks be traveled in one journey without retracing an arc?

4. Which of the following networks (Fig. 10) can be traveled in one journey without retracing any line?

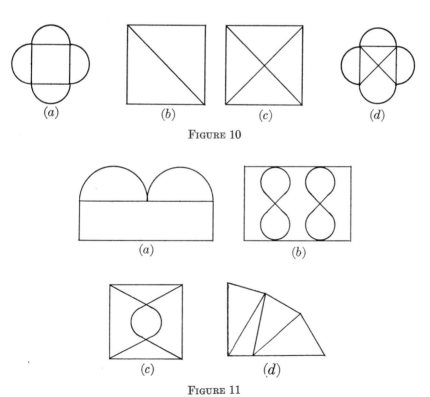

FIGURE 10

FIGURE 11

Encircle the vertices which are odd vertices in each figure. How many odd vertices does each network have? Have many even vertices does each network have?

5. Which of the networks in Figure 11 can be traveled in one journey without retracing any line?

Encircle the vertices which are odd vertices in each figure. How many odd vertices does each network have? How many even vertices does each network have?

6. Make a record of each of the networks in Examples 3, 4, and 5 in Table I below.

TABLE I

Network	Total Number of Vertices	Number of Even Vertices	Number of Odd Vertices	Can It Be Traveled Without Retracing an Arc?
3a	4	4	0	yes
3b				
3c				
3d				
4a				
4b				
4c				
4d				
5a				
5b				
5c				
5d				

7. From your results in Example 6, answer these questions:

(a) Can a network be traveled in one journey without retracing an arc if all the vertices are even?

(b) Can a network be traveled if it has more than two odd vertices?

(c) Can a network be traveled if it has only two odd vertices?

(d) Do you notice something special about the number of odd vertices (always even)? Can you see why?

Extending the Lesson

Explore a variety of more complex networks to test these conclusions.

SAMPLE LESSON 3

Aim

To discover Euler's formula for the closed networks of topology.

Background and Materials

For the Teacher: A knowledge of Euler's formula, $V + R = A + 2$, where A represents the number of arcs, R the number of regions, and V the number of vertices of a closed network.

References: Johnson and Glenn, *Topology, the Rubber Sheet Geometry.* St. Louis: Webster Publishing Co., 1960. 40 pp. Kasner and Newman, *Mathematics and the Imagination.* New York: Simon and Schuster, 1940. 380 pp.

For the Pupils: The pupil should know what a closed network is and the parts of a network, namely: vertex, arc, and region.

Development of the Lesson

1. Make sketches of networks such as shown in Figure 12, to become familiar with the terms vertices, arcs, and regions.

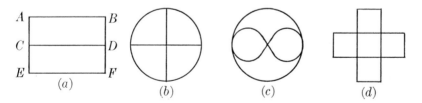

FIGURE 12

How many regions are formed by each network in Figure 12. Be sure to count the outside region as one region. Thus, network (a) has three regions. How many vertices does each network have? Network (a) has six vertices. How many arcs does each network have? Network (a) has seven, namely: AB, BD, DF, FE, EC, CA, and CD.

2. Count the arcs, vertices, and regions for the networks in Figure 13. Tabulate the results in Table II.

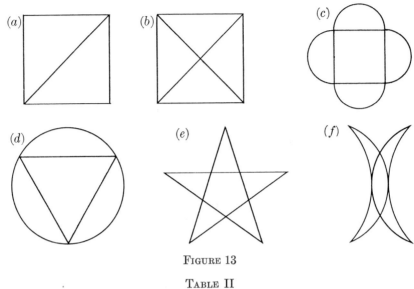

FIGURE 13

TABLE II

Network	Number of Vertices (V)	Number of Arcs (A)	Number of Regions (R)	V + R	A + 2
(a)					
(b)					
(c)					
(d)					
(e)					
(f)					

What formula expresses the relationship between vertices, arcs, and regions of these networks? Does this formula also work for the networks of Example 1?

3. Draw several other networks. Count the arcs, vertices, and regions. Check to see if the relationship discovered in Example 2 is true for other networks.

4. Draw a large circle. Divide this circle into regions with lines drawn across the circle. No three lines may pass through the same point. What is the largest number of regions formed for each number of lines? Complete Table III.

TABLE III

Number of Lines	Drawing	Number of Regions Inside Circle	Increase in Number of Regions
0		1	
			1
1		2	
			2
2		4	
3		7	
4		11	
5		16	

How do the differences in the number of regions change for each line added? How many regions would you predict for six lines? Seven lines? Eight lines?

Extending the Lesson

Consider the four-color mapping problem. Color the map in Figure 14 so that no states with common borders have the same color. What is the smallest number of colors you can use?

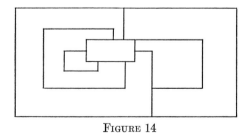

FIGURE 14

Try to draw a map that needs more than four colors but which shows all states with common borders having different colors.

SAMPLE LESSON 4

Aim

To develop an appreciation of the unusual aspects of topology.

Background and Materials

For the Teacher: A knowledge of Euler's formula and the Moebius strip.

References: Johnson and Glenn, *Topology, The Rubber Sheet Geometry.* St. Louis: Webster Publishing Co., 1960. 40 pp. Gardner, *Mathematical Puzzles and Diversions.* New York: Simon and Schuster.

Materials: Models of tetrahedron, cube, octahedron, dodecahedron, icosahedron, several large Moebius strips, patterns for polyhedrons.

For the Pupils: Experience with Euler's formula for plane figures and understanding of single closed curves.

Materials: Paper strips about 2 inches by 24 inches, cardboard, shears, crayon, cellophane tape, rubber cement.

Development of the Lesson

1. Have the pupils make Moebius strips. Cut two strips of paper about 2 inches wide and 24 inches long from wrapping paper. Use rubber cement to glue these strips together to discover the magic of Moebius strips. Glue the strips together with a half twist to form Moebius strips as shown in Figure 15.

   (a) Use a crayon to color one side of the Moebius band. How many sides does a Moebius strip have?

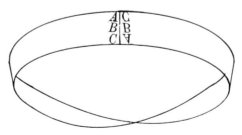

FIGURE 15

(b) Cut one band lengthwise along a line in the center of the strip. What is the result?

(c) Cut one band lengthwise along a line one-third of the way in from one edge of the band. What is the result?

2. Demonstrate the five regular polyhedrons, illustrating the meaning of edges, faces, and vertices. Emphasize the equality of the faces, edges, and polyhedral angles. Have the pupils trace the patterns of polyhedrons on cardboard. Cut the patterns out of cardboard and fold on the dotted lines. Fasten the edges with cellophane tape. Check to see if Euler's formula applies to these objects. For each polyhedron, count the number of faces, edges, and vertices and complete Table IV. See patterns on page 164.

TABLE IV

Name of Polyhedron	Number of Faces (F)	Number of Edges (E)	Number of Vertices (V)	V + F	E + 2
tetrahedron					
hexahedron					
octahedron					
dodecahedron					
icosahedron					

What formula describes the pattern of numbers in this table?

3. Which of these containers are simple closed surfaces?

(a) unopened can of fruit juice

(b) a basketball

(c) a sheet of paper

(d) a cup.

Extending the Lesson

Make a double Moebius strip by placing two strips of paper together. Give them a half twist as if they were one strip. Join their ends in the same way as you made the single Moebius strip. Run your finger between and around the bands until you return to the starting point. Are the strips separate? Make a mark on the bottom strip and travel the inside loop again. When you return to the starting point, is the mark on the bottom or top? Separate the strips to see if you have one or two bands.

Patterns of Polyhedrons: (Fold on dotted lines)

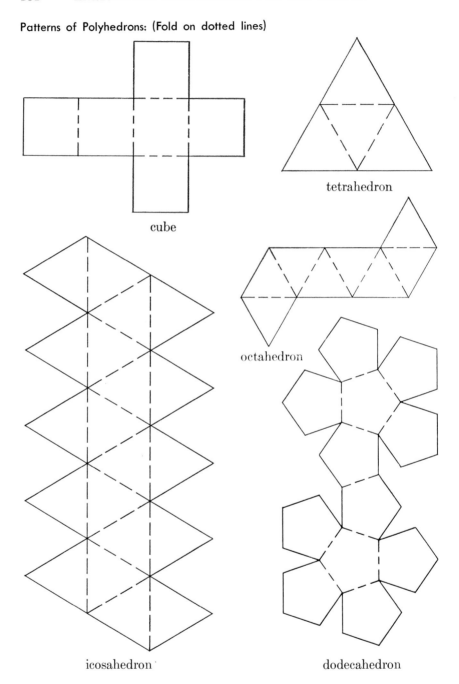

tetrahedron

cube

octahedron

icosahedron

dodecahedron

11

SOME SIMPLE
LAWS OF PHYSICS*

W. W. SAWYER

Wesleyan University
Middletown, Connecticut

Although there is some justification for devoting a fraction of the elementary school mathematics programs for the talented to a study of abstract patterns, there is overwhelming agreement among mathematicians and mathematics educators that the school mathematics program, elementary and secondary, is most significant when it is tied very directly to scientific applications and to other social situations.

This article provides the elementary school teacher with several applications of mathematics to problems in science. The teacher should familiarize himself with the problems, and obtain the needed laboratory materials. Other problems relating mathematics to science, industry, and business situations can be readily found.

THE PLATFORM ON ROLLERS

This experiment requires that you make a piece of apparatus. However, only junk is needed to make the apparatus. A study of the pictures (Fig. 1) will show how the junk is put together.

For the rollers you can use cotton spools or doweling. The platform and base are simply flat pieces of wood.

Put the rollers and the platform in the starting position as shown in the picture. Both pointers are opposite zero.

With your hand flat on the platform, push forward until the pointer

* By special permission of Wesleyan University, Middletown, Connecticut, publishers, as reprinted from the soft-bound booklet, *Math Patterns in Science*, by W. W. Sawyer. Available for purchase in quantities of ten or more from American Education Publications, Education Center, Columbus 16, Ohio.

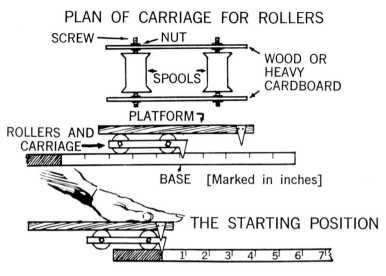

FIGURE 1

fastened to the rollers is opposite 1. Where is the other pointer? The other pointer shows how far the platform has gone.

Push the rollers another inch forward, and see where the platform is now. Continue in this way for several more inches and enter your results in Table I in the right-hand column.

As you push the platform forward, you find that it moves faster than the rollers.

In this experiment, we shall use r to stand for the number of inches the rollers have gone. We shall use p to stand for the number of inches the platform has gone.

What law or rule can you see in the numbers in Table I? Fill in the space in the sentence below. Just a number is needed.

TABLE I

Number of Inches Rollers Have Gone (r for short)	Number of Inches Platform Has Gone (p for short)
0	0
1	
2	
3	

The number of inches the platform goes is _____ times the number of inches the rollers go.

If you write the same number in the space below, you have this law in the shorthand of algebra.

$$p = \underline{\hspace{1cm}} r$$

THE PULLEY

This is another very simple experiment. Mark a chalkboard with numbers as shown in Figure 2. The divisions could be, say, 4 inches each.

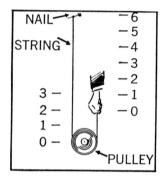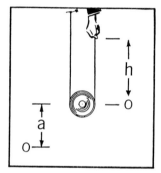

FIGURE 2

The actual length does not matter so long as the divisions are all the same.

If you do not have a pulley, you can use your thumb for the string to go around. Equally well you could use a pencil or a bottle—any small object that the string can slip around. You will have to use your left hand to keep the object you use for a pulley from falling.

As the hand pulls up on the string, both the hand and the pulley rise. But they do not rise through equal distances. Try it and see. Checking by the numbers on the chalkboard, fill in the spaces in Table II.

TABLE II

Number of Divisions the Pulley Has Risen (a for short)	Number of Divisions the Hand Has Risen (h for short)
0	0
1	
2	
3	

The rule is $h = \underline{\hspace{1cm}} a.$

Note: You may wonder why *a* was used to show how far the pulley went. The reason is this. In the question before, *p* was used for the distance the platform went. As "pulley" also begins with "P", we might have used *p* again in this question for the number of divisions the pulley has risen. Some readers might find it confusing to use the same letter with two different meanings so close together. We use *a* because the pulley can rise any number you like to mention.

HOMEMADE WEIGHING MACHINE

The support can be made in any way you like. Wood or a metal bracket can be used. The bar must turn freely about the rail.

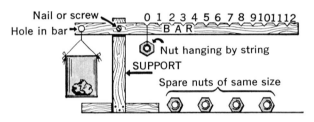

FIGURE 3

If you use a long bar, you can have more notches than the picture (Fig. 3) shows. A convenient spacing is to have the notches 1 inch apart and the hole in the bar 4 inches from the rail.

It is important that the can should hang from the bar by a single string as shown. There is a package of old nails, screws, bolts, etc., in the can. You must put just enough in this package to make the bar balance with the nut in notch 0. Then put one nut in the can. Find how many notches the hanging nut must slide along the bar to restore balance. Next, put

TABLE III

Number of Nuts in Can (*w* for short)	Number of the Notch Giving Balance (*n* for short)
0	0
1	
2	
3	
4	

another nut in the can, and so on. Of course all the nuts must be identical with the nut that hangs by a string. Fill in the spaces in Table III.

What is the rule? $n =$ _____ w.

THE BOUNCING BALL

I stood near a brick wall and dropped a new tennis ball. I used the thickness of the bricks as a measure. I noticed the height (h bricks) from which the ball was dropped, and the height (b bricks) to which it bounced. I did not expect to get very exact results. Probably I did not catch the ball when it was exactly at the top of its bounce.

Anyhow, here are the results I got.

h	b
0	0
4	$2\frac{1}{4}$
8	$4\frac{1}{4}$
9	5
12	$6\frac{1}{2}$
20	$10\frac{1}{2}$
28	14
30	15

The results for heights 0, 28, and 30 suggest that the bounce is just half the original height: $b = \frac{1}{2}h$. Look at the other numbers and see if you think that this rule is fairly near to the truth.

Nearly always, when we make measurements, errors creep in. Do not expect to find a simple law that fits all observations exactly.

See if you get results like these with a bouncing ball. You may get different laws with a new ball and with an old one. The hardness of the ground will also make a difference.

TWO NOTES ON ALGEBRA

NOTE 1. In arithmetic, we use the sign $\times$ for multiplication. In algebra, this sign is not convenient because it can easily be mistaken for x.

If you will look back through our work in algebra, you will find that *we have never used any multiplication sign at all.*

For example, we used $3x$ for "three times the number we thought of." We did not put any sign at all for *times.*

This was natural. We used "a bag" to picture "the number of things thought of." For three times this number, we drew a picture of three bags. We say "three bags"; we do not say "three times a bag." In our shorthand we do the same thing. We write $3x$. We do not write $3 \times x$, as a rule.

In algebra, when we write $5a$ or $2h$ or $7x$, with no sign at all between the number and the letter, remember that each of these represents multiplication. If a stands for any number, $5a$ stands for 5 times that number.

We have used this way of writing in our experiments above. The platform goes 2 *times* as far as the rollers; $p = 2r$. If the pulley rises any number of divisions, the hand rises twice as many; $h = 2a$. The number of the notch is 4 *times* the number of nuts being weighed; $n = 4w$.

Sometimes a dot is used to show multiplication. The formula $y = 10 \cdot x$, then, means y is 10 times x.

NOTE 2. *The simplest law of all*—it is a strange thing that some questions are hard to answer because they are too simple. In the guessing game, you may be able to write the laws $y = 2x$ and $y = x + 3$, and yet find difficulty with the following:

Number Called	Number Answered
x	y
5	5
2	2
7	7
3	3
11	————
4	————

You will be able to fill in the spaces above. The number answered is the *same as* the number called. But how shall we write this law in our shorthand? We write $y = x$.

THE LAW OF FLOATING

This law is used in the design and making of ships, rafts, pontoon bridges, and balloons.

Get some tin cans of different sizes. Fill each can with water, weigh it, and write down the weight.

Empty the cans and let them dry. Put old nails or scrap metal into one of the cans, and float the can in a bowl of water. Keep adding metal (without tipping the can) until the can is on the point of sinking, as shown in the picture.

Remove the can, dry it, and weigh it with the metal still in it.

Do the same for another can. Try the same experiment with several cans.

What do you learn about the two weights—the weight of the can full of water, and the weight of the can with enough metal to make it almost sink?

Code:

W = the number of ounces the can full of water weighs.

M = the number of ounces the can with its load of metal weighs.

Which of the following laws agrees best with the results of your experiments? (Of course, you expect some errors of measurement.) None of these laws will fit your results exactly.

(1) $W = M + 5$ (2) $W = 2M$ (3) $M = 2W$
(4) $M = W$ (5) $W = 2M + 3$

The law seems to be _____.

The law you have just discovered here is known as the Principle of Archimedes. Archimedes (är′kĭ-mē′dēz) was a famous Greek mathematician who lived more than 2,000 years ago.

POWER FOR AIRPLANE ENGINES

The following information about airplane engines was taken from the *Encyclopaedia Britannica*. It tells us the weight (in thousands of pounds) that various airplanes can carry, and it also tells us the power of the engine in each plane (in hundreds of horsepower). We should use P for the power and W for the weight.

Airplane	Power P in Hundreds of Horsepower	Total Weight W in Thousands of Pounds
DC-3	24	24
DC-6	84	95
Convair 240	42	40
Grumman G73	12	$12\frac{1}{2}$
Beech Model k85	9	9
Lockheed Constellation	88	94
Boeing Stratocruiser	140	$142\frac{1}{2}$

There is no simple law that connects these numbers *exactly*, but they do come quite close to a very simple law. This law tells you about how much power to provide for an airplane of known weight.

Which of the laws below do you think is most suitable for this purpose?

(1) $P = 2W$ (2) $W = 2P$ (3) $P = W - 10$

(4) $P = W + 10$ (5) $P = W$ (6) $P = 2W + 5$

The most suitable law is _____.

If you were designing an airplane similar to one of the types above, and this plane was to weigh 70,000 pounds fully loaded, about how much power would you expect it to require? _____

_____.

12

TRICKS AND
WHY THEY WORK*

W. W. SAWYER

Wesleyan University
Middletown, Connecticut

One of the most recent uses of mathematics is its application to games and puzzles. This article provides for the talented elementary school child a simple introduction to the use of mathematics for "explaining" why a trick works. The teacher who is successful with the following material may wish to extend it further: by using tricks with more steps, by providing a set of "shorthand" expressions and requiring the child to write the "word" expression, or by having the child make up his own "tricks" for use with the other children in the class.

A CLASSIC TRICK

Probably you have often heard tricks like this one:

Think of a number.
Add three to it.
Double.
Take away four.
Halve.
Take away the number you first thought of.

Whatever number you think of at the start of the trick, provided you made no mistakes, you will end up with the answer: 1.

Why does this work? We can see by thinking in pictures. When you have thought of your numbers, suppose you put that many stones in a bag. I can see the bag, but I do not know how many stones are inside it.

When I say, "Add three," you put three more stones next to the bag.

---

* By special permission of American Education Publications, Columbus, Ohio: Reprinted from *Math Patterns in Science* by W. W. Sawyer, 1960. 32 pp.

I can now see a bag and three stones. I tell you, "Double." You bring up exactly the same things again. I now see two bags and six stones. "Take away four." You remove four stones. I see two bags and two stones. "Halve." You do so. I now see one bag and one stone. "Take away the number you first thought of." The number you first thought of was the number of stones in the bag. So you remove the bag. That leaves just one stone. The final answer of this trick is always 1—regardless of the number you may have chosen.

Think of a number	🛍	
Add 3	🛍	0 0 0
Double	🛍	0 0 0
	🛍	0 0 0
Take away 4	🛍	0
	🛍	0
Halve	🛍	0
Take away the number you first thought of		0

In C1 through C4 below, some are tricks and some are not. In the tricks you can tell a person the answer, because whatever number he thinks of, he will always get the same answer. In the others, this does not happen.

Can you find out, by drawing bags, which are the tricks? It is easy to make mistakes, so test your answers; think of numbers and see whether the same answer always comes.

C1. Think of a number   _5_ B
    Add 5   _10_ B+5
    Multiply by 3   _30_ 3B+15
    Subtract 9   _21_ 3B+6
    Divide by 3   _7_ B+2
    Take away the number you first thought of   _2_ 2

C3. Think of a number   _5_ B
    Add 1   _6_ B+1
    Double   _12_ 2B+2
    Add 10   _22_ 2B+
    Divide by 4   _5½_ ½B +
    Take away the number you first thought of   _½_ No

C2. Think of a number   _5_ *B*   C4. Think of a number   _5_ *B*
     Add 3   _8_ *B+B*   Add 3   _8_ *B+3*
     Double   _16_ *2B+6*   Double   _16_ *2B+6*
     Subtract 2   _14_ *2B-4*   Add the number you
     Take away the num-   *B+4*   first thought of   _21_ *3B+6*
     ber you first thought of _11_ *No*   Divide by 3   _7_ *B+2*
                                                         Take away the num-
                                                         ber you first thought of _2_ *Yes*

MAKING UP YOUR OWN TRICKS

You will see that there are very many tricks of this kind. How would you make one up for yourself? Before you read further, you may like to see if you can invent your own trick.

What makes a trick of this kind work?

Suppose you say to a friend, "Think of a number. Add 2." You certainly cannot tell him the answer. If he thought of 5, his answer would be 7. If he thought of 10, his answer would be 12. If he thought of 1,000,000, his answer would be 1,000,002. You have no way of telling what his answer is.

In pictures it would look like this:

Think of a number	🛍️
Add 2	🛍️ 0 0

You do not know how many stones are in the bag. That is why you cannot tell the answer. To make a trick, you have to arrange things so that no bag is left at the end.

Suppose you say, "Think of a number. Add two. Take away the number you first thought of." The first two steps are shown in the picture above.

In the last step—"Take away the number you first thought of"— you remove the bag. Two stones are left and the answer is always 2. This is a trick, but a rather feeble one. Many people would see right through it, and would not be surprised at all. To surprise your friend, you have to make the trick a bit more complicated to hide what you are really doing.

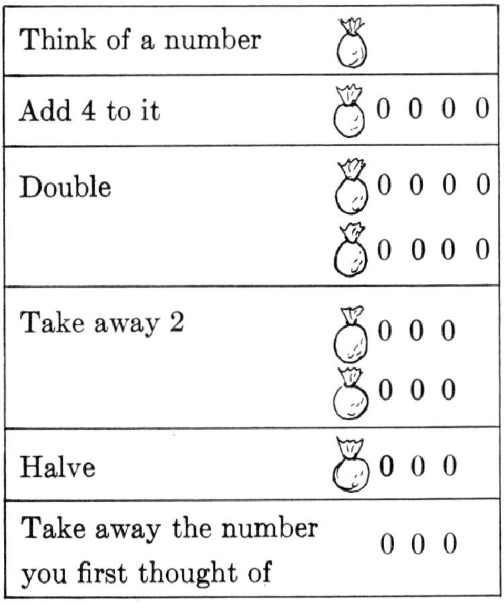

Think of a number	
Add 4 to it	0 0 0 0
Double	0 0 0 0 0 0 0 0
Take away 2	0 0 0 0 0 0
Halve	0 0 0
Take away the number you first thought of	0 0 0

FIGURE 1

Figure 1 is an example.

The answer is 3. Here, as you see, no bag remains at the end. The step before the end should leave just one bag. When you say, "Take away the number you first thought of," that gets rid of this bag.

A Plan to Save Labor

You may get tired of drawing bags and stones, particularly if fairly large numbers appear in a trick. It is easy to avoid most of this labor.

Instead of drawing two bags and six stones, you can write 2  + 6.

It would be most wearisome to draw a hundred bags and fifty stones. It is easy to write 100  + 50. (Be careful not to let your bag look like the figure 8.) In this shortened way, write:

Five bags and three stones_____

Ten bags and one stone_____

Three bags and four stones_____

Three bags_____

   Four stones_____

Two bags_____

   Five stones_____

Two bags and five stones_____

Finding Endings for Tricks

It does not matter how a trick begins, it can always be finished successfully. Here is the beginning of a trick in words and pictures.

Think of a number	$\bigcirc$
Add one	$\bigcirc + 1$
Double	$2\,\bigcirc + 2$
Add one	$2\,\bigcirc + 3$
Double	$4\,\bigcirc + 6$

How are we to finish this trick? We aim to get to a statement that has but one bag, and then to take that bag away. This is how all our earlier tricks ended.

At present we have four bags in the picture. If we divide by 4, that would give one bag. But it is awkward to divide 6 stones by 4. Before we divide, we might add or subtract some number that would make the division by 4 easy to carry out.

There are many ways of finishing this trick. Here is one of them:

Think of a number	$4\,\bigcirc + 6$
Add 6	$4\,\bigcirc + 12$
Divide by 4	$\bigcirc + 3$
Take away the number you first thought of	$\bigcirc$
The answer is	3.

It is wise to make sure that we have not made any slips. Try thinking of different numbers, and check to see that the answer 3 does come whatever number you choose.

TRY IT YOURSELF

Some beginnings of tricks are given below. Find a good ending for each of them. There are many different ways in which these tricks could end. If you get a trick that works, your answer is right.

Endings wanted for these tricks!

D1.	Think of a number	8	D4.	Think of a number	___
	Add 3	$8+3$		Add 1	___
	Double	$28+6$		Double	___
	Subtract 2	$28+4$		Add 1	___
D2.	Think of a number	___		Multiply by 3	___
	Add 5	___		Add 3	___
	Double	___	D5.	Think of a number	___
	Subtract 6	___		Add 1	___
D3.	Think of a number	___		Double	___
	Add 1	___		Add 1	___
	Double	___		Double	___
	Add 4	___		Subtract 5	___
				Double	___

FROM PICTURES TO SHORTHAND

We have been using the picture of a bag to help us imagine the number somebody thought of. Suppose we erase the top and bottom of the bag. This will change to $\times$, which looks much like the letter X.

It is much quicker to write X than to draw a bag.

Instead of 4 + 6, we now write $4X + 6$.

This is a very convenient shorthand. We now have four different ways of showing the same thing:

1. *Words:* Four times the number somebody thonght of with 6 added.

2. *Picture:* + 6

3. *Shortened Picture:* 4 + 6

4. *Shorthand:* $4X + 6$

The shorthand form, Number 4 above, is very quick and easy to write. It says just as much as all the words in Number 1. A scientist will often use Number 4, hardly ever Number 1. The pictures are useful to help you imagine what is happening, and to see why the trick works.

SOME SHORTHAND EXERCISES

Figure 2 shows the three ways we may tell about the same trick.

The following exercises are given so that you may get used to reading and writing in our shorthand. We take an idea and express it in each

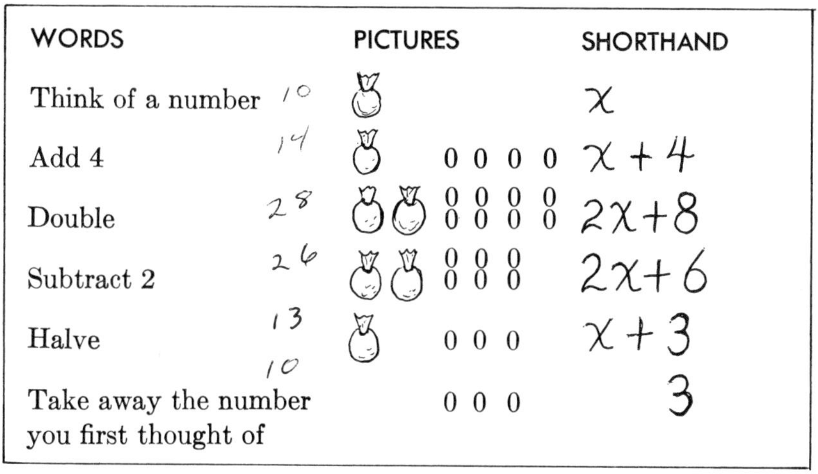

WORDS	PICTURES	SHORTHAND
Think of a number *(10)*	[bag]	x
Add 4 *(14)*	[bag] 0 0 0 0	$x + 4$
Double *(28)*	[bag][bag] 0000 0000	$2x+8$
Subtract 2 *(26)*	[bag][bag] 000 000	$2x+6$
Halve *(13)*	[bag] 0 0 0	$x+3$
Take away the number you first thought of *(10)*	0 0 0	3

FIGURE 2

of the forms which we shall call ⚹1, ⚹2, ⚹3, and ⚹4. The first exercise has the answers given. You should be able to fill the spaces in the other exercises.

    I. ⚹1 Three times the number thought of.

       ⚹2 [bag][bag][bag]

       ⚹3 3 [bag]

       ⚹4 3 X

  II. ⚹1 Twice the number thought of.

       ⚹2 _____

       ⚹3 _____

       ⚹4 2 X

 III. ⚹1 Four times the number thought of.

       ⚹2 _____

       ⚹3 _____

       ⚹4 _____

 IV. ⚹1 Three times the number thought of with 2 added.

       ⚹2 _____

       ⚹3 _____

       ⚹4 _____

  V. ⚹1 Twice the number thought of with 3 added.

       ⚹2 _____

       ⚹3 _____

       ⚹4 _____

13

ARITHMETIC FOR
THE FAST LEARNER
IN ENGLISH SCHOOLS

ANGELA PACE*

*State University of New York
Cortland, New York*

Angela Pace had the unique opportunity to study first-hand the arithmetic programs of England and Italy. Here she reports informally one phase of her study—that of the provisions being made for the talented in content, methodology, and administrative groupings in England. The American teacher can find suggestions for vitalizing the arithmetic program in both content and methodology, while maintaining a self-contained classroom.

ARITHMETIC FOR THE FAST LEARNER IN THE INFANT SCHOOL

The idea that all children should make progress in arithmetic at a rate commensurate with their ability seems to permeate both the infant school and the junior schools in England.

In the infant school, which is attended by children from age five to seven, teachers attempt to care for the needs of children of different ability in arithmetic by having children work in small groups, or else individually at their own rates. A variety of carefully planned activities and materials are then made available to care for the needs of all from the slowest to the brightest.

Five-Year-Olds

In one school, for instance, in a class of 5-year-olds, the slowest children were having experiences in recognizing figures up to 10 and

---

* Dr. Pace wrote this paper during a year's study (1960–61) in England and Italy under a Special International Scholarship of the Delta Kappa Gamma Society.

their corresponding number groups by means of bead threading. A box of large beads of various colors and number cards were provided. The children began by threading one bead along with the appropriate number card. Two beads of different color were threaded next, together with the correct card. This procedure was continued until a group of ten was reached.

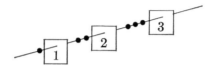

In the meantime, other children in the same class, who were more advanced, were busily engaged in such activities as the following:

1. Matching domino patterns with correct number cards.

Card *A*

2. Putting groups together to find the total. Materials for this consisted of cards, like Card A, and smaller number cards representing totals.

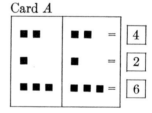

Card *B*

3. Finding totals by partial counting, for which work cards, like Card B, were provided, along with the number cards representing the totals.

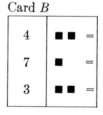

At the same time, the brighter children in the group were engaged in such activities as the following:
1. Individual games involving the recording and addition of two scores, each less than 10.
2. Solving of problems requiring both addition and subtraction, the problems being related to the play store in the classroom. With a card, like Card C to guide him, the child went to the store with 9

pence, bought two items, returned with them to his desk, recorded the cost of each, and then computed the total cost of both items and the amount of change from 9 pence.

Card C

9d. Buy two things. How much change?

In another school, the most immature were having experiences in simple sorting of materials like milk tops, acorns, matchboxes, and buttons. Children who were more advanced were sorting materials into groups, drawing around them, counting them, and writing the number to show how many in each group. The faster learners, using the same materials, were busy forming groups of various sizes and making up such equations as 3 buttons + 2 buttons = 5 buttons. One child was busy with such equations as $10 \times 2 = 20$, $10 \times 5 = 50$, and $10 + 10 + 10 + 10 + 10 + 10 + 10 = 70$.

In a third school, while the most immature children were still learning to recognize numbers, and developing basic ideas of big and little, thick and thin, heavy and light, long and short, the faster learners, supplied with various kinds of counters, were making discoveries by themselves, and recording such findings as $3 + 2 + 2 + 1 = 8$ and $5 - 2 = 3$. The teacher, rather than providing them with specific examples to be solved, allowed these children complete freedom to form groups of random size, to find totals, and to summarize their discoveries in written form. No ceiling was set to their thinking.

Six-Year-Olds

Teachers of 6-year-olds in the infant school also provide carefully prepared activities so that every child can make progress according to his ability. Through the use of well-constructed "job" cards, the child is encouraged to do his own finding, to make his own observations, and to draw his own conclusions. The job cards make it possible for the child to progress from one stage to the next, with the bright child going faster and farther than the slow one. Examples of job cards related to activities with weight follow.

First step: balancing—

How many beans balance four shells?

Second step: weighing of various amounts requiring ounce weights—

> Weigh 5 ounces of rice.
> Use 2 ounces.
> How much do you have left?

> Weigh 2 ounces of peas.
> Share them between two children.
> How many will each get?

Third step: finding the total weight of more than one object, comparison of weights, and more difficult problems—

> Weigh the big parcel.
> Weigh the small parcel.
> How much do they weigh together?

> Weigh the long parcel.
> Weigh the square parcel.
> Which is heavier?
> By how much?

> How many 3-ounce bags of peas can
> you make from 1 pound?

Many types of activities and materials, other than job cards, are prepared by teachers in order to provide for varying needs of 6-year-olds. Examples of a few of the activities seen in a single classroom follow.

1. Using sticks representing whole and halves to find answers to

$$2\tfrac{1}{2} + 1\tfrac{1}{2} + \tfrac{1}{2} = \qquad 4\tfrac{1}{2} \times 2 =$$
$$4\tfrac{1}{2} + 2\tfrac{1}{2} - 1 = \qquad 2 - 1\tfrac{1}{2} =$$

2. Buying three items at a time from a shopping list consisting of some 20 items with prices given, and then computing the cost for each group of three items chosen.

> Shopping List
> Toothpaste   1s.0d.
> Matches   2d.1
> Pepper   9½d.
>   etc.

3. Making up problems and solving them. Data for the problems
consisted of a number of cards on each of which was a picture of a
toy with its price. The child using the material was recording his
findings thus:

Ball, 3d. each.   Buy 4 = 1s.
Book, 6d. each.   Buy 4 = 2s.

4. Having experiences with multiplication. Five small glasses with
tiddley-winks were provided. By using these materials the child
was able to make up multiplication examples for himself and to
find the answers. For example, the child began by putting four
tiddley-winks in each of three glasses, and then proceeded to find
the total. With the materials, the child provided himself with many
experiences in combining equal-sized groups and in recording his
findings in equation form.

Carefully graded activities to develop understandings relative to
capacity, length, time, and shapes are also provided by teachers to
challenge the bright 6-year-old. A bright child, not satisfied with merely
measuring the length of each side of a rectangle ($3'' \times 7\frac{1}{2}''$) proceeded
to find its perimeter. He did this by adding $3''$, $3''$, and $6''$ (from the
$7\frac{1}{2}''$) to make one foot, and then by counting on the remaining inches to
obtain the correct perimeter. Other bright children in the class, provided
with boxes of squares and half squares and with cards like Card D to
guide their thinking, were making greater discoveries.

Card D

How many squares are there?
How many half squares are there?
Put two half-squares together to make a whole
    square.
How many whole squares do they make?
How many squares are there altogether in the box?

Other children, by manipulating squares and half-squares, were making
discoveries on their own and recording such findings as:

3 whole squares + 5 half-squares = $5\frac{1}{2}$ squares.
2 half squares + 3 whole squares = 4 squares.

In some schools children are provided with concrete materials. They
are then encouraged to manipulate the materials, make discoveries, note
relationships, and record their findings. There is a complete shift in
emphasis from "telling and showing" by the teacher to "discovery"
of meanings and important relationships by the child himself. There

are *no* questions to channel the children's thinking. Such a procedure allows the bright child complete freedom to forge ahead in discovering important relationships and in gaining deeper understandings of number.

Excerpts from children's notebooks give an idea of the type of discoveries being made by 6-year-olds in schools where this new approach is being used.

   1. Discoveries made by filling various types of containers.
     First child:
     "Today I have water.
     2 pints fill up 1 quart ∴ the pint is half the quart."
     Second child:
     "4 gills fill 1 pint ∴ the gill is $\frac{1}{4}$ of the pint.
     4 half pints = 1 quart ∴ my half pint is $\frac{1}{4}$ of my quart.
     8 gills = 1 quart ∴ my gill is $\frac{1}{8}$ of my quart."
     Third child:
     "I have two pints. I have a long shape and a wide shape and they both are worth the same. I tectid (tested) it."
     (This little child was becoming aware by experiment that differently shaped containers may hold the same amount of liquid.)
   2. Discoveries made through manipulating sticks of various lengths.
     "Today I have sticks:

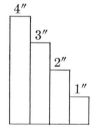

                  All of these numbers = my big stick and
                                          my little sticks.
           4 (1″) sticks   = 1 (4″) stick.
           ∴ my 1″ stick   = $\frac{1}{4}$ of my 4″ stick."

   3. Discoveries made by manipulating wholes, halves, thirds, quarters, sixths, and twelfths.
     "Today I have wholes and parts:
     1 whole   = 1 whole
     2 halves   = 1 whole
     3 thirds   = 1 whole
     4 quarters   = 1 whole
     6 sixths   = 1 whole
     12 twelfths = 1 whole."

4. Discoveries made by manipulating wholes and halves.

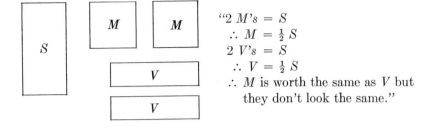

"2 M's = S
∴ $M = \frac{1}{2} S$
2 V's = S
∴ $V = \frac{1}{2} S$
∴ M is worth the same as V but they don't look the same."

5. Discoveries made by manipulating wholes and parts of wholes. While slow and average children were finding the lengths of the sides of the shapes below, bright children were exploring the problem of equal areas with different shapes, as their notes indicate.

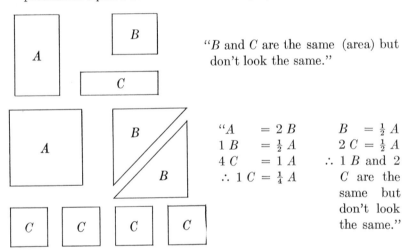

"B and C are the same (area) but don't look the same."

"A = 2 B B = $\frac{1}{2} A$
1 B = $\frac{1}{2} A$ 2 C = $\frac{1}{2} A$
4 C = 1 A ∴ 1 B and 2
∴ 1 C = $\frac{1}{4} A$ C are the same but don't look the same."

PROBLEM SOLVING

Infant schools are also adapting the work in verbal problem solving to the needs of the slow, average, and bright. In one school, each child had a booklet of problems matched to his ability, experience, and interest. In some cases, the problems had been constructed by the teacher. In other cases, the child had dictated the problems to the teacher. Problems and solutions taken from the children's notebooks will give an idea of the type of work being done.

1. Peter has lots of soldiers: 29 are English, 18 are Canadian, and 32 are American. How many soldiers has he altogether?

   SOLUTION: 29
              18
              32
              79

2. Mummie needs 74 cakes for the party. She has made 28 already How many more must she make?

   SOLUTION: 74
              28
              46

3. This is a box of biscuits. One quarter of the number are creams, one quarter are chocolate biscuits, one quarter are curranty biscuits, and all the rest are plain biscuits. There are 36 plain biscuits. How many biscuits are there altogether in the tin?

   SOLUTION:

$c.\ \frac{1}{4}$ 36	$ch.\ \frac{1}{4}$ 36	$\begin{array}{r} 36 \\ \times\ 4 \\ \hline 144 \end{array}$
$cu.\ \dfrac{1}{4}$ 36	$pl.\ \dfrac{1}{4}$ 36	

4. I have a big tin of sweets. Half of them are wrapped in green paper (*gp*). One quarter of those left are wrapped in blue paper (*bp*) and all the rest are wrapped in silver paper (*sp*). Sixteen are wrapped in blue paper. How many sweets are there altogether?

   SOLUTION:

gp	*bp* 16	*sp*	$\begin{array}{r} 16 \\ \times\ 8 \\ \hline 128 \end{array}$
	sp	*sp*	

5. I am sorting my beads. They are all red or yellow. Now I find that for each one red bead there are two yellow beads. There are 45 beads altogether.

   (a) How many are red?

   (b) How many are yellow?

Solution: 45

r	y	y
.	.	.
15	15	15

group of 3

$a.\ 3\overline{)45}$ $b.\ \begin{array}{r} 15 \\ \times\ 2 \\ \hline 30 \end{array}$

(with $\frac{15}{3\overline{)45}}$)

6. Jill has 18 red balloons 9 more green than red, and 14 blue balloons. These are all. How many balloons has Jill got altogether?

Solution: $18 + (9 + 18) + 14$ $\begin{array}{r} 18 \\ 27 \\ 14 \\ \hline 59 \end{array}$

7. I have 28 toffees and 18 creams, and these are just exactly half my number of sweets. How many sweets have I?

Solution:

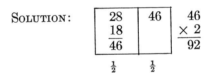

$\begin{array}{r} 28 \\ 18 \\ \hline 46 \end{array}$ | 46 | $\begin{array}{r} 46 \\ \times\ 2 \\ \hline 92 \end{array}$

$\frac{1}{2}$ $\frac{1}{2}$

8. One third of my class are away ill. Oh dear! Oh dear! Sixteen children are away ill. How many children are there in my class altogether?

Solution:

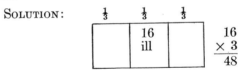

$\frac{1}{3}$ $\frac{1}{3}$ $\frac{1}{3}$

| | 16 ill | |

$\begin{array}{r} 16 \\ \times\ 3 \\ \hline 48 \end{array}$

9. I spend one third of all my money on Monday. I use one sixth of all my money on Tuesday. I have exactly 9 shillings left. How much money did I have in the beginning?

Monday	Tuesday	
	$\frac{1}{6}$	
$\frac{1}{3}$		

$\frac{1}{2}$ of all my money = 9/—
∴ all my money = 18/—

10. I give my friend one-fourth of my marbles. I give him exactly 25 marbles. How many will I have left for myself?

SOLUTION:

25	25
25	25

$$\begin{array}{r} 25 \\ \times\ 3 \\ \hline 75 \end{array}$$

Bright children, in infant schools, constantly challenged with carefully planned experiences and materials and working in small groups or individually at their own rates, tend not only to cover more ground than the slow or the average children, but also to delve more deeply and to gain deeper insights.

ARITHMETIC FOR FAST LEARNERS IN THE JUNIOR SCHOOL

In junior schools, attended by children from 7 to 12 years of age, the head teachers also adhere to the principle that each child ought to be able to progress in arithmetic according to his ability. They attempt to provide for varying abilities largely through variation in the content and through various forms of class organization, such as "streams," "sets," grouping within the class, and individual work. Rapid promotion of the very bright and the allocation of the better teachers to the bright classes also help to take care of the needs of the bright.

Class Organizations to Provide for Varying Abilities

STREAMING.[1] Since each head teacher in England has complete freedom to organize his school in the way which he thinks best for the children, different kinds of organization for work are found. In the great majority of junior schools, once the number of children in an age group is more than can be contained in a single class, head teachers decide to stream them, so as to keep the bright, the average, and the slow children in separate classes. The allocation of children in a particular age group to A, B, C, or even D streams, or classes, depends on the results of a test of ability, or attainment. Thus, the children with the highest scores on the intelligence tests go into the A stream. The others, depending upon scores obtained, are placed in B, C, or D streams. The size of the classes is roughly the same, although C and D streams tend to be slightly smaller. If the results of intelligence tests are not available, children are then streamed according to proficiency in basic skills. Some schools

---

[1] "Streaming" and "Sets" are terms commonly used in English educational terminology.

make use of both intelligence tests and tests in basic skills to separate the children into streams.

The different streams are then each taught at a different pace and at a different level. By streaming, head teachers feel that the able child, working among his peers, can be taught more easily and successfully, and that he can be "stretched" without discouraging the slower pupils. Thus each stream can move forward at its own pace. The abler ones are allowed to press on rapidly with a view to passing at 11 years of age the selection examination — the "open sesame" to a grammar school education and to the universities.

SETS.[2] Some head teachers feel that they can best provide for varying abilities by grouping together children of similar arithmetical ability into "sets" for their arithmetic work. In this case, children throughout the junior school, or perhaps only part of the school, are tested for arithmetical ability and attainment. They are then divided into sets according to the results of the testing. This form of organization means that the arithmetic period for all classes occurs at the same time so children can go to their particular set in arithmetic of mixed-age range but similar level of attainment. The set in which a particular child is placed may not bear much relation to the class in which he studies subjects other than arithmetic. Thus, an 11-year-old may be found in a set where most of the children are age 9. Similarly, a 9-year-old may be doing arithmetic with older children. Teachers feel that sets are more homogeneous arithmetically, and this greater homogeneity enables children to proceed at more suitable rates. Bright children can move ahead rapidly while slower children can be given work and taught at a pace which suits them.

In many schools, A streams and sets are taught by the most successful teachers, with the young and the weaker teachers being assigned to B and C streams or sets. It seems to be part of the pattern of thinking of teachers that a certain amount of prestige is associated with "promotion" to the teaching of the bright classes.

GROUPING WITHIN THE CLASS. Where the school is not organized in streams or sets, provision for the needs of children of varying ability is often made by means of grouping within a given classroom, with each of several groups working at its own pace and level.

In some schools, the arithmetic program is organized on an individual basis with each child progressing at his own pace.

_____

[2] This administrative device is found in the United States in several plans for grouping children for instruction, notably the Dual Progress Plan mentioned in Chapter 1.

RAPID PROMOTION. Some head teachers feel that they can best provide for the gifted children by rapid promotion. Thus, in many schools, a small number of 9-year-old children may be found working along with 10- and 11-year-olds in the final year of the junior school.

VARIATION IN CONTENT IN ARITHMETIC
FOR CHILDREN OF VARYING ABILITIES

One of the outstanding characteristics of the educational system in England is that it not only allows head teachers freedom to organize their schools in the way that they deem best for the children but also to frame syllabuses in order to care for varying needs. In the arithmetic syllabus, many head teachers provide for variation in the content to be covered by A, B, and C streams or sets. They do not feel that they must get all children in any given age-group to the same stage in learning. Thus the extent of work in arithmetic for the average and slow learners tends to be reduced. Bright children, making more rapid progress, cover more ground, and before leaving the junior school may be introduced to such topics as multiplication and division of a decimal by a decimal, profit and loss, factoring to find the least common multiple, and proportion. As an example, children in A streams or sets in a school using the arithmetic text, *Top of the Form* by A. Keith, complete both Book One and Book Two of the series during their first year in the junior school; Book Three, in the second year; Book Four, in the third year; and Book Five, in the fourth year. Children in B streams or sets, who are able to complete only one book each year, in most cases do not reach Book Five.

Further experiences for bright children may include such things as: a study of the history of number, use of the slide rule, old methods of multiplication (lattice method, Napier's bones), an introduction to algebra and to geometry.

Well-constructed "job" cards, used in junior schools as well as in the infant schools, also provide means for the bright child to progress at his own rate. Job cards being used in an A stream of 9- and 10-year-olds contained problems related to such topics as:
1. Furnishing a house
2. Measuring rainfall
3. Postal money orders
4. January sales
5. Travel
6. Telegrams
7. Staying at the Grasmere Hotel
8. Graphs.

Examples of job cards follow.

> You are going to visit your aunt in Worcester for one week. Consult the railway time-table and find a convenient morning train from London. Send a telegram to your aunt to tell her the time of your arrival. Then withdraw an adequate amount of money from the bank to pay for your fares and to allow you a suitable amount of pocket money.

> School Meals
> 1. How much money would you bring to school during the week if you pay one shilling daily for your dinner?
> 2. Your two brothers get their dinners at school also. Write the weekly amount each one would bring.
> 3. Next week there is no school on Wednesday. How much money will you need to bring to school next week to pay for the dinners?
> 4. How many dinners do you eat at school in a month and how much do you need to pay for them?

> Areas
> Find the areas of these shapes.

(This card is accompanied by a number of irregular shapes and inch squares.)

> Use *Whitaker's Almanac* to make a graph showing:
> (a) Lighting-up times during the months of June and January
> (b) Length of day during these months.

> Record the length of the "shadow stick" at hourly intervals during the day.
> Show this on a graph.
> Record the length of the stick at noon every day for a month and show your results on a graph.

In some junior schools, greater use of inductive methods is being made in at least one arithmetic period during the week. Activities are planned for children which will enable pupils to discover the important relationships for themselves. Below are a few of the activities[3] through which children are being encouraged to discover relationships.

1. Make and calibrate: (a) a water clock, (b) a candle clock.

   Time the water as it flows from the water clock and see if you can find whether a law is obeyed or not.

2. Make a pendulum by fastening a 42-inch weighted string to a shelf edge with a drawing pin or a thumb tack. Set it swinging through a small angle.

   Use a watch with second hand to time 50 beats.

   Repeat, gradually shortening the length.

   Represent your results on a graph.

   Can you say from this if the relationship between the time and the length obeys a law?

3. Make a beam balance with a ruler balanced on a matchbox. Can you find the "law of the lever" from the beam balance?

   (a) Place four 3d. pieces at the third unit from the center of the ruler. See how many you need on the other side to restore balance at 1, 2, 4, 6 units from the center on the other side.

   (b) Place six nails on one side, two units from the center. See how many nails you need to place on the other side to restore balance at 1, 2, . . . units from the center. Represent your results on a graph. Does your graph seem to represent a law?

4. You have a yard of string with which to make rectangles of different shapes. Starting with a width of 1″ (hence length of 17″) write down the dimensions of all the rectangles you can make, increasing the width by 1″ each time.

   Calculate the area of each rectangle and arrange your results—

Width	Length	Area

   Draw a graph showing how the area grows with the width. Does this obey a law? State any interesting features of your graph.

5. (a) Find how squares grow, using $\frac{1}{4}$″ squared paper. Make a square whose sides are—twice, three times, four times, etc. of the small square.

   (b) Find how cubes grow. Make a cube whose edges are twice, three times, four times . . . those of the small cube.

---

[3] These activities were supplied by J. Walls, Headmaster of The Redbridge Junior Mixed School, Ilford, Essex.

Arrange your results—

Length of side	Perimeter	Area	Volume

Using a large sheet of graph paper and taking 1″ for the unit of length and a convenient unit for the other axis, represent your results graphically.

Can you say, from your graphs, whether the growth obeys laws or not?

6. On $\frac{1}{4}$″ squared paper, draw circles of radius 1″, 2″ . . . . 10″. By counting squares, find the area of each and tabulate your results. Draw a graph to show how the area varies with the radius. Does this represent a law?

In England, while people may tend to criticize the infant and junior schools for paying too little attention to its weaker members, the more able members are far from neglected. Teachers and heads of schools pay a great deal of attention to the fast learners, and do all that they can to see that they progress as far as their ability permits.

A GRADED LIST OF READERS

To Enrich the Elementary School Mathematics Program

VINCENT J. GLENNON
Syracuse University

KINDERGARTEN TO GRADE 3

AMBLER, C. GIFFORD. *Ten Little Foxhounds.* New York: Grosset & Dunlap, 1958. Unpaged. (K–1)

BAER, HOWARD. *Now This, Now That.* New York: Holiday House, 1957. Unpaged. (1–3)

*BANIGAN, SHARON. *One by One.* Chicago: Hampton Publishing Co., 1953. (1–2)

*BEEBE, CATHERINE. *The Calendar.* New York: Oxford University Press, 1940. 63 pp. (1–2)

BEHN, HARRY. *All Kinds of Time.* New York: Harcourt, Brace and Co., 1950. (K–3)

BEIM, JERROLD. *The Smallest Boy in the Class.* New York: William Morrow and Co. 47 pp. (1–3)

BENDICK, JEANNE. *All Around You.* New York: McGraw-Hill Book Co., 1951. 48 pp. (K–4)

*BERKLEY, ETHEL S. *The Size of It.* New York: William R. Scott, 1950. (1–3)

*BERKLEY, ETHEL S. *Ups and Downs: A First Book about Space.* New York: William R. Scott, 1951. Unpaged. (1–2)

BIANCO, PAMELA. *The Doll in the Window.* Toronto: Oxford, 1953. 32 pp. (1–2)

BISHOP, CLAIRE. *The Five Chinese Brothers.* New York: Coward-McCann, 1938. 44 pp. (1–3)

BLOUGH, GLENN O. *Wait for the Sunshine.* New York: McGraw-Hill Book Co., 1954. 47 pp. (1–5)

BRANN, ESTHER. *Five Puppies for Sale.* New York: Macmillan Co., 1955. 78 pp. (1–3)

*BRODY, VIRGINIA. *Round the Clock Book.* Irvington-On-Hudson, N. Y.: Harvey House, 1956. (K–2)

BROWN, MARGARET WISE. *Two Little Trains.* New York: William R. Scott, 1949. 32 pp. (K–3)

BUDNEY, BLOSSOM. *A Kiss Is Round.* New York: Lothrop, Lee & Shepard Co., 1954. (K–3)

CAMERON, POLLY. *The Dog Who Grew Too Much.* New York: Coward-McCann, 1958. Unpaged. (K–3)

CHALMERS, AUDREY. *Hundreds and Hundreds of Pancakes.* New York: Viking Press, 1941. 38 pp. (K–2)

CORCOS, LUCILLE. *Joel Spends His Money.* New York: Abelard-Schuman, 1954. 40 pp. (2–3)

D'AULAIRE, INGRI, and PARIN, EDGAR. *Don't Count Your Chickens.* New York: Doubleday & Co., 1943. 40 pp. (preschool)

---

* Denotes books not included in *Books in Print,* 1960 edition. Readers may be able to obtain these out-of-print books in local and state libraries.

*Doisheau, Robert. *1, 2, 3, 4, 5*. New York: J. B. Lippincott Co., 1956. (K–2)

*Duvoisin, Roger. *Counting Book*. New York: Alfred A. Knopf, 1955. (1–3)

Duvoisin, Roger. *Two Lonely Ducks*. New York: Alfred A. Knopf, 1955. 36 pp. (K–1)

Eichenberg, Fritz. *Dancing in the Moon: Counting Rhymes*. New York: Harcourt, Brace and World, 1955. 20 pp. (K–3)

Elkin, Benjamin. *Six Foolish Fishermen*. Chicago: Children's Press, 1957. (K–3)

*Fraser, Phyllis. *Counting Rhymes*. New York: Simon & Schuster, 1947. 22 pp. (K–2)

Friskey, Margaret. *Chicken Little Count to Ten*. Chicago: Children's Press, 1946. 28 pp. (K–1)

Gag, Wanda. *Millions of Cats*. New York: Coward-McCann, 1945. 30 pp. (K–2)

Geisel, Theodor Seuss. *McElligot's Pool*. New York: Random House, 1947. 56 pp. (K–3)

Geisel, Theodor Seuss. *Yertle the Turtle*. New York: Random House, 1958. Unpaged. (K–3)

Grimm Brothers. *The Wolf and the Seven Little Kids*. New York: Harcourt, Brace and Co., 1959. (K–2)

Hall, William. *Telltime the Rabbit*. New York: Thomas Y. Crowell Co., 1955. 28 pp. (K–1)

Hall, William, and Robin. *Telltime Goes A'Counting*. New York: Thomas Y. Crowell Co., 1956. (preschool)

Hengesbaugh, Jane. *I Live in So Many Places*. New York: Grosset & Dunlap, 1956. Unpaged. (1–3)

Hoberman, Mary Ann and Norman. *All Shoes Come in Twos*. Little, Brown & Co., 1957. 48 pp. (1–3)

Hogan, Inez. *Twin Lambs*. New York: E. P. Dutton & Co., 1951. 44 pp. (K–2)

Ipcar, Dahlov. *Ten Big Farms*. New York: Alfred A. Knopf, 1958. Unpaged. (K–3)

Ipcar, Dahlov. *The Wonderful Egg*. New York: Doubleday & Co., 1958. (K–2)

Ipcar, Dahlov. *Brown Cow Farm: A Counting Book*. New York: Doubleday & Co., 1959. (preschool)

*Jackson, Kathryn. *School Days*. New York: Simon & Schuster, 1954. 64 pp. (K–2) (–K)

Kahl, Virginia. *The Habits of Rabbits*. New York: Charles Scribner's Sons. 1957. (K–3)

Karasz, Ilonka. *The Twelve Days of Christmas*. New York: Harper Brothers, 1949. (K–1)

Kaula, Edna M. *One, Two, Buckle My Shoe*. Chicago: Albert Whitman & Co., 1951 (K). Unpaged. (1–3)

Kay, Helen. *One Mitten Lewis*. New York: Lothrop, Lee & Shepard Co., 1955. Unpaged. (1–3)

*King, Dorothy. *Set the Clock*. New York: Franklin Watts, 1946. (K)

Krasilovsky, Phyllis. *The Very Little Girl*. New York: Doubleday & Co., 1953. Unpaged. (K–1)

KRAUSS, RUTH. *The Growing Story.* New York: Harper Brothers, 1947. 32 pp. (preschool)

LANGSTAFF, JOHN, and ROJANKOVSKY, FEODOR. *Over in the Meadow.* New York: Harcourt, Brace and Co., 1957. 32 pp. (K–3)

LANSDOWN, BRENDA. *Arithmetic for Beginners.* New York: Grosset & Dunlap, 1959. 69 pp. (K–1)

LEAF, MONRO. *Arithmetic Can Be Fun.* Philadelphia: J. B. Lippincott Co., 1949. 64 pp. (K–3)

*LEWELL, BETTY LOU. *Tooooot! A Train Whistle.* Chicago: Melmont Publishers, 1958. 26 pp. (K–3)

LEWELLEN, JOHN. *The True Book of Airports and Airplanes.* New York: Grosset & Dunlap, 1956. 46 pp. (1–4)

MARINO, DOROTHY. *Edward and the Boxes.* Philadelphia: J. B. Lippincott Co., 1957. Unpaged. (1–3)

McCULLOUGH, JOHN G., and KESSLER, LEONARD. *Farther and Faster.* New York: Thomas Y. Crowell Co., 1954. Unpaged. (1–4)

McLEOD, EMILIE WARREN. *The Seven Remarkable Bears.* New York: Houghton Mifflin Co., 46 pp. (K–3)

MEEKS, ESTHER K. *One Is the Engine.* Chicago: Follett Publishing Co., 1956. 24 pp. (K–3)

MERWIN, DECIE. *Time for Tamie?* New York: H. Z. Walck, 1946. 40 pp. (K–2)

*MONRAD, JEAN. *How Many Kisses Goodnight?* New York: William R. Scott, 1949. 20 pp. (K or younger)

MOORE, LILLIAN. *Count to Ten.* New York: Golden Press, 1957. 20 pp. (K–2)

MOORE, LILLIAN. *My Big Golden Counting Book.* New York: Golden Press, 1957. 32 pp. (1–3)

MOORE, LILLIAN. *My First Counting Book.* New York: Golden Press, 1956. 22 pp. (K)

NORLING, JO and ERNEST. *Pogo's Sea Trip: A Story of Boats.* New York: Henry Holt & Co., 1949. 50 pp. (1–3)

OSMOND, EDWARD. *Animals of the World.* Vol. 3. New York: H. Z. Walck, 1958. 129 pp. (2–3)

PETER, JOHN. *The Counting Book.* New York: Wonder Books, 1957. 20 pp. (1–3)

*PETER, JOHN. *What Time Is It?* New York: Wonder Books, 1954. 22 pp. (1–3)

PINE, TILLIE S., and LAVINE, JOSEPH. *The Chinese Knew.* New York: McGraw-Hill Book Co., 1958. 32 pp. (K–4)

PODENDORF, ILLA. *The True Book of Space.* New York: Gosset & Dunlap, 1959. 48 pp. (2–4)

*REED, MARY. *Counting Rhymes.* New York: Simon and Schuster, 1947. 22 pp. (K–2)

REED, MARY and OSSWALD. *The Golden Picture Book of Numbers: What They Look Like and What They Do.* New York: Golden Press, 1954. 80 pp. (1–3)

ROTHSCHILD, ALICE. *Bad Trouble in Miss Alcorn's Class.* New York: William R. Scott, 1959. 101 pp. (1–4)

RUSSELL, BETTY. *Big Store, Funny Door.* Chicago: Albert Whitman Co., 1955. Unpaged. (K–3)

SCHLEIN, MIRIAM. *Fast Is Not a Ladybug.* New York: William R. Scott, 1953. Unpaged. (K–3)

SCHLEIN, MIRIAM. *Heavy Is a Hippopotamus.* New York: William R. Scott, 1954. 30 pp. (K–4)

SCHLEIN, MIRIAM. *Shapes.* New York: William R. Scott, 1952. 40 pp. (K–2)

SCHLEIN, MIRIAM. *It's About Time.* William R. Scott, 1955. 48 pp. (K–4)

SCHLEIN, MIRIAM. *City Boy, Country Boy.* New York: Grosset & Dunlap, 1955. Unpaged. (2–3)

*SCHLEIN, MIRIAM. *The Four Little Foxes.* New York: William R. Scott, 1953. 34 pp. (1–2)

SCHLOAT, G. WARREN. *Adventures of a Letter.* New York: Charles Scribner's Sons, 1949. 48 pp. (1–4)

SCHNEIDER, HERMAN and NINA. *How Big Is Big?* New York: William R. Scott, 1946. (K-4)

SCHWARTZ, JULIUS. *I Know a Magic House.* New York: McGraw-Hill Book Co., 1956. 32 pp. (K–4)

SEIGNOBOSE, FRANCOISE. *Jeanne Marie Counts Her Sheep.* New York: Charles Scribner's Sons, 1951. 32 pp. (K–3)

SHAPP, CHARLES and MARTHA. *Let's Find Out What's Big and What's Small.* Franklin Watts, 1959. Unpaged. (1–2)

SKAAR, GRACE. *The Very Little Dog.* New York: William R. Scott, 1949. 19 pp. (preschool)

SLOBODKIN, LOUIS. *Millions and Millions.* New York: Vanguard Press, 1955. Unpaged. (K–1)

SLOBODKIN, LOUIS. *One Is Good.* New York: Vanguard Press, 1956. 25 pp. (K or younger)

SOOTIN, LAURA. *Let's Go to an Airport.* New York: G. P. Putnam's Sons, 1957. 48 pp. (2–4)

*STEINER, CHARLOTTE. *The Big Laughing Book.* New York: Grosset & Dunlap, 1949. 56 pp. (1–3)

THOMAS, JOAN GALE. *One Little Baby.* New York: Lothrop, Lee & Shepard Co., 1956. (K–2)

TODD, MARY FIDELIS. *ABC and 1 2 3.* New York: McGraw-Hill Book Co., 1955. 27 pp. (K–3)

TRESSELT, ALVIN. *Follow the Road.* New York: Lothrop, Lee & Shepard Co., 1953. 26 pp. (1–2)

TRUE, LOUISE, and OWNES, LILLIAN. *Number Men.* Chicago: Children's Press, 1948. 32 pp. (1–2)

TUDOR, TASHA. *1 Is One.* New York: H. Z. Walck, 1956. (K–1)

TUDOR, TASHA. *Around the Year.* New York: H. Z. Walck, 1957. 56 pp. (K–3)

WALLER, LESLIE. *Time.* New York: Holt, Rinehart and Winston, 1959. (1–2)

WATSON, NANCY DIGMAN. *What Is One?* New York: Alfred A. Knopf, 1954. 42 pp. (K–1)

WATSON, NANCY DIGMAN. *When Is Tomorrow?* New York: Alfred A. Knopf, 1955. (K–1)

WATSON, NANCY DIGMAN. *Annie's Spending Spree.* New York: Viking Press, 1957. 45 pp. (K-3)

WEBBER, IRMA E. *It Looks Like This.* Chicago: Scott, Foresman and Co. Unpaged. (1–3)

*WEISSENBOR, H. *Counting.* London: Acorn Press, 1948. 26 pp. (1–2)

*WITHERS, CARL. *1, 2, 3.* New York: Grosset & Dunlap, 1958. 20 pp. (K or younger)

WITHERS, CARL. *Counting Out.* New York: H. Z. Walck, 1946. (1–4)

WONDRISKA, WILLIAM. *1, 2, 3, A Book to See.* New York: Pantheon Books, 1959. (K–2)

*WOOLEY, CATHERINE. *Two Hundred Pennies.* New York: William Morrow and Co., 128 pp. (1–3)

ZINER, FEENIE. *The True Book of Time.* Chicago: Children's Press, 1956. (1–3)

ZOLOTOW, CHARLOTTE. *One Step, Two.* New York: Lothrop, Lee & Shepard Co., 1955. (K)

*ZOLOTOW, CHARLOTTE. *The True-to-Life ABC Book.* New York: Grosset & Dunlap, 1952. (K)

ZOLOTOW, CHARLOTTE, *Over and Over.* New York: Harper & Brothers, 1957. Unpaged. (preschool)

GRADE 3 THROUGH GRADE 8

ABBOTT, E. A. *Flatland.* New York: Dover Publications, 1950. 103 pp. (7–8)

ADLER, IRVING. *Mathematics: The Story of Numbers, Symbols and Space.* New York: Golden Press, 1958. 55 pp. (7–8)

ADLER, IRVING. *The New Mathematics.* New York: John Day Co., 1958. 187 pp. (8)

ADLER, IRVING. *The Tools of Science: From Yardstick to Cyclotron.* New York: John Day Co., 1958. 128 pp. (7–11)

ADLER, IRVING. *Time in Your Life.* New York: John Day Co., 1955. 127 pp. (6–9)

ADLER, IRVING. *The Giant Golden Book of Mathematics.* New York: Golden Press, 1958. 92 pp. (6 and up)

ADLER, IRVING. *Magic House of Numbers.* New York: John Day Co., 1957. 128 pp. (6–9)

ANDERSON, RAYMOND. *Romping Through Mathematics.* New York: Alfred A. Knopf, 1947. 152 pp. (7 and up)

*ANDREWS, F. E. *New Numbers.* New York: Harcourt, Brace and Co., 1935. 168 pp. (6)

ASIMOV, ISAAC. *The Realm of Numbers.* New York: Houghton Mifflin Co., 1959. 200 up. (6–11)

ASIMOV, ISAAC. *The Clock We Live On.* New York: Abelard-Schuman, 1959. 160 pp. (7–8)

ASIMOV, ISAAC. *The Realm of Measure.* New York: Houghton Mifflin Co., 1960. 186 pp. (5 and up)

BAKST, AARON. *Mathematics: Its Magic and Mastery.* New York: D. Van Nostrand Co., 1952. 790 pp. (7–8)

BECKHARD, ARTHUR. *Albert Einstein.* New York: G. P. Putnam's Sons, 1959. 126 pp. (5–9)

BELL, ERIC T. *Men of Mathematics.* New York: Simon & Schuster, 1937. 592 pp. (7–8)

---

* Denotes books not included in *Books in Print,* 1960 edition. Readers may be able to obtain these out-of-print books in local and state libraries.

BELL, THELMA. *Snow.* New York: Viking Press, 1954. 56 pp. (3–7)

BENDICK, JEANNE. *How Much and How Many.* New York: McGraw-Hill Book Co., 1947. 188 pp. (5–8)

BETZ, BETTY. *The Betty Betz Teen-Age Cookbook.* New York, Holt, Rinehart and Winston, 1953. 182 pp. (7–8)

*BOEHM, DAVID ALFRED, and REINFELD, FRED. *Coinmetry.* New York: Sterling Publishing Co., 1952. 93 pp. (4–8)

BOEHM, GEORGE A. W., and THE EDITORS OF FORTUNE. *The World of Mathematics.* New York: Dial Press, 1959. 128 pp. (7–8)

BOLTON, SARAH. *Famous Men of Science.* New York: Thomas Y. Crowell Co., 1946. 308 pp. (7–11)

*BOWERS, HENRY and JOAN. *Arithmetical Excursions.* New York: Dover Publications, 1961. 320 pp. (6 and up)

BRAGDON, L. J. *Tell Me the Time, Please.* New York: J. B. Lippincott Co., 1946. (7–9)

*BRADES, LOUIS GRANT. *Math Can Be Fun.* Portland, Maine: J. Weston Walch, 1956. 200 pp. (5 and up)

BRINDZE, RUTH. *The Story of Our Calendar.* New York: Vanguard Press, 1949. 64 pp. (4–7)

BRINDZE, RUTH. *Johnny Get Your Money's Worth.* New York: Vanguard Press, 1938. 230 pp. (4–6)

BUFF, MARY and CONRAD. *Big Tree.* New York: Viking Press, 1946. 80 pp. (5–9)

CARLSON, BERNICE W. *Make It and Use It.* Nashville: Abingdon Press, 1958. 160 pp. (3–5)

COURT, NATHAN A. *Mathematics in Fun and Earnest.* New York: Dial Press, 1958. 250 pp. (7–8)

COWAN, HARRISON J. *Time and Its Measurement from the Stone Age to the Nuclear Age.* Cleveland: World Publishing Co., 1958. 160 pp. (6 and up)

CROCKER, BETTY. *Betty Crocker's Cook Book for Boys and Girls.* New York: Golden Press, 1957. 191 pp. (4–6)

EPSTEIN, SAMUEL and BERYL. *First Book of Maps and Globes.* New York: Franklin Watts, 1959. (6–9)

FENTON, CARROLL L. and MILDRED A. *Worlds in the Sky.* New York: John Day Co., 1950. (4–7)

*FLYNN, HARRY EUGENE, and LUND, CHESTER BENFORD. *Tick-Tock, A Story of Time.* Boston: D. C. Heath and Co., 1938. 234 pp. (3–4)

*FOSTER, CONSTANCE J. *The Story of Money.* New York: McBride Co., 1950. 205 pp. (5–8)

FOWLER, H. WALTER, JR. *Kites.* New York: Ronald Press Co., 1953. 92 pp. (5–6)

FREEMAN, MAE and IRA. *Fun with Figures.* New York: Random House, 1946. 60 pp. (10–12)

FREEMAN, MAE and IRA. *Fun with Astronomy.* New York: Random House, 1953. 57 pp. (4–6)

FREEMAN, MAE BLACKER. *The Story of Albert Einstein: The Scientist Who Searched Out the Secrets of the Universe.* New York: Random House, 1958. 178 pp. (5–9)

FRIEND, NEWTON. *Numbers: Fun and Facts.* New York: Charles Scribner's Sons. 208 pp. (7 and up)

GALT, THOMAS FRANKLIN. *Seven Days from Sunday.* New York: Thomas Y. Crowell Co. 215 pp. (5–9)

GAMOW, GEORGE, and STERN, MARVIN. *Puzzle-Math.* New York: Viking Press, 1958. 119 pp. (7–8)

GARDNER, MARTIN. *Mathematics, Magic and Mystery.* New York: Dover Publications, 1955. 176 pp. (8)

GARDNER, MARTIN. *The Scientific American Book of Mathematical Puzzles and Diversions.* New York: Simon and Schuster, 1959. 178 pp. (7–8)

*GILLES, WILLIAM F. *The Magic and Oddities of Numbers.* New York: Vantage Press, 1953. 65 pp. (7 and up)

*HEATH, ROYAL VALE. *Math & Magic.* New York: Dover Publications, 1953. 126 pp. (6 and up)

*HENRY, THOMAS. *Charles Steinmetz.* New York: G. P. Putnam's Sons, 1959. 126 pp. (5–9)

*HIGGINS, LOYTA. *Let's Save Money.* New York: Golden Press, 1958. 18 pp. (3–4)

HOGBEN, LANCELOT. *The Wonderful World of Mathematics.* New York: Doubleday & Co., 1955. 69 pp. (5–9)

HOGBEN, LANCELOT. *The Wonderful World of Energy.* New York: Doubleday & Co., 1957. 69 pp. (5–9)

HOOPER, ALFRED. *Makers of Mathematics.* New York: Random House, 1948. 402 pp. (7–8)

*HUFFMAN, PEGGY. *Miss B's First Cookbook.* Columbus, Ohio: Charles E. Merrill Co., 1950. 43 pp. (3–4)

KIENE, JULIA. *The Step-by-Step Cookbook for Girls and Boys.* New York: Golden Press, 1950. 125 pp. (4–6)

KLINE, MORRIS. *Mathematics and the Physical World.* New York: Thomas Y. Crowell Co., 1959. 482 pp. (9–12)

KRAITCHIK, MAURICE. *Mathematical Recreations.* New York: Dover Publications, 1953. 330 pp. (7–8)

*KUOJIMA, TAKASHI. *The Japanese Abacus.* Rutland, Vt.: Charles E. Tuttle Co., 1959. 102 pp. (7 and up)

*LACH, ALMA S. *A Child's First Cookbook.* New York: Hart Publishing Co., 1956. 96 pp. (3–4)

*LARSEN, HAROLD D. *Enrichment Program for Arithmetic.* Evanston, Ill.: Row, Peterson and Co., 1956. (3–8)

LATHAM, JEAN LEE. *Carry On, Mr. Bowditch.* New York: Houghton Mifflin Co., 1955. 252 pp. (6–12)

LATHAM, JEAN LEE. *Trail Blazer of the Seas.* New York: Houghton Mifflin Co., 1956. 245 pp. (5–9)

LAUBER, PATRICIA. *The Quest of Galileo.* New York: Doubleday & Co., 1959. 56 pp. (2–6)

LEE, RECTOR. *Gil's Discovery in the Mine.* Boston: Little, Brown and Co., 1957. 202 pp. (7 and up)

LEEMING, JOSEPH. *Fun with Puzzles.* Philadelphia: J. B. Lippincott Co., 1946. 128 pp. (4 and up)

LEEMING, JOSEPH. *More Fun with Puzzles.* Philadelphia: J. B. Lippincott Co., 1947. 149 pp. (3 and up)

*LEEMING, JOSEPH. *From Barter to Banking.* New York: Appleton-Century-Crofts, 1940. 131 pp. (5–6)

LEVINGER, ELMA EHRLICH. *Albert Einstein.* New York: Julian Messner, 1959. 174 pp. (8–12)

LIEBER, HUGH G. and LILLIAN R. *The Education of T. C. Mits.* New York: W. W. Norton & Co., 1944. 230 pp. (7–8)

LIEBER, LILLIAN R. *Infinity.* New York: Holt, Rinehart and Winston, 1953. 359 pp. (7–8)

LIEBER, LILLIAN R. *Take a Number: Mathematics for the Two Billion.* New York: Ronald Press Co., 1946. 221 pp. (7–8)

MALONEY, TERRY. *The Story of Maps.* Buffalo: Sterling Publishing Co., 1959. 48 pp. (3–7)

MALTER, MORTON S. *Our Largest Animals.* Chicago: Albert Whitman & Co., 1958. 31 pp. (3–6)

MALTER, MORTON S. *Our Tiniest Animals.* Chicago: Albert Whitman & Co., 1955. 32 pp. (3–4)

*MARSHAK, ILIN. *What Time Is It?* Philadelphia: J. B. Lippincott Co., 1932. (5–6)

MASSOGLIA, ELINOR. *Fun-Time Paper Folding.* New York: Grosset & Dunlap, 1959. 31 pp. (4–6)

MAYALL, NEWTON, and WYCOFF, MARGARET and JEROME. *The Sky Observers' Guide.* New York: Golden Press, 1959. 125 pp. (6 and up)

McCLOSKEY, ROBERT. *Time of Wonder.* New York: Viking Press, 1957. 63 pp. (4–6)

MERRILL, HELEN A. *Mathematical Excursions.* New York: Dover Publications, 1957. 145 pp. (7–8)

MEYER, JEROME. *Fun with Mathematics.* New York: World Publishing Co., 1952. 176 pp. (8–12)

MOORE, LILLIAN. *The Important Pockets of Paul.* New York: David McKay Co., 1954. 73 pp. (3–6)

MOORE, PATRICK. *Isaac Newton.* New York: G. P. Putnam's Sons, 1958. 124 pp. (4–7)

MOTT-SMITH, GEOFFREY. *Mathematical Puzzles for Beginners and Enthusiasts.* New York: Dover Publications, 1954. 176 pp. (7–8)

NEAL, HARRY E. *The Story of the Kite.* New York: Vanguard Press, 1954. 61 pp. (4–6)

NEURATH, MARIE. *Too Small to See.* Buffalo: Sterling Publishing Co., 1957. 36 pp. (4–6)

NEWELL, HOMER E., JR. *Space Book for Young People.* New York: McGraw-Hill Book Co., 1958. 114 pp. (3–7)

*NORMAN, GERTRUDE. *The First Book of Music.* New York: Franklin Watts, 1954. 65 pp. (3–6)

PARKER, BERTHA MORRIS. *Golden Book of Science.* New York: Golden Press, 1956. 98 pp. (5–7)

*PARKER, BERTHA MORRIS. *Heat.* Evanston, Ill.: Row, Peterson and Co., 1942. 36 pp. (5–6)

PATTON, PRICE A. and MARTHA. *Money in Your Pocket.* New York: David McKay Co., 1959. 181 pp. (9–12)

PERKINS, WILMA LORD. *Fannie Farmer Junior Cook Book.* Boston: Little Brown and Co., 1957. 208 pp. (5–9)

PLOTZ, HELEN. *Imagination's Other Place.* New York: Thomas Y. Crowell Co., 1955. 200 pp. (7 and up)

RAVIELLI, ANTHONY. *An Adventure in Geometry.* New York: Viking Press, 1957. 117 pp. (10–12)

REID, CONSTANCE. *From Zero to Infinity.* 2nd revised edition. New York: Thomas Y. Crowell Co., 1960. 145 pp. (7–8)

REINFELD, FRED. *The Story of Paper Money.* Buffalo: Sterling Publishing Co., 1957. 128 pp. (7–8)

ROMBAUER, IRMA. *Cookbook for Girls and Boys.* Indianapolis: Bobbs-Merrill Co., 1952. 243 pp. (4–8)

ROSEN, SIDNEY. *Galileo and the Magic Numbers.* Boston: Little, Brown and Co., 1958. 212 pp. (7–11)

ROSS, FRANK, JR. *The World of Engineering.* New York: Lothrop, Lee & Shepard Co., 1957. 186 pp. (8–12)

*RUCHLIS, HYMAN, and ENGELHARDT, JACK. *The Story of Mathematics.* Irvington-on-Hudson, N. Y.: Harvey House, 1958. 149 pp. (7–8)

SANFORD, VERA. *A Short History of Mathematics.* New York: Houghton Mifflin Co., 1930. 402 pp. (7–8)

*SAXON, G. R. *How Fast?* New York: Thomas Y. Crowell Co., 1954. Unpaged. (5–6)

SAWYER, W. W. *Mathematician's Delight.* Baltimore: Penguin Books, 1943. 238 pp. (7–8)

SAWYER, W. W. *Prelude to Mathematics.* Baltimore: Penguin Books, 1955. 214 pp. (7–8)

*SAWYER, W. W., and SRAWLEY, L. G. *Designing and Making.* New York: Oxford University Press, 1952. 192 pp. (7 and up)

SHACKLE, G. L. S. *Mathematics at the Fireside.* Cambridge: University Press, 1952. 155 pp. (7 and up)

SHARP, ELIZABETH N. *Simple Machines and How They Work.* New York: Random House, 1959. 96 pp. (3–5)

SMITH, DAVID EUGENE. *History of Mathematics.* 2 vol. Boston: Ginn & Co., 1951–53. (7–8)

SMITH, DAVID EUGENE. *Number Stories of Long Ago.* New York: Scripta Mathematica, 1955. (5 and up)

*SMITH, DAVID EUGENE, and GINSBERG, J. *Numbers and Numerals.* Washington, D. C.: National Council of Teachers of Mathematics, a department of the National Education Association, 1937. 62 pp. (5 and up)

SMITH, DAVID EUGENE. *The Wonderful Wonders of One-Two-Three.* New York: Scripta Mathematica, 1937. 47 pp. (3 and up)

SOONG, MAYLING. *The Art of Chinese Paper Folding.* New York: Harcourt, Brace and World, 1948. (6 and up)

SOOTIN, HARRY. *Isaac Newton.* New York: Julian Messner, 1955. 191 pp. (8–12)

SOOTIN, LAURA. *Let's Go to a Bank.* New York: G. P. Putnam's Sons, 1957. 47 pp. (4–6)

STEINHAUS, H. *Mathematical Snapshots.* New York: Oxford University Press, 1950. 266 pp. (7–8)

STICKER, HENRY. *How to Calculate Quickly.* New York: Dover Publications, 1956. 256 pp. (7–8)

STRADER, WILLIAM W. *Five Little Stories.* Washington, D. C.: National Council of Teachers of Mathematics, a department of the National Education Association, 1960. 16 pp. (7–8)

TANNENBAUM, BEULAH, and STILLMAN, MYRA. *Isaac Newton: Pioneer of Space Mathematics.* New York: McGraw-Hill Book Co., 1959. 128 pp. (7–8)

TANNENBAUM, BEULAH, and STILLMAN, MYRA. *Understanding Time.* New York: McGraw-Hill Book Co., 1958. (8–12)

TANNENBAUM, BEULAH, and STILLMAN, MYRA. *Understanding Maps.* New York: McGraw-Hill Book Co., 1957. 144 pp. (5–8)

THURBER, JAMES. *The Great Quillow.* New York: Harcourt, Brace and World, 1944. 54 pp. (3–7)

THURBER, JAMES. *Many Moons.* New York: Harcourt, Brace and World, 1943. 42 pp. (3–6)

*TOWNSEND, HERBERT. *Our Wonderful Earth.* Englewood Cliffs, N. J.: Allyn and Bacon, 1950. 152 pp. (3 and up)

VALENS, EVANS G. *Me and Frumpet: An Adventure with Size and Science.* New York: E. P. Dutton & Co., 1958. 128 pp. (4–6)

VERGARA, WILLIAM C. *Mathematics in Everyday Things.* New York: Harper & Brothers, 1959. 301 pp. (7–8)

WATSON, JANE WERNER. *The World of Science.* New York: Golden Press, 1958. 216 pp. (7–8)

*WEEKS, RAYMOND. *Boys' Own Arithmetic.* New York: E. P. Dutton & Co., 1924. 188 up. (6 and up)

WERNER, ELSA JANE. *The Golden Geography Book.* New York: Golden Press, 1952. 96 pp. (4–6)

WEYL, PETER K. *Men, Ants, and Elephants: Size in the Animal World.* New York: Viking Press, 1959. 103 pp. (5–9)

WILCOX, LOUISE K., and BURKS, GORDON E. *What Is Money?* Austin: Steck Co., 1959. 48 pp. (3–7)

WILKINS, H. PERCY, and MOORE, PATRICK. *How to Make and Use a Telescope.* New York: W. W. Norton & Co., 1956. 195 pp. (7–8)

*WINTERS, MARY. *Teach Me Numbers.* New York: Hart Publishing Co., 1957. (5–8)

WYLER, ROSE, and AMES, GEROLD. *The Golden Book of Astronomy.* Revised edition. New York: Golden Press, 1959. 97 pp. (5–7)

WYLIE, C. R., JR. *101 Puzzles in Thought and Logic.* New York: Dover Publications, 1957. 128 pp. (6 and up)

ZARCHY, HARRY. *Wheel of Time.* New York: Thomas Y. Crowell Co., 1957. 144 pp. (7 and up)

ZARCHY, HARRY. *Let's Make a Lot of Things.* New York: Alfred A. Knopf, 1948. 156 pp. (5–9)

ZARCHY, HARRY. *Let's Make Something.* New York: Alfred A. Knopf, 1941. 158 pp. (3–7)

ZIM, HERBERT S. *Codes and Secret Writing.* New York: William Morrow and Co., 1948. 154 pp. (7–9)

ZIM, HERBERT S. *The Sun.* New York: William Morrow and Co., 1953. Unpaged. (3–7)

SECTION II | THE JUNIOR HIGH SCHOOL YEARS

SECTION II / The Junior High School Years

INTRODUCTION

ALICE M. HACH

Ann Arbor Public Schools
Ann Arbor, Michigan

PROVIDING CHALLENGE FOR STUDENTS

One of the challenges of education today is to encourage every child to work to the limit of his ability. If the teacher is in a small school there may be only one or two students in each class who need challenge beyond the regular work of the classroom. If the teacher is in a larger school, it may be that classes are sectioned into ability groups. However, regardless of whether there is grouping, there will always be a few students who have ability to do work beyond that offered to the group. Although there is recognition of need for providing challenge for these students, teachers are often handicapped because they do not have material available for this purpose.

It is with these thoughts in mind that sample enrichment units to be used by pupils in the junior high school are provided in this section. Of even greater importance to the teacher is the possible use of these materials as guides in developing additional units. There are many topics suitable for enrichment, and there is no attempt to exhaust the possibilities in this section. The bibliography which follows this introduction will give the reader a guide to further exploration.

No one type of unit will serve the needs and interests of all children. For this reason a variety in difficulty-level and type of unit is suggested. However, greater emphasis has been placed on developing materials which require a high level of ability, since such materials are not generally available.

It is important that students be guided into units which measure up to their abilities and provide challenge. Some students might be directed into the high school level and then again some might be encouraged to work in the upper elementary level. In most cases it is expected that this section will offer enough variety so that suitable enrichment material can be found for nearly all junior high school students.

SUGGESTIONS FOR USE OF ENRICHMENT UNITS

As for the method of using the units, there is no one best way. Instead, the material may be used in a variety of ways and adapted to the needs of the school, the teacher, and type of class.

Guided Selection

If a class is heterogeneous, there may be only one or two pupils who show a need for additional work beyond the regular assignments. It is these pupils who should be encouraged to work beyond the level of the group and at a level commensurate with their ability. Time might be taken to discuss with these pupils the possible direction to take in order to develop further interest and experience in mathematics. If a child reveals any particular interests, a teacher might recommend units from this yearbook relating to his interests. In order to do this effectively, it is important that the teacher familiarize himself with the enrichment materials to the extent that he knows the emphasis and range of difficulty. Children should neither work below their potential nor become discouraged because they are required to work at a level beyond their ability.

Self Selection and Pacing

Another possible way to stimulate the interest of pupils is through self-selection of units and pacing, whereby a child looks over all possible choices and then selects the ones that interest him. The number of units worked will depend upon each child's ability and initiative. With this method, it is expected that a pupil will seek units at his own level of ability.

If the entire class is of high ability, a list of the units might be posted in the room and all the pupils encouraged to select units to work as time and ability permit. Some teachers might require each pupil to select at least one unit as a special project and do as much of it as possible. This would encourage all talented students to participate in independent work, and might possibly stimulate interest that might never have been stimulated in any other way. Evidences of individual differences would become apparent from the level of accomplishment of each student.

Group Work

There is also the possibility of committees or groups working on a unit. Several students might find working together a profitable experience. In this way they might benefit from an exchange of ideas on various approaches, and have the advantage of group discussion on the topic. Students might even choose to work in pairs but independently on the same unit. They could then compare work and discuss the unit after both had completed the unit. Exchange of opinions and discussion among students can often be stimulating and thought-provoking.

There may be students who prefer to pursue one topic for a semester or possibly longer. Such students might complete a unit and then follow this by further investigation and study on the topic. They might refer to various resource books and texts, to an engineer, or to a mathematician for further insight and guidance.

The units may be used in still another way. A teacher might be in need of enrichment material for an entire group of high ability students. From the units in this volume, material might be selected which would relate to the level and needs of the group as a whole.

Stimulus for Exploration

Naturally teachers cannot be expected to have answers to all questions. This should not discourage a teacher from using the materials. The fact that a pupil raises questions is an indication that he has been challenged to explore further. It is at this point that a teacher can, through his interest and questioning, encourage the pupil to explore far beyond the ideas given in the unit.

Although the teacher expects independent work on the part of the students, he should at all times be aware of the work that is being done by individuals in the class. At the same time the teacher needs to avoid getting in the way of the pupil and blocking his originality. It is the skill of the teacher that determines when a child needs to be left completely alone, when he needs words of encouragement, when suggestions and leads on the topic are advantageous to the pupil, and when he needs to be referred to a mathematician in the community or nearby college.

There is no one best way to use these units, but many ways. The units are flexible as to their use. The more nearly a teacher relates this activity to the needs of the group the more effective will be the outcome.

At all times it is hoped that the enrichment units will open doors for the pupils and stimulate further exploration. No unit is thought of as an end in itself, but rather as a start on an idea. Some students, no doubt, will be challenged to go far beyond the ideas developed in the unit.

JUNIOR HIGH SCHOOL YEARS SUBCOMMITTEE

The subcommittee which gathered, appraised, and organized the materials for the Junior High School Years Section consisted of the following.

ALICE M. HACH, Ann Arbor Public Schools, Ann Arbor, Michigan

EUGENE P. SMITH, Wayne State University, Detroit, Michigan

BRUCE R. VOGELI, Bowling Green State University, Bowling Green, Ohio

LAUREN G. WOODBY, U. S. Department of Health, Education and Welfare, Washington, D. C.

JOSEPH N. PAYNE, *Chairman*

University of Michigan, Ann Arbor, Michigan

14

SOME PUZZLERS
FOR THINKERS

EUGENE P. SMITH

Wayne State University
Detroit, Michigan

Puzzlers which do not place heavy demands on previously learned skills and which students can reason out add spice to many mathematics class. The puzzlers may be used individually as *Puzzle for the Day* (or *Week*) for the entire class or for selected students.

The set presented in this chapter contains many ancient puzzlers, some of which are given in a modern setting. The section at the end on magic squares combines the puzzle element with an analysis of reasons for making such squares.

Students who like puzzle problems may wish to study those chapters in the Elementary School Years Section of this yearbook entitled, "Short Cuts and Why They Work," "A Method of Front-End Arithmetic," and "Tricks and Why They Work."

More advanced problems and puzzlers will be found in The High School Years Section of the Twenty-Eighth Yearbook in the following chapters: "Problem Solving," "More Non-Routine Problems," and "What Every Young Mathlete Should Know."

TEST YOUR SKILL ON THESE

Mathematical puzzles have fascinated the best of thinkers for centuries. Do you consider yourself a good problem solver? If so, perhaps this chapter will tickle your imagination. The following problems will serve as a good warm-up. They may even help to sweep away some of the cobwebs that tend to collect in almost any "upper story."

We should warn you to "keep your thinking cap on" because an answer which seems obvious may not be the correct one.

1. The Bookworm

Two volumes of *Life on the Moon* stand side by side in order on a book-shelf with Volume I left of Volume II and the bindings facing you. Each cover is $\frac{1}{4}$ inch thick, and each book without its covers is $1\frac{1}{2}$ inches thick. If a bookworm should eat his way directly from page one, Volume I to the last page of Volume II how far would he travel?

2. Making a Chain

Assume you have six sections of chain, each consisting of four links. If the cost of cutting open one link is 10¢, and welding it together again 25¢, what is the *least* it should cost to have the six pieces joined into one chain?

3. Socks

In your bureau drawer there are 10 blue socks and 16 red socks. You reach into the drawer in the dark to get a pair of socks. What is the smallest number of socks you must take out to make sure of getting a pair that match?

4. Age and Month

Write down your age in years. Multiply the number you have written by 10 and add 5. Multiply this sum by 10 again. Add the number of the month in which you were born, counting January as 1, February as 2, and so on. Subtract 50. The first two numerals on the left will be your age. The next two will be the number of the month in which you were born. Can you figure out why this works?

5. The Bear Truth

A hunter walks due *south* 3 miles from a given point, then walks due *east* for another 3 miles, and there shoots a bear. He drags the bear 3 miles *north* from the spot where he shot it and finds he is at his original starting point. What is the color of the bear?

Now that you have been successful on the first question, you may be interested to know that the outstanding success of this hunter "went to his head." He decided to go to the opposite side of the earth to hunt. There he shot an animal of a much smaller size. What kind of animal was this? Where did he go? How many starting points are possible in this region?

6. Long Division

$$
\begin{array}{r}
EDB \\
2E5\overline{\smash{)}B7E9J} \\
\underline{HEJ} \\
EJA9 \\
\underline{E5K5} \\
A4J \\
\underline{A4J}
\end{array}
$$

What numeral does each of the letters in this problem stand for?

7. Double or Nothing

Suppose you were offered a job and were promised one cent for the first day's work, double that or two cents for the second day, four for the third, and so on, getting double the previous day's pay for every day in the month. Would you take the job? Now figure out how much you would have made had you taken the job for the month of March.

8. Spending

A man goes into a store and says to the proprietor: "Give me as much money as I have with me and I will spend $10 with you." It is done. The man repeats the operation in a second and a third store, after which he has no money left. How much did he start with?

9. Help Needed

A young college student, Mr. Kantstand Prospearuty, wishing to be subtle about how much money he needed, sent the following telegram to his father:

$$
\begin{array}{r}
\$SE.ND \\
\underline{MO.RE} \\
\$MON.EY
\end{array}
$$

If each of these letters stands for a digit, how much money did Kantstand want?

10. Plugging a Hole

How can you cut the board into two equal pieces to cover the hole completely?

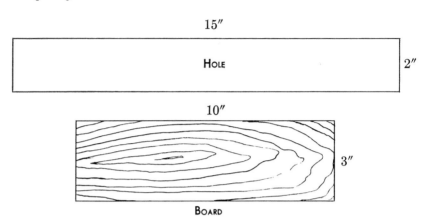

The answers to the above questions, or some hints on how to find the answers, may be found at the end of Section II.

There are two important skills that help to solve problems such as those presented at the beginning of this article—the ability to think logically and the ability to see patterns and relationships among the parts of a puzzle. How good are you at seeing and rearranging patterns? Each of the following puzzles will give you a chance to test your skill. Some toothpicks are needed for aids. For example, make diagrams with toothpicks, placing them so that there are five squares in each diagram as shown in Figures 1 and 2.

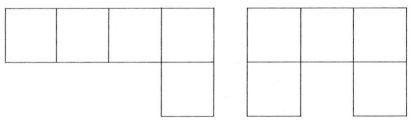

FIGURE 1   FIGURE 2

11. The Disappearing Square

Move three toothpicks (Fig. 1) to new positions and leave only four squares of equal area. All 16 of the toothpicks must be used in the final diagram and each toothpick must be a part of a square.

12. Another Disappearing Square

Move three toothpicks (Fig. 2) to new positions so that only four squares of equal area are formed.

Now that you have solved the first two toothpick puzzles, try your logic on the following one with 10 of the toothpicks.

13. The Double-Cross

Lay ten toothpicks in a row with each one about one inch apart. Cross these ten toothpicks in the following manner: take one pick at a time, jump two picks with it, and then place the jumping pick on top of the third pick to form an X. You should finish with five X's. Caution: When you jump a crossed pair, you are jumping two toothpicks.

PUZZLE GAMES

The next two puzzles are games played by two people. Twenty-one toothpicks are required for the first game.

14. Twenty-one

Lay 21 toothpicks in a row as in Puzzle 13. The object of the game is to make your opponent pick up the last toothpick by following these rules:

1. Players take alternate turns.
2. Each turn a player must pick up 1, 2, or 3 toothpicks as he chooses.
3. The player who must pick up the the last pick loses.

Is this just a game of chance or can it be played scientifically? We leave it up to you to find out.

15. Nim

There is a very famous game called *Nim* which is interesting and somewhat like the game above. Take as many toothpicks as you please and place them in several different piles. The number of piles does not matter. Again the object of the game is to make your opponent pick up the last toothpick by following the rules below:

1. Players take alternate turns.
2. Each player at his turn picks up as many toothpicks (at least one) as he pleases, as long as they *all* come from the *same* pile.
3. The player who picks up the last pick loses the game.

This game is much more intriguing than the previous one. It is also more difficult to analyze. Play the game to see if you can develop a winning strategy. If you look up the game of *Nim* in books of mathematical recreations, you may become a real expert at the game. Then you can always win from your unsuspecting opponents. Incidentally, we should tell you that the analysis involves binary (base two) arithmetic.

REASONING PLUS TRIAL AND ERROR

Solutions to the following puzzlers are dependent on intelligent trial and error and logical reasoning. (We might add that a bit of luck here and there may help.)

16. The Huntsmen and Noblemen

In a faraway land lived two and only two groups of people, huntsmen and noblemen. The huntsmen were incurable liars who never told the truth. The noblemen were incapable of telling a lie. A stranger, visiting in this land, asked three natives of the country to which groups they belonged. The first man said something the visitor did not hear. The second said, "He said he is a nobleman." The third man immediately retorted, "The first man is a liar because he is a huntsman." How many huntsmen and how many noblemen are there among the three natives? Can you tell which ones are noblemen and which are huntsmen?

17. A Fair Problem

A 158-pound farmer was going to the fair with his prize 39-pound fox, 19-pound goose, and a 40-pound sack of corn. On his way he came to a wide stream which he had to cross. It was too wide to swim across and there were no bridges. The only available means of crossing was a water-soaked rowboat. On it was this sign: "Beware—do not load with more than 200 pounds."

Now, if he took over the fox, the goose would eat the corn. If he took over the corn, the fox would eat the goose. (This would also happen, of course, if the two were left on the other side by themselves.)

Can you help this farmer solve his problem? If so, how?

18. The Cannibals and the Missionaries

Solving the farmer's problem may help you to give a helping hand to three distressed missionaries who hope to take three cannibals across a wide river. They are restricted, however, by the following conditions:

1. They have only one boat.
2. Only two people can ride in the boat for each crossing.
3. All missionaries but *only one* cannibal can row the boat.
4. There must *never* be fewer missionaries on one side than cannibals, for the cannibals will feast on the missionaries when they

outnumber them. (Count any people in the boat when it is on a given side as members of the group on that side.)

This problem can be solved. Can you do it? The real test, if you solve it once, is to be able to solve it again and again.

(To aid you in solving this problem, it may be helpful to use one set of coins of the same denomination to represent the missionaries and another set of three coins of a different denomination to represent the cannibals. Be sure to designate clearly which cannibal can row.)

19. The Three Jealous Men and Their Wives

Three men, traveling with their wives, came to a river which they wished to cross. The one available boat would accommodate only two people. Since the husbands were very jealous, no woman could be with a man unless her own husband was present. Under these severe handicaps, how can they get across the river using the one boat?

20. A Refreshing Problem

If you are still with us, refresh yourself by considering the problem of the 8-pint container full of chocolate milk shake. The story is that two boys have agreed to split the milk shake evenly between them. Unfortunately, they have only one 8-pint, one 5-pint, and one 3-pint container. None of the three containers has subdivision marks on it. How can these boys, using only these three containers, divide the 8 pints of milk shake *evenly* between them? Can you find two different solutions for this problem?

21. An Odd Problem

To end this set we shall propose this simple conjecture: "Every even number greater than 4 may be expressed as the sum of two odd prime numbers." For example, $6 = 3 + 3$, $8 = 5 + 3$, $\cdots$, $116 = 113 + 3$. Gather all of the evidence you can to support the conjecture. Do you believe this statement to be true? Can you find a counter example to disprove the conjecture? If not, can you prove the conjecture?

This problem is a very old and famous one. You can find out more about it by looking up the Goldbach Conjecture.

MAGIC SQUARES

A magic square is an array of numbers arranged in the pattern of a square in such a manner that the sums of the columns, rows, and diagonals of the array are all equal.

22. A Square to Complete

Suppose that the cells of the square array in Figure 3 are to be filled with the integers from 1 to 16.

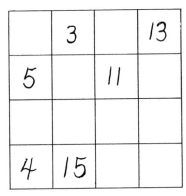

FIGURE 3

The table has been partially filled in. Copy the square as it is shown and fill in the empty cells to complete the magic square. These questions should help you complete the job:

1. What is the sum of all natural numbers from 1 to 16?
2. What should be the sum of each row, column, and diagonal of the magic square?
3. After you have filled in all of the cells, check your magic square by finding the sums of the rows, columns, and diagonals. Are they equal?
4. Add 16 to the number in each cell. Do you get another magic square?
5. Multiply the number in each cell of your original magic square by 20. Do you get another magic square?
6. In what other ways could you get new magic squares from your original one?

23. Magic Squares with Letters and Numerals

Let's try another magic square. Make a 3 × 3 square such as that shown in Figure 4. Using the numerals 1 through 9, inclusive, fill in the cells of your magic square.

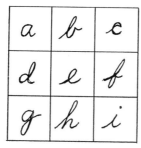

FIGURE 4

1. Find the sum of the numbers 1 through 9 (i.e., $1 + 2 + 3 + \cdots + 9$).

2. What should be the sum of each row, column, and diagonal? Hints: From here on, you have two alternatives—start guessing and solve the puzzle in this way; or, if you know some algebra, take the following steps.

(a) Let the letters from a through i represent the numerals to be placed in each cell as indicated in Figure 3.

(b) Complete the following mathematical sentences and you will get a start.

$$a + e + i = \underline{\hspace{1cm}}?$$
$$b + e + h = \underline{\hspace{1cm}}?$$
$$c + e + g = \underline{\hspace{1cm}}?$$
$$(a + e + i) + (b + e + h) + (c + e + g) = \underline{\hspace{1cm}}?$$

(c) By regrouping, using the associative and commutative properties in this last equation, we get:

$$(a + b + c) + (g + h + i) + 3e = \underline{\hspace{1cm}}?$$

(d) Can you find the numeral represented by e?

(e) After placing the numeral for e in the center cell, try placing 1 in a cell and following the consequences! Then try 2 in another cell. If you reach an impossible situation, start over by placing 1 in another cell. Keep trying and you will soon solve this puzzle.

Making Your Own Magic Squares

This is not the end of magic squares at all. You may like to try your hand at a 5×5 or higher-order magic square. Once you have one magic square, you can make numerous other magic squares of that same order from the first one. The squares with which we have worked have been magic-sum squares. There are also magic-product squares in which the products rather than the sums of the rows, columns, and diagonals are equal. Can you discover a magic-product square?

FURTHER EXPLORATION

In conclusion, if you have found this section on puzzles interesting, we invite you to explore the topic further by reading the references suggested below.

REFERENCES

1. GARDNER, MARTIN. "Mathematical Games." *Scientific American*. New York. (The column appears in each month's issue.)
2. GARDNER, MARTIN. *The Scientific American Book of Mathematical Puzzles & Diversions*. New York: Simon and Schuster, 1959.
3. GARDNER, MARTIN. *The Second Scientific American Book of Mathematical Puzzles & Diversions*. New York: Simon and Schuster, 1961.
4. KASNER, EDWARD, and NEWMAN, JAMES. *Mathematics & The Imagination*. New York: Simon and Schuster, 1940.
5. MADACHY, JOSEPH F. *Recreational Mathematics Magazine*. Idaho Falls. (Published bi-monthly.)
6. SCHAAF, WILLIAM L. *Recreational Mathematics: A Guide to the Literature*. (2nd ed.) Washington, D.C.: National Council of Teachers of Mathematics, a department of the National Education Association, 1958.

UNIT FRACTIONS

JULIA ADKINS

Central Michigan University
Mount Pleasant, Michigan

Using Egyptian fractions as a unifying theme, this short article helps students discover some of the fascinating things that can be done with fractions which have a numerator of 1. Only a background with fractions is needed for study of this historically interesting topic.

EGYPTIAN FRACTIONS

Ancient Egyptians (2000–1800 B.C.) did not know as much about fractions as we do. With the exception of $\frac{2}{3}$, they used only unit fractions, fractions with numerators of 1, for computation. Neither did they write their fractions as we do. They had special symbols for $\frac{2}{3}$ and $\frac{1}{2}$; but for the other fractions they used a fraction symbol, $\bigcirc$, under which they placed a symbol representing the value of the denominator. Here are a few illustrations of the way they wrote their fractions:

Our Fractions	Egyptian Fractions
$\frac{2}{3}$	
$\frac{1}{2}$	
$\frac{1}{3}$	
$\frac{1}{4}$	
$\frac{1}{5}$	
$\frac{1}{10}$	
$\frac{1}{30}$	

221

Information about the Egyptians' use of unit fractions is given in the oldest known mathematical book in the world. This book (*The Rhind Mathematical Papyrus*) was written about 1650 B.C., but contained the mathematical knowledge of an earlier period.

For some unknown reason, the Egyptians expressed fractions as the sum of *unlike* unit fractions. For example (using our symbols),

$$\tfrac{3}{8} \text{ would be expressed as } \tfrac{1}{4} + \tfrac{1}{8}.$$

If you had lived in those early times, you would have learned to express any fraction as the sum of unlike unit fractions. Try a few. At first you will probably have to guess, but later you should discover systematic methods.

EXERCISES

Exercise 1:

(a) Express each of these fractions as the sum of unlike unit fractions.

$$\tfrac{1}{2} =$$
$$\tfrac{1}{3} =$$
$$\tfrac{1}{4} =$$

(b) Can you get two different sets of answers for these fractions?
A leather scroll, said to have been found with the Rhind papyrus, contained the following fundamental formulas:

(1) $$\tfrac{1}{6} + \tfrac{1}{6} = \tfrac{1}{3}$$

(2) $$\tfrac{1}{6} + \tfrac{1}{6} + \tfrac{1}{6} = \tfrac{1}{2}$$

(c) Using these two formulas, how would you obtain the following formula?

(3) $$\tfrac{1}{3} + \tfrac{1}{6} = \tfrac{1}{2}$$

The following list of formulas was also found in the papyrus scroll. These were obtained from formula (3) upon division by 3, by 4, by 5, and so on.

$$\tfrac{1}{9} + \tfrac{1}{18} = \tfrac{1}{6}$$

$$\tfrac{1}{12} + \tfrac{1}{24} = \tfrac{1}{8}$$

$$\tfrac{1}{15} + \tfrac{1}{30} = \tfrac{1}{10}$$

$$\tfrac{1}{18} + \tfrac{1}{36} = \tfrac{1}{12}$$

$$\tfrac{1}{21} + \tfrac{1}{42} = \tfrac{1}{14}$$

$$\tfrac{1}{24} + \tfrac{1}{48} = \tfrac{1}{16}$$

$$\tfrac{1}{30} + \tfrac{1}{60} = \tfrac{1}{20}$$

$$\tfrac{1}{45} + \tfrac{1}{90} = \tfrac{1}{30}$$

$$\tfrac{1}{48} + \tfrac{1}{96} = \tfrac{1}{32}$$

(d) Were any omitted in the above list?

(e) By what number would you divide formula (3) to get the following formula?

$$\tfrac{1}{24} + \tfrac{1}{48} = \tfrac{1}{16}$$

Exercise 2:

(a) Can you find two unit fractions (like the ones in the above list) whose sum is $\tfrac{1}{18}$?

Examine the list carefully. Can you discover a relationship between the denominators of each pair of fractions on the left of the equal sign? Do you also see a relationship between the denominator of the fraction on the left-hand side of each formula and the denominator of the fraction immediately to the left of the equal sign? Do you see a short-cut rule for writing the sum of two fractions having this relationship? If so, write your rule.

Now let's see if you can write your short-cut rule as a formula. Here is a start and you finish the formula (see *Note A*, page 224, if you need help).

$$\frac{1}{n} + \frac{1}{-} = \frac{1}{-},$$

where n is any positive multiple of 3.

Are you wondering why the sum of two unit fractions of this type equals another unit fraction? See if you can determine mathematically why this is true. (See *Note B*, page 225, if you need help.)

(b) If you discovered the correct formula, you should now be able to complete the formulas that are started. Try to determine the fraction on the right-hand side without having to add the two fractions on the left. Then check your answer by addition.

$$\tfrac{1}{33} + \tfrac{1}{-} = \tfrac{1}{-}$$

$$\tfrac{1}{60} + \tfrac{1}{-} = \tfrac{1}{-}$$

$$\tfrac{1}{75} + \tfrac{1}{-} = \tfrac{1}{-}$$

(c) Using the general formula you discovered, try to develop another general formula that you could use to express a unit fraction having an *even* number in the denominator, as the sum of two unlike unit fractions. For example,

$$\tfrac{1}{2} = \tfrac{1}{3} + \tfrac{1}{6}$$

$$\tfrac{1}{4} = \tfrac{1}{6} + \tfrac{1}{12}$$

$$\tfrac{1}{6} = \tfrac{1}{9} + \tfrac{1}{18}.$$

Exercise 3:

A technique used by the Egyptians for expressing *any* fraction as the sum of unlike unit fractions is shown below.

Express $\tfrac{7}{8}$ as the sum of unlike unit fractions.
1 divided by 8 gives $\tfrac{1}{8}$
2 divided by 8 gives $\tfrac{1}{4}$
3 divided by 8 gives $\tfrac{1}{4} + \tfrac{1}{8}$
4 divided by 8 gives $\tfrac{1}{2}$
5 divided by 8 gives $\tfrac{1}{2} + \tfrac{1}{8}$
6 divided by 8 gives $\tfrac{1}{2} + \tfrac{1}{4}$
7 divided by 8 gives $\tfrac{1}{2} + \tfrac{1}{4} + \tfrac{1}{8}$

Exercise 4:

How was the above table developed? Using this technique in combination with any other, express each of the following fractions as the sum of unlike unit fractions:

$$\tfrac{3}{5}; \quad \tfrac{9}{10}; \quad \tfrac{4}{15}; \quad \tfrac{7}{15}.$$

You have become acquainted with the types of fractions with which the early Egyptians worked. Perhaps you would like to learn about the kinds of fractions with which the ancient people of other countries worked. See what you can find about the fractions of the early Greeks, Chinese, Romans, or Babylonians.

NOTES

References were made to two notes in order to provide additional help. These notes are listed below.

Note A. $\dfrac{1}{n} + \dfrac{1}{2n} = \dfrac{3}{2n} = \dfrac{1}{\dfrac{2n}{3}},$ where n is any positive multiple of 3.

If the above formula is written

$$\frac{3}{2n} = \frac{1}{n} + \frac{1}{2n},$$

where n is any integer greater than one,

additional relationships may be determined.

Note B. $\dfrac{1}{n} + \dfrac{1}{2n} = \dfrac{3}{2(3p)} = \dfrac{1}{2p}$ Let $n = 3p$, where p is an integer greater than zero.

ADDITIONAL FORMULAS

Many additional formulas could be developed. A few of them are the following ones:

(1) $\dfrac{1}{n} + \dfrac{1}{3n} = \dfrac{4}{3n}$ where n is any integer greater than 1.

(2) $\dfrac{1}{n} + \dfrac{1}{4n} = \dfrac{5}{4n}$ where n is any integer greater than 1.

(3) $\dfrac{1}{n} + \dfrac{1}{5n} = \dfrac{6}{5n}$ where n is any integer greater than 1.

(4) $\dfrac{1}{n} + \dfrac{1}{xn} = \dfrac{x + 1}{xn}$ where x and n are any integers greater than 1.

(5) $\dfrac{1}{n} = \dfrac{1}{n + 1} + \dfrac{1}{n(n + 1)}$ where n is any integer greater than 1.

Suggestion to teachers: Prepare a guide sheet in which the students are led to discover formulas such as (1), (2), and (3), above. With this type of work as a background, they could be guided to discover the general formula (4).

The following formula will give more than one pair of unit fractions:

(6) $\dfrac{1}{n} = \dfrac{n + x}{n(n + x)}$

$\qquad = \dfrac{1}{n + x} + \dfrac{x}{n(n + x)}$ where x and n are any integers greater than 1, and x is a factor of n which is less than n.

The formula listed and illustrated below will also give more than one pair of unlike unit fractions whose sum is equal to another unit fraction:

(7) $\dfrac{1}{n} = \dfrac{1}{x} + \dfrac{1}{y}$ where n, x, and y are integers greater than 1.

Illustration:

$$\frac{1}{8} = \frac{1}{x} + \frac{1}{y}$$

$$xy = 8y + 8x$$

$$xy - 8y = 8x$$

$$(x - 8)y = 8x$$

$$y = \frac{8x}{x - 8}$$

$$= 8 + \frac{64}{x - 8}$$

Make a table of values for x and y such that:
  x is any integer greater than 8 and
  $(x - 8)$ divides 64.

x	y
9	72
10	40
12	24

Therefore,

$$\tfrac{1}{9} + \tfrac{1}{72} = \tfrac{1}{8}$$

$$\tfrac{1}{10} + \tfrac{1}{40} = \tfrac{1}{8}$$

$$\tfrac{1}{12} + \tfrac{1}{24} = \tfrac{1}{8}.$$

REFERENCES

NEUGEBAUER, OTTO. *The Exact Sciences in Antiquity*. Princeton: Princeton University Press, 1952. Chapter 4.

RANSOM, WILLIAM R. "One Over– ", *Mathematics Teacher* 54: 100–01; 1961.

SMITH, DAVID EUGENE. *History of Mathematics*. Boston: Ginn and Co., 1925. 2: 209–12.

VAN DER WAERDEN, B. L. *Science Awakening*. Groningen, Holland: P. Noordhoff Ltd., 1954. pp. 16–31.

16

ON DIVISIBILITY RULES

HAROLD TINNAPPEL

Bowling Green State University
Bowling Green, Ohio

This article presents a study of divisibility rules for 2, 3, 4, 5, 7, 9, and 11, when the number is expressed in base ten.

This study is appropriate for students who have an understanding of natural number and zero exponents. Although algebraic notation is used, no extensive background in algebra is needed.

DEVELOPING THE RULES

Is 654321789 divisible by 11? If so, there will be a remainder of 0 when 654321789 is divided by 11.

Frequently when we are faced with such a problem whose solution requires some laborious computation, we suspect that a little knowledge of the properties of the numbers involved would reduce our work. Suppose we wish to determine if the fraction $\dfrac{1111}{654321789}$ can be reduced. We see at a glance that the numerator is divisible by 11, for $1111 = 11 \cdot 101$. But what do we know about the denominator? In this section we will develop some rules which will tell whether or not the large number 654321789 is or is not divisible by 11, without actually performing this division. In fact, we will devise tests in this section which will tell whether or not a given number is divisible by 2, 3, 4, 5, 7, 8, 9 or 11.

In agreement with our conventional decimal representation, a number n with $k + 1$ digits is written in the following manner:

$$n = A \cdot 10^k + B \cdot 10^{k-1} + \cdots + M \cdot 10^3 + H \cdot 10^2 + T \cdot 10^1 + U \cdot 10^0,$$

where each coefficient $A, B, \cdots, M, H, T, U$ is one of the digits 0, 1, 2, $\cdots$, 9. You can easily surmise why we have chosen the symbols M, H, T, U for the particular places they occupy. We will refer to U as the units' digit, or the last digit, of n.

227

Divisibility by 2

We recall that an even number is one which is divisible by 2, and so each even number will be represented by the expression:

$$2n = 2(A \cdot 10^k + B \cdot 10^{k-1} + \cdots + T \cdot 10 + U)$$
$$= 2A \cdot 10^k + 2B \cdot 10^{k-1} + \cdots + 2T \cdot 10 + 2U.$$

Now we note that the last digit of any even number must be one of the following five possibilities:

$$2 \cdot 0 = 0, \quad 2 \cdot 1 = 2, \quad 2 \cdot 2 = 4, \quad 2 \cdot 3 = 6, \quad 2 \cdot 4 = 8.$$

Even multiples of the remaining five digits, 5, 6, 7, 8, 9, give two-digit products, but only the second digit (underlined here) will appear in the unit's place of the number $2n$ and so yield the same five digits already noted: $2 \cdot 5 = 1\underline{0}$, $2 \cdot 6 = 1\underline{2}$, $2 \cdot 7 = 1\underline{4}$, $2 \cdot 8 = 1\underline{6}$, $2 \cdot 9 = 1\underline{8}$. Summarizing then, any number whose last digit is one of the five possibilities: 0, 2, 4, 6, 8 (zero or an even digit) will be divisible by 2.

Exercise 1:

What units' digits are possible for an odd number?

Since we wish to use the technique in later cases, let us now prove the proposition just found in a different way. We say that the number e is even if, when divided by 2, there is a zero remainder. In the demonstrations that follow we will work with four-digit numbers for the sake of convenience. Of course, the proofs for numbers with a greater number of digits than four will not differ materially from that which we give. We will first illustrate with a numerical example. Note in particular that our division by 2 does not involve the first three coefficients, but is affected by the value of the units' digit; for example,

$$\frac{4756}{2} = \frac{4 \cdot 10^3 + 7 \cdot 10^2 + 5 \cdot 10 + 6}{2}$$

$$= 4 \cdot 500 + 7 \cdot 50 + 5 \cdot 5 + \frac{6}{2} = 2375 + 3.$$

Now, performing this same division on the even number e, we have

$$\frac{e}{2} = \frac{M \cdot 10^3 + H \cdot 10^2 + T \cdot 10 + U}{2}$$

$$= M \cdot 500 + H \cdot 50 + T \cdot 5 + \frac{U}{2} = q.$$

For the quotient, q, to be an integer it is necessary and sufficient for $\dfrac{U}{2}$ to be an integer. Therefore U must be equal to 0, 2, 4, 6, or 8, which is in agreement with our previous result.

Divisibility by 3

A number is divisible by 3 if and only if the sum of its digits is divisible by 3. This fact is demonstrated in the following manner (we note that $10^3 = 999 + 1$, $10^2 = 99 + 1$, $10 = 9 + 1$):

$$\frac{n}{3} = \frac{M \cdot 10^3 + H \cdot 10^2 + T \cdot 10 + U}{3}$$

$$= \frac{M(999 + 1) + H(99 + 1) + T(9 + 1) + U}{3}$$

$$= 333M + 33H + 3T + \frac{M + H + T + U}{3}.$$

To demonstrate the use of this rule, let us see if 37,984,268 is divisible by 3. The sum of the digits is $3 + 7 + 9 + 8 + 4 + 2 + 6 + 8 = 47$. Now we must decide if 47 is divisible by 3. The sum of the digits of 47 is $4 + 7 = 11$. We must decide if 11 is divisible by 3. The sum of the digits of 11 is $1 + 1 = 2$. But 2 is not divisible by 3, so 11 is not divisible by 3, 47 is not divisible by 3; and so finally, the original number 37,984,268 is also not divisible by 3. Is 18,437,827 divisible by 3?

Divisibility by 4

Let us determine the conditions under which a number is divisible by 4. If we examine the quotient q we obtain

$$\frac{n}{4} = \frac{M \cdot 10^3 + H \cdot 10^2 + T \cdot 10 + U}{4} = 250M + 25H$$

$$+ \frac{T \cdot 10 + U}{4} = q.$$

The number q will be an integer if and only if $T \cdot 10 + U$ is 00 or divisible by 4. That is, in testing for the divisibility by 4 we can ignore all digits of the given number except the last two.

Exercise 2:

Tell which of the following are leap years (dates divisible by 4, except century years in which case they must be divisible by 400): 1592, 1816, 1968, 2242, 2634, 2856, 3000.

Exercise 3:

(a) If a number is divisible by 4, must it be even? (b) If a number is even, is it necessarily divisible by 4? (c) If the last digit of a number is either 0, 4, 8, must it be divisible by 4? Why? Can you prove that you are right?

Exercise 4:

Adapt the above demonstration to find the rule for divisibility by 8.

Divisibility by 5

Let us discover when a number is divisible by 5. We set

$$\frac{n}{5} = \frac{M \cdot 10^3 + H \cdot 10^2 + T \cdot 10 + U}{5} = 200M + 20H + 2T + \frac{U}{5} = q.$$

Hence q is an integer if and only if $U = 0$ or 5. Therefore, a number which is divisible by 5 ends in 0 or 5.

Divisibility by 9

The derivation of the rule for divisibility by 9 is similar to that used in finding the rule for divisibility by 3:

$$\frac{n}{9} = \frac{M \cdot 10^3 + H \cdot 10^2 + T \cdot 10 + U}{9}$$

$$= \frac{M(999 + 1) + H(99 + 1) + T(9 + 1) + U}{9}$$

$$= 111M + 11H + T + \frac{M + H + T + U}{9} = q.$$

If q is an integer, the sum of the digits must be divisible by 9.

Exercise 5:

(a) If a number is divisible by 9, must it be divisible by 3? (b) Is the converse true (that is, if a number is divisible by 3 must it also be divisible by 9)?

Exercise 6:

If we reverse the digits of the number 2365, we obtain 5632. If we reverse the digits of $n = M \cdot 10^3 + H \cdot 10^2 + T \cdot 10 + U$, we obtain $n' = U \cdot 10^3 + T \cdot 10^2 + H \cdot 10 + M$. Prove that if n is divisible by 9, then n' is also divisible by 9, and n' is not divisible by 9 whenever n is not. Prove that if n is divisible by 9, then so is

$$n'' = H \cdot 10^3 + U \cdot 10^2 + M \cdot 10 + T.$$

In fact, any number obtained by rearranging the order of the digits for n will also be divisible by 9 if n is. This comment helps explain why the checking method of "casting out nines" fails to reveal an error of an interchange (transposition) of digits.

Exercise 7:

Suppose that $a = M \cdot 10^3 + H \cdot 10^2 + T \cdot 10 + U$, and the number with digits reversed is $a' = U \cdot 10^3 + T \cdot 10^2 + H \cdot 10 + M$. Prove that $a - a'$ is divisible by 9.

Exercise 8:

Fill in the blank with the missing digit which makes the number divisible by 9: 7843__; 30__7; 1156__91; 811__623. Is there a unique answer in each case?

Divisibility by 11

In deriving the rule for the divisibility by 11, we use a technique similar to that used for deriving the rules for the divisibility by 3, except this time we write each power of 10 as a multiple of 11 increased or decreased by 1. We note that $10^3 = 1001 - 1$; $10^2 = 99 + 1$; $10 = 11 - 1$. We set

$$\frac{n}{11} = \frac{M \cdot 10^3 + H \cdot 10^2 + T \cdot 10 + U}{11}$$

$$= \frac{M(1001 - 1) + H(99 + 1) + T(11 - 1) + U}{11}$$

$$= 91M + 9H + T + \frac{(H + U) - (M + T)}{11} = q.$$

Hence q is an integer if and only if $(H + U) - (M + T)$ is 0 or is divisible by 11. Verbally we can state the rule in the following manner: Form the sum of the odd-ordered digits (starting with the units' digit, add every other one) and from this sum subtract the sum of the even-ordered digits (starting with the tens' digit, add every other one). If this difference is 0 or divisible by 11 (it may be a negative number or zero) then n is also divisible by 11. To illustrate, let $n = 89157486$. Then $(6 + 4 + 5 + 9) - (8 + 7 + 1 + 8) = 24 - 24 = 0$. Hence n is divisible by 11.

Exercise 9:

If possible, fill in each blank with the missing digit so that the resulting number is divisible by 11: 51__1; 34769__; 2015493__7; 78__93486.

Divisibility by 7

The last divisibility rule we give is the rule to determine if a number is divisible by 7. Since the rule is somewhat more involved than the preceding rules, it may not be used as frequently as the other rules. However, it so well demonstrates another instance in which the form of the number is changed without altering its value that it merits careful study. To illustrate the rule, suppose we test 6895 to see if it is divisible by 7. We first drop the last digit 5, leaving 689 from which we subtract $2 \cdot 5$, leaving 679. Repeating the rule, we drop the last digit 9, leaving 67; and then subtract $2 \cdot 9$ from 67, leaving 49. Since 49 is divisible by 7, then 6895 is also divisible by 7.

The proof of this rule may be given in the following way:

$$\frac{n}{7} = \frac{M \cdot 10^3 + H \cdot 10^2 + T \cdot 10 + U}{7} = q.$$

Now $q' = q - 3U$ is an integer if and only if q is an integer.

$$q - 3U = \frac{M \cdot 10^3 + H \cdot 10^2 + T \cdot 10 + U}{7} - 3U$$

$$= \frac{M \cdot 10^3 + H \cdot 10^2 + T \cdot 10 + U - 21U}{7}$$

$$= \frac{M \cdot 10^3 + H \cdot 10^2 + T \cdot 10 - 20U}{7}$$

$$= 10 \frac{M \cdot 10^2 + H \cdot 10 + T - 2U}{7}.$$

From the last expression we observe that n will be divisible by 7 whenever the difference between the number obtained by dropping the units' digit from n, $(M \cdot 10^2 + H \cdot 10 + T)$, and twice the units' digit, $(2U)$, is divisible by 7.

We might note here that this test may be adapted to also test for the divisibility by 3. That is, n is divisible by 3 if and only if

$$\frac{n - U}{10} - 2U$$

is divisible by 3.

Exercise 10:

Using the above rule, test the following numbers for divisibility by 7:
830456; 91362; 5873; 31029; 69146; 142857.

Exercise 11:

A test for divisibility by 7 which is faster for large numbers than the
one given is this: the number n is divisible by 7 if

$$m = \frac{n - U}{10} - 9U$$

is divisible by 7. Can you prove this result? Incidentally, the same
test can be used for divisibility by 13; that is, n is divisible by 13 if
and only if m is divisible by 13. In looking at the two tests for divisibility by 7, can you detect a pattern? Does this suggest to you ways
of inventing tests for divisibility?

Exercise 12:

To illustrate the application of the above rules of divisibility, for
each of the following find an equivalent fraction in which the numerator and denominator do not have common factors.

(a) $\frac{117}{156}$ (e) $\frac{111}{10101}$

(b) $\frac{156}{8580}$ (f) $\frac{7364}{23331}$

(c) $\frac{97}{907}$ (g) $\frac{781}{111111}$

(d) $\frac{1043}{7063}$ (h) $\frac{1234}{4321}$

NUMERATION SYSTEMS

JOSEPH N. PAYNE

The University of Michigan
Ann Arbor, Michigan

The first part of this article deals with numeration systems employing place value but using a base other than 10. A study of divisibility by 2, when numbers are expressed in a different base, points up the meaning of divisibility as a property of numbers and not of the way in which they are represented.

The last section, on a prime numeration system, gives an example of a system with characteristics in sharp contrast to our own notational system. Some familiarity with exponents is needed for this last section.

Many books now contain chapters on systems of numeration. This article contains: (a) a brief review of numeration systems with various bases as presented in many texts; (b) a study of divisibility by 2, using various bases; and (c) an introduction to a way of expressing numbers, using a system that is quite unlike our own base-ten numeration system.

DIFFERENT NUMBER BASES

Our own decimal system has a base of ten. To find the number of cross-marks in this set $\{\times\times\times\times\times\ \ \times\times\times\times\times\ \ \times\times\times\}$ we "pull out" one group of ten cross marks as shown below

$$\{(\times\times\times\times\times\ \ \times\times\times\times\times)\ \ \times\times\times\}$$

and have three single cross marks left. The symbol 13 is, of course, used to designate the one group of ten and three "singles."

Using a base of seven, we have a group of seven and six singles,

$$\{(\times\times\times\times\times\ \ \times\times)\ \ \times\times\times\times\times\times\}$$

and we write the numeral 16_{seven} to show how many cross marks are in the set. The "seven" written below and to the right shows the base.

In a seven system, when you get seven groups of seven, you regroup getting one group of "seven sevens." The cross marks below have been grouped to show one group of "seven sevens," two groups of seven, and five singles. We write the numeral 125_{seven} to show the number of marks.

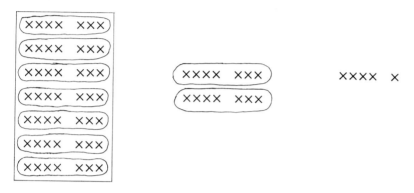

Exercises A:

1. Use base ten and write the number of cross marks above.
2. Tom is 24_{five} years old. Express his age in base ten; in base seven.
3. There are 100_{ten} pages in a book. Express this number in base nine; in base five.
4. 342_{five} can be expressed as $3 \times (5 \times 5) + 4 \times 5 + 2 \times 1$, or as $3 \times 5^2 + 4 \times 5 + 2 \times 1$. This is called expanded notation. Express 473_{eight} in expanded notation. Do the same for 265_{twelve}.
5. To add 34_{five} and 23_{five}, you think
   - $4 + 3 = 1$ group of five and 2 singles
   - Regrouping the fives, 3 fives and 2 fives $+ 1$ five $= 1$ group of "five fives" and 1 five; two singles are left over
   - The sum is 112_{five}.

   You might write it like this:

   $34_{\text{five}} = 3$ fives $+ 4$ ones
   $23_{\text{five}} = 2$ fives $+ 3$ ones

   5 fives $+ 7$ ones $= 6$ fives $+ 2$ ones (using "5", "7" and "6" as in base ten)
   $= 1$ "five fives" $+ 1$ five $+ 2$ ones or
   112_{five}.

   Now, find the following sums:
   (a) $12_{\text{five}} + 22_{\text{five}}$
   (b) $34_{\text{seven}} + 25_{\text{seven}}$

(c) $38_{twelve} + 11_{twelve}$

(d) $246_{seven} + 642_{seven}$

(e) $1463_{seven} + 2641_{seven} + 3456_{seven}$.

6. To subtract 241_{six} from 431_{six} you could write:

$$431_{six} = 4 \text{ "six sixes"} + 3 \text{ sixes} + 1 \text{ one}$$
$$241_{six} = 2 \text{ "six sixes"} + 4 \text{ sixes} + 1 \text{ one}$$

or

$$431_{six} = 3 \text{ "six sixes"} + 9 \text{ sixes} + 1 \text{ one}$$
$$241_{six} = 2 \text{ "six sixes"} + 4 \text{ sixes} + 1 \text{ one}$$

$$1 \text{ "six sixes"} + 5 \text{ sixes} + 0 \text{ one}$$

or

$$150_{six} .$$

(Note that the "9" is used as in base ten. One group containing "six sixes" was "broken apart" or "decomposed" in the second step.)

Find the following differences:

(a) $42_{six} - 13_{six}$

(b) $243_{six} - 125_{six}$

(c) $345_{seven} - 126_{seven}$

(d) $736_{eight} - 457_{eight}$.

7. To find the product 3×5 in a base seven system, you may think "$3 \times 5 = 15_{ten}$." This is 2 groups of seven and one left over. Hence, $3 \times 5 = 21_{seven}$.

Find these products and express in the given base:

(a) $5 \times 6_{seven}$ (d) $8 \times 8_{nine}$

(b) $4 \times 5_{eight}$ (e) $4 \times 12_{six}$

(c) $7 \times 9_{ten}$ (f) $14_{seven} \times 12_{seven}$.

Our Decimal System

There are five major characteristics of our decimal system:

1. SYMBOLS OR DIGITS: 0, 1, 2, 3, 4, 5, 6, 7, 8, 9. These digits have evolved historically. They name the number of elements in sets with fewer elements than the base.

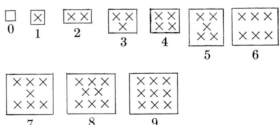

2. BASE OF TEN.

3. PLACE VALUE, INCREASING RIGHT TO LEFT. When the digit 2 is in the second place from the right, it designates two groups of ten. In the third place from the right it designates two groups of "ten 10's." The *place* the digit occurs determines the value it represents. The value of each place is $\frac{1}{10}$ the value of the place to its left. It is necessary to say that value increases right to left; otherwise 34 might mean $3 \cdot 1 + 4 \cdot 10$ instead of $3 \cdot 10 + 4 \cdot 1$.

4. VALUES DESIGNATED BY THE VARIOUS DIGITS ARE ADDED. For example, 245 means $2 \cdot (10^2) + 4 \cdot 10 + 5 \cdot 1$. Otherwise it would be conceivable that 24 could mean $2 \cdot 10 - 4 \cdot 1$ instead of $2 \cdot 10 + 4 \cdot 1$.

5. AGREED-UPON STARTING PLACE. For a whole number, the digit farthest to the right designates the ones' place. In a numeral such as 234.7 the decimal point designates the starting place. The digit to the left of the decimal point designates the ones' place.

So far in studying systems with other bases we have assumed these same characteristics except for the base and, of course, the number of digits. For a seven system, seven digits—0, 1, 2, 3, 4, 5, and 6—are needed.

Exercises B:

1. How many digits are needed for a base-twelve system?

2. Use the digits 0–9 in their base-ten sense and t and e as digits for "ten" and "eleven." Then $4t_{\text{twelve}}$ means $4 \times 12_{\text{ten}} + 10 \times 1_{\text{ten}}$ or 58_{ten} .

Express each of the following in base ten:

(a) $5e_{\text{twelve}}$ (b) $2t_{\text{twelve}}$ (c) 241_{twelve} (d) 4256_{twelve}
(e) $12_{\text{twelve}} + 19_{\text{twelve}}$ (f) $241_{\text{twelve}} + 599_{\text{twelve}}$ (g) $46_{\text{twelve}} - 19_{\text{twelve}}$
(h) $365_{\text{twelve}} - 1t4_{\text{twelve}}$ (i) $7 \times 5_{\text{twelve}}$ (j) $t \times t_{\text{twelve}}$

3. Suppose that in a base-five system we use the digits $\emptyset, /, \lrcorner, \sqcup, \boxslash$ to name the number of marks in each set shown below.

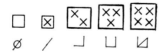

Complete the addition and multiplication shown below.

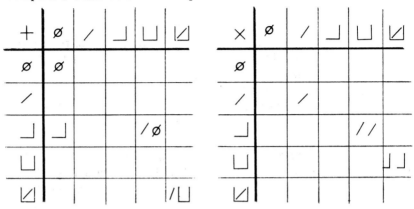

EXAMPLES: The following show how tables are used to subtract and divide.

1. $/\emptyset - \sqcup = ?$ Since $\sqcup + \lrcorner = /\emptyset$, from the addition table the answer is $\lrcorner$.

2. $// \div \sqcup = ?$ Since $\sqcup \times \lrcorner = //$, from the multiplication table, the answer is $\lrcorner$.

(a) $/\lrcorner - \sqcup = ?$ (c) $\lrcorner\lrcorner \div \sqcup = ?$

(b) $/\sqcup - ⊠ = ?$ (d) $\lrcorner\lrcorner \div ⊠ = ?$

4. Suppose place value increased left-to-right with the first digit on the left designating the ones' place, then 245 would mean
$$2 \cdot 1 + 4 \cdot 10 + 5 \cdot 10^2.$$
Find the following in base ten and write the answers with place value increasing left to right.

(a) $46 + 25$ (d) $342 - 941$ (g) 342×149

(b) $267 + 349$ (e) 4×32 (h) $42 \div 2$

(c) $64 - 13$ (f) 17×98 (i) $27 \div 8$

DIVISIBILITY BY TWO

A number is divisible by 2 if and only if there is a remainder of 0 when the number is divided by 2. For a number as small as 7, for example, you can draw a picture showing three groups of two and a remainder of 1: (× ×)(× ×)(× ×) × . There is an extra cross mark which does not have a "mate." Hence, 7 is not divisible by 2. For larger numbers such as 563, you can divide, find a remainder of 1, and conclude that it is not divisible by 2. To draw cross marks would be tedious. But the essential idea is that if you have an extra one with no mate, the number is not divisible by 2. If there is not an extra cross mark, the number is divisible by 2.

Of course, you know by now that you can look at a numeral in base ten and decide by looking at the last digit on the right if the number is divisible by 2. If the last digit is 0, 2, 4, 6, or 8 the number is divisible by 2; if the last digit is 1, 3, 5, 7, or 9 it is not divisible by 2. Let's see why this works.

First examine the numbers 10; 100; 1,000; 10,000 and so on—the values of each place in a base-ten system. To show that 10 is divisible by 2, it is easy to draw a picture showing five groups of 2 and a remainder of

0. $\left(\times\times\right)\left(\times\times\right)\left(\times\times\right)\left(\times\times\right)\left(\times\times\right)$. Now the value of the next

place to the left is 10×10 or 100. This would be 10 groups of "five twos." Then 100 is divisible by 2. Likewise 1,000 is 100 groups of "five twos" and is divisible by 2. Then each of the numbers 10, 100, 1,000, 10,000, etc., is divisible by 2. From this you can see that numbers such as 40; 500; 6,000 are divisible by 2. For example, $40 = 4\cdot10$; since 10 is divisible by 2, $4\cdot10$ is divisible by 2.

Now you can see why a number such as 5,658 is divisible by 2. First express it in expanded notation: $5642 = 5\cdot1000 + 6\cdot100 + 4\cdot10 + 8$. Each of the numbers $5\cdot1000$, $6\cdot100$, and $4\cdot10$ is divisible by 2. Thus, the only number left to check is 8. Since 8 is divisible by 2, the number is divisible by 2. In general, if a is the thousand's digit, b the hundred's digit, c the ten's digit, and d the unit's digit of a four-digit numeral, then the number can be expressed as: $a\cdot1000 + b\cdot100 + c\cdot10 + d$. Each of the numbers $a\cdot1000$, $b\cdot100$ and $c\cdot10$ is divisible by 2. Hence, if d is divisible by 2, the number is divisible by 2. This same reasoning applies to any number of digits.

Numbers Expressed in Other Bases

How would you decide if numbers such as 42_{six}, 35_{seven}, 142_{nine}, and 34275_{eleven} are divisible by 2? See if you can find a way before reading further. The questions in these next exercises should help you see when a number expressed with an even base is divisible by 2.

Exercises C:

1. $42_{six}\quad = 4\cdot6 + 2$. Is $4\cdot6$ divisible by 2? Is 2?

2. $84_{twelve} = 8\cdot12 + 4$. Is $8\cdot12$ divisible by 2? Is 4?

3. $342_{eight} = 3\cdot64 + 4\cdot8 + 2$. Are $3\cdot64$ and $4\cdot8$ divisible by 2? Is 2?

4. $231_{four} = 2 \cdot 16 + 3 \cdot 4 + 1$. Are $2 \cdot 16$ and $3 \cdot 4$ divisible by 2? Is 1?

5. $67_{eight} = 6 \cdot 8 + 7$. Is $6 \cdot 8$ divisible by 2? Is 7?

6. If the base is even, will the value represented by every place except the ones' place be divisible by 2?

7. How can you tell if a number is divisible by 2 when it is written in an even-base notation?

You are correct if you concluded that a number is divisible by 2 when expressed in an even base if the number for the ones' place is divisible by 2.

The Generalization for an Odd Base

The generalization for an odd base is a little more difficult. Work through the following exercises and see if you can discover the generalization.

Exercises D:

1. $40_{seven} = 4 \cdot 7$.   Is $4 \cdot 7$ divisible by 2?

2. $50_{seven} = 5 \cdot 7$.   Is $5 \cdot 7$ divisible by 2?

3. $60_{seven} = 6 \cdot 7$.   Is $6 \cdot 7$ divisible by 2?

4. $70_{nine} = 7 \cdot 9$.   Is $7 \cdot 9$ divisible by 2?

5. $80_{nine} = 8 \cdot 9$.   Is $8 \cdot 9$ divisible by 2?

6. If a two-digit numeral ends in 0, and the base is odd, when is the number divisible by 2?

7. $42_{seven} = 4 \cdot 7 + 2$. Is $4 \cdot 7$ divisible by 2? Is 2? Is $4 \cdot 7 + 2$? Is $4 + 2$?

8. $52_{seven} = 5 \cdot 7 + 2$. Is $5 \cdot 7$ divisible by 2? Is 2? Is $5 \cdot 7 + 2$? Is $5 + 2$?

9. $43_{seven} = 4 \cdot 7 + 3$. Is $4 \cdot 7$ divisible by 2? Is 3? Is $4 \cdot 7 + 3$? Is $4 + 3$?

10. $53_{seven} = 5 \cdot 7 + 3$. Is $5 \cdot 7$ divisible by 2? Is 3? Is $5 \cdot 7 + 3$? Is $5 + 3$?

11. $64_{nine} = 6 \cdot 9 + 4$. Is $6 \cdot 9$ divisible by 2? Is 4? Is $6 \cdot 9 + 4$? Is $6 + 4$?

12. $74_{nine} = 7 \cdot 9 + 4$. Is $7 \cdot 9$ divisible by 2? Is 4? Is $7 \cdot 9 + 4$? Is $7 + 4$?

13. $65_{nine} = 6 \cdot 9 + 5$. Is $6 \cdot 9$ divisible by 2? Is 5? Is $6 \cdot 9 + 5$? Is $6 + 5$?

14. $75_{nine} = 7 \cdot 9 + 5$. Is $7 \cdot 9$ divisible by 2? Is 5? Is $7 \cdot 9 + 5$? Is $7 + 5$?

15. In a two-place numeral, if both digits represent even numbers, is the number divisible by 2? What if both digits represent odd numbers? What if one is odd and one is even?

16. For a two-digit numeral with an odd base, when is the number divisible by 2?

You are correct if you concluded that a number is divisible by 2 when expressed with an odd base if the sum of all the numbers for the digits (disregarding place value) is divisible by 2. This generalization is true no matter how many digits it takes to express the number.

Now you should understand that divisibility is a property of numbers, and not of the numerals used to name it. However, we do look at the numeral to make it easy to recognize whether or not the number it represents is divisible by 2.

SUGGESTIONS FOR FURTHER STUDY ON DIVISIBILITY

1. If a number is expressed in base ten, when is it divisible by 5? By 4? By 8? By 3? By 6? By 9?
2. Devise tests for divisibility by 3, 4, and 5 in base six.
3. Devise tests for divisibility by 3, 4, 5, and 6 in base eight and in base seven.

AN UNUSUAL WAY TO EXPRESS NUMBERS

Can you imagine a numeration system where these statements are true? $5 \cdot 4 = 9$; $12 \cdot 24 = 36$; $8 \div 2 = 6$; $52 \div 21 = 31$. There *is* such a system. But as you will find out, this system is quite different from the systems studied earlier.

In this strange new system, *there is no base.* The value of each place is a consecutive prime, beginning right to left. (A prime number is a counting number greater than 1 that has only itself and 1 as factors. The first ten prime numbers are 2, 3, 5, 7, 11, 13, 17, 19, 23, and 27.) For the moment, we will use a dash for each place and will write the value represented by each place below the dash.

$$\underline{\quad}\ \ \underline{\quad}\ \ \underline{\quad}\ \ \underline{\quad}\ \ \underline{\quad}\ \ \underline{\quad}\ \ \underline{\quad}\ \ \underline{\quad}\ \ \underline{\quad}\ \ \underline{\quad}$$

| 29 | 23 | 19 | 17 | 13 | 11 | 7 | 5 | 3 | 2 |

Continuing to the left, the value of each place is the next prime number. As with our own base-ten system, this new system continues indefinitely to the left. Let's call our new system the *prime system.*

We will write 4_p, for example, to mean 4 in the prime system. In this prime system, the numerals in any place do not tell you how many groups of that place you have. *The numerals in any place show the exponent for the place!* For example, 4_p means 2^4; 8_p means 2^8.

In our previous systems we added the values of the places. In our prime system, we *multiply* the values of the places. For example 45_p

means $3^4 \cdot 2^5$. You may, if you wish, express this in base ten. $45_p = 3^4 \cdot 2^5 = 81 \cdot 32_{\text{ten}} = 2592_{\text{ten}}$. You can see that 45_p is a large number.

You may not yet have studied exponents of 0. For example, we define 4^0 as 1, 25^0 as 1, 267^0 as 1. In general, we agree that any non-zero number with an exponent of 0 is 1. (Can you find out why we choose this definition?)

Then 40_p means $3^4 \cdot 2^0{}_{\text{ten}} = 81 \cdot 1_{\text{ten}} = 81_{\text{ten}}$.

EXAMPLE: $2140_p = 7^2 \cdot 5^1 \cdot 3^4 \cdot 2^0{}_{\text{ten}} = 49 \cdot 5 \cdot 81 \cdot 1_{\text{ten}} = 19{,}845_{\text{ten}}$.

Exercises E:

Express each of the following numbers in a base ten numeration system.

1. 31_p	**5.** 221_p	**9.** 1000_p
2. 24_p	**6.** 1234_p	**10.** $1{,}000{,}000_p$
3. 15_p	**7.** 201_p	**11.** $1{,}000{,}001_p$
4. 41_p	**8.** 1201_p	**12.** $1{,}000{,}000{,}111_p$

To be able to express numbers from base ten in the prime system, you must be able to find the prime factors of a number. Look at this example which shows one way to get the prime factors of 460_{ten},

$$460_{\text{ten}} = 2 \cdot 230 = 2 \cdot 2 \cdot 115 = 2 \cdot 2 \cdot 5 \cdot 23 = 2^2 \cdot 5 \cdot 23.$$

You get the same factors no matter what steps you take; $460 = 20 \cdot 23 = 4 \cdot 5 \cdot 23 = 2 \cdot 2 \cdot 5 \cdot 23$. Perhaps you already know that any whole number greater than 1 can be expressed as the product of prime factors in one and only one way.

Exercises F:

Express each of the following numbers (already in base ten) as the product of prime factors.

1. 50	**3.** 125	**5.** 1000	**7.** 350	**9.** 2000
2. 100	**4.** 400	**6.** 260	**8.** 135	**10.** 630

Express each of the numbers below in the prime system, as in the following examples: (a) $7^2 \cdot 5^4 \cdot 3^0 \cdot 2^3 = 2403_p$; (b) $960_{\text{ten}} = 5^1 \cdot 3^1 \cdot 2^6 = 116_p$.

11. $7^4 \cdot 5^2 \cdot 3^4 \cdot 2^0 =$	**19.** 125_{ten}
12. $5^4 \cdot 3^5 \cdot 2^7 =$	**20.** 400_{ten}
13. $17^2 \cdot 13^0 \cdot 11^0 \cdot 7^5 \cdot 5^4 \cdot 3^2 \cdot 2^4 =$	**21.** 1000_{ten}
14. $7^4 = 7^4 \cdot 5^0 \cdot 3^0 \cdot 2^0 =$	**22.** 260_{ten}
15. 13^2	**23.** 350_{ten}
16. 29^5	**24.** 135_{ten}
17. 50_{ten}	**25.** 2000_{ten}
18. 100_{ten}	**26.** 630_{ten}

27. Count from 1 to 100_{ten} in the prime system. Is there any pattern to the way you write numbers in the prime system?

MULTIPLICATION AND DIVISION

To find the product $2^4 \cdot 2^5$, you add the exponents, $2^4 \cdot 2^5 = 2^{4+5} = 2^9$. Now do you see why $4_p \cdot 5_p = 9_p$? Also $(3^4 \cdot 2^5) \cdot (3^2 \cdot 2^3) = (3^{4+2})(2^{5+3}) = 3^6 \cdot 2^8$. In the prime system, $45_p \cdot 23_p = 68_p$. You add the values of the respective places.

The quotient $\dfrac{2^8}{2^3} = 2^{8-3} = 2^5$. You subtract the exponents. Then $8_p \div 3_p = (8 - 3)_p = 5_p$. Also, $\dfrac{5^7 \cdot 3^4 \cdot 2^3}{5^6 \cdot 3^1 \cdot 2^2} = (5^{7-6})(3^{4-1})(2^{3-2}) = 5 \cdot 3^3 \cdot 2$. Then $743_p \div 612_p = 131_p$. You subtract the numbers in the respective places.

By now you must have realized that you can have more than one digit in a place. For example, to express 2^{12} in the prime system, you must put 12 in the "two" place. We can keep our place values in order by putting multi-digit numerals for any given place in brackets. Thus $4\,[12]_p$ would mean $3^4 \cdot 2^{12}$

Exercises G:

Find these products and quotients in a prime system.
1. $423_p \cdot 516_p$
6. $87_p \cdot 62_p$
2. $4\,[12]_p \cdot 32_p$
7. $494_p \div 162_p$
3. $[16]\,2_p \cdot 14_p$
8. $79624_p \div 40213_p$
4. $96_p \cdot 22_p$
9. $9876543_p \div 8765432_p$
5. $42756_p \cdot 56192_p$
10. $[12][42]\,6_p \div 982_p$
11. In the prime system when is a number divisible by 2? By 3? By 4? By 5? By 6? By 7? By 8? By 9?

Perhaps such a system doesn't have much practical value. But it should help you gain a deeper understanding of prime numbers. Also it should help you to understand better our own system of numeration.

SUGGESTIONS FOR FURTHER EXPLORATION

1. Would it be easy to add and subtract in the prime system? Explain.

2. Knowing that $\dfrac{1}{2^3} = 2^{-3}$, how would you express $\dfrac{1}{8_{ten}}$ in the prime system if we allow negative numbers in any given place? Express the fractions $\frac{1}{8}$, $\frac{2}{8}$, $\frac{3}{8}$, $\frac{4}{8}$, $\frac{5}{8}$, $\frac{6}{8}$, $\frac{7}{8}$, and $\frac{8}{8}$ in this system. Express the fractions between 0 and 2 with denominators of 10 in the prime system.

3. Does the prime system have a symbol for the number zero? Explain.

4. Investigate the early systems of notation used by the Egyptians, Greeks, Romans, and Indians. Books by Jones and Mueller listed in the references below will be helpful. Encyclopedias will be helpful also.

5. Investigate the possibility of a numeration system, with all the five characteristics or properties listed on pages 236–37 with base ten (2nd characteristic) changed to a base of -10. The values of the places, from right to left would be 1, -10, 100, -1000, 10,000, etc. Try to count from 1 to 25 in this system. How would you add, subtract, multiply, and divide?

REFERENCES

JONES, P. S., *Understanding Numbers: Their History and Use.* Ann Arbor: Ulrich's Bookstore, 1954.

MUELLER, FRANCIS J., *Arithmetic—Its Structure and Concepts.* Englewood Cliffs, N. J.: Prentice-Hall, 1956.

SCHOOL MATHEMATICS STUDY GROUP. *Mathematics for Junior High School,* Volume I. New Haven, Conn.: Yale University Press, 1961.

SWAIN, ROBERT L. *Understanding Arithmetic.* New York: Rinehart & Co., 1957.

NUMBERS AND GAMES

CHARLES F. BRUMFIEL

University of Michigan
Ann Arbor, Michigan

Here are five short but separate topics dealing with numbers and games.

"Repeating Decimals" requires only arithmetic as background. It should help students see, for example, that $0.9\bar{9}$ (the bar shows nine repeats) is a perfectly good way to represent the number 1.

"Continued Fractions," an interesting topic from number theory, ranges in difficulty. The first two sets of exercises can be done by students with no algebra at all. The rest of the unit will be easier after pupils have had some study of radicals and algebra.

While it may appear that substantial algebra is needed for "Algebra with Finite Sets," students with only a slight background in algebra can undertake this unit. The use of finite sets will help students begin to understand some basic concepts of algebra.

In "Ticktacktoe," there is an interesting analysis of the strategy for playing the familiar game. No special skills are required except a clear head and an interest in getting a look behind the game to see how and why you can win.

Substantial background in algebra is needed to understand and appreciate "Number Line Games" fully. The games in the first set of number line exercises can be done by beginning algebra students, or they can be adapted by the teacher for classroom use prior to algebra. These games provide an interesting application of the function concept.

REPEATING DECIMALS

It is easy to see that every rational number has a repeating decimal representation. For example, if you divide 73 by 291 to get the decimal representation for $\frac{73}{291}$ only 291 remainders are possible, namely: 0, 1, 2, $\cdots$, 290. If the zero remainder is never obtained, eventually some nonzero remainder repeats, and the decimal repeats. If a zero remainder occurs, we say that the decimal repeats zeros. Thus,

$$\frac{3}{25} = .12\bar{0}\bar{0} \cdots \text{ (the zeros repeat)}$$
$$\frac{3}{11} = .27\overline{27} \cdots \text{ (the block of digits "27" repeats)}$$

The proof that every repeating decimal represents a rational number is rather difficult. Strictly speaking, the proof depends upon sophisticated limit concepts, but an intuitive "proof" can be built around a little computation.

Students readily admit that the decimals $.33\bar{3}\cdots$ and $.11\bar{1}\cdots$, where the 3's and 1's repeat, respectively represent $\frac{1}{3}$ and $\frac{1}{9}$. Students have a division technique for generating these decimals from their fractions. However, nearly all students reject $.99\bar{9}\cdots$ as a repeating decimal representation of 1. Acceptance for this must be won, and perhaps this can be managed with the following sequence of problems.

Exercises A:

1. Add the following, and interpret your answers.

(a) $0.33\bar{3}\cdots$ (b) $0.33\bar{3}\cdots$ (c) $0.99\bar{9}\cdots$
 $0.33\bar{3}\cdots$ $0.66\bar{6}\cdots$ $0.99\bar{9}\cdots$

2. Subtract and interpret the results.

(a) $0.66\bar{6}\cdots$ (b) $0.99\bar{9}\cdots$ (c) $1.00\bar{0}\cdots$
 $0.33\bar{3}\cdots$ $0.66\bar{6}\cdots$ $0.99\bar{9}\cdots$

3. Multiply and interpret.

(a) $3 \times (0.33\bar{3}\cdots)$
(b) $2 \times (0.99\bar{9}\cdots)$
(c) $3 \times (0.66\bar{6}\cdots)$
(d) $9 \times (0.11\bar{1}\cdots)$
(e) $99 \times (0.01\overline{01}\cdots)$
(f) $999 \times (0.001\overline{001}\cdots)$
(g) $10 \times (0.033\bar{3}\cdots)$
(h) $100 \times (0.0011\bar{1}\cdots)$

4. What rational numbers are represented by the following?

(a) $0.\overline{111}\cdots$ (f) $0.\overline{001}\cdots$
(b) $0.\overline{01}\cdots$ (g) $0.\overline{005}\cdots$
(c) $0.\bar{2}\cdots$ (h) $0.\overline{214}\cdots$
(d) $0.\overline{02}\cdots$ (i) $1.\overline{25}\cdots$
(e) $0.\overline{14}\cdots$ (j) $0.0\bar{1}\cdots$

From the exercises, the student should learn that $0.\overline{99}\cdots$ does represent 1, and since $0.\overline{01}\cdots$ represents $\frac{1}{99}$, $0.\overline{34}\cdots$ is $\frac{34}{99}$, etc. Now the student has a technique for computing the rational number represented by a repeating decimal. For example,

$$0.321\overline{515}\cdots = 0.32 + 0.00\overline{15}\cdots$$
$$= \tfrac{32}{100} + \tfrac{1}{100}(0.1\overline{515}\cdots)$$
$$= \tfrac{32}{100} + \tfrac{1}{100}\cdot\tfrac{15}{99} \text{ etc.}$$

CONTINUED FRACTIONS

Continued fractions are compound fractions of a special type, quite important in number theory. Here are examples:

$$2 + \cfrac{1}{3 + \cfrac{1}{2 + \cfrac{1}{4}}}, \qquad 0 + \cfrac{1}{5 + \cfrac{1}{2 + \cfrac{1}{4 + \cfrac{1}{7}}}}.$$

The first represents $\frac{71}{31}$, and the second $\frac{65}{354}$. Each "numerator" in a continued fraction is 1. These two continued fractions are represented by the symbols,

$$(2; 3, 2, 4) \qquad (0; 5, 2, 4, 7).$$

We compute the continued fraction for $\frac{31}{73}$ below. This example illustrates the method for calculating the continued fraction for any real number.

$$\frac{31}{73} = \cfrac{1}{\frac{73}{31}} = \cfrac{1}{2 + \cfrac{11}{31}} = \cfrac{1}{2 + \cfrac{1}{\frac{31}{11}}} = \cfrac{1}{2 + \cfrac{1}{2 + \cfrac{9}{11}}} =$$

$$\cfrac{1}{2 + \cfrac{1}{2 + \cfrac{1}{1 + \cfrac{2}{9}}}} = \cfrac{1}{2 + \cfrac{1}{2 + \cfrac{1}{1 + \cfrac{1}{4 + \cfrac{1}{2}}}}}$$

Hence $\frac{31}{73} = (0; 2, 2, 1, 4, 2)$.

Exercises B:

1. Show that $\frac{42}{29} = (1; 2, 4, 3)$ and $\frac{43}{33} = (1; 3, 3, 3)$.

2. Write $\frac{11}{4}$ as a continued fraction.

3. Compute backwards and verify that the continued fraction above represents $\frac{31}{73}$.

Consider the continued fraction $r = (0; 1, 2, 3, 2, 7, 4, 3)$. We construct the set of "approximating" continued fractions.

$r_0 = 0; \qquad r_1 = (0; 1); \qquad r_2 = (0; 1, 2); \qquad r_3 = (0; 1, 2, 3);$

$r_4 = (0; 1, 2, 3, 2); \qquad r_5 = (0; 1, 2, 3, 2, 7);$

$r_6 = (0; 1, 2, 3, 2, 7, 4); \qquad r_7 = r.$

Computing these approximations we get, for the sequence $r_0, r_1, r_2, \cdots,$ r_7, the following:

$$\frac{0}{1}, \frac{1}{1}, \frac{2}{3}, \frac{7}{10}, \frac{16}{23}, \frac{119}{171}, \frac{492}{707}, \frac{1595}{2292}.$$

The sequence of numbers above has many interesting properties. Each is a better approximation to r than the number preceding it. The numbers r_1, r_3, and r_5 are greater than r and r_2, r_4, r_6 are less than r. Hence, the number $r = \frac{1595}{2292}$ lies between each pair of consecutive r's. The difference between any two consecutive r's is a fraction with numerator 1 whose denominator is the product of the two denominators. For example:

$$r_5 - r_6 = \frac{119}{171} - \frac{492}{707} = \frac{1}{171 \cdot 707},$$

$$r_7 - r_6 = \frac{1595}{2292} - \frac{492}{707} = \frac{1}{707 \cdot 2292}.$$

The number r_6 is very close to r. The difference is less than one-millionth. In a certain sense the r's are as close together as possible. They are the "best possible" approximations to r, considering their denominators.

When the symbol for a continued fraction is given, there is a simple rule for calculating this approximating sequence. We illustrate for the example above:

$$r = (0; 1, 2, 3, 2, 7, 4, 3).$$

We write the first two approximations, $r_0 = \frac{0}{1}$ and $r_1 = \frac{1}{1}$. Then r_2 is computed from these two. The numerator and denominator of $\frac{1}{1}$ are multiplied by the 2, and numerators and denominators are added. Then r_3 is computed from r_1 and r_2, using the 3 as a multiplier:

$$r_2 = \frac{0 + 1 \cdot 2}{1 + 1 \cdot 2} = \frac{2}{3}; r_3 = \frac{1 + 3 \cdot 2}{1 + 3 \cdot 3} = \frac{7}{10}.$$

Continue the computation, getting:

$$r_4 = \frac{2 + 2 \cdot 7}{3 + 2 \cdot 10} = \frac{16}{23},$$

$$r_5 = \frac{7 + 7 \cdot 16}{10 + 7 \cdot 23} = \frac{119}{171},$$

$$r_6 = \frac{16 + 4 \cdot 119}{23 + 4 \cdot 171} = \frac{492}{707},$$

$$r_7 = \frac{119 + 3 \cdot 492}{171 + 3 \cdot 707} = \frac{1595}{2292}.$$

Exercises C:

    1. Using the rule described above, compute the sequence of approximating fractions for $(1; 2, 2, 2, 2, 2, 2)$.

    2. Square each number of the sequence in Exercise 1, and compare each result with the number 2.

    3. The continued fraction for π begins $(3; 7, 16, \cdots)$. Compute the first three approximations. Convert the second and third approximations to decimals and compare with the decimal representation

$$\pi = 3.1415926535 \cdots .$$

Continued Fractions for Irrational Numbers

    The continued fraction for each rational number terminates. (Can you prove this?) Those for irrational numbers continue forever. (Prove this.) We illustrate this last remark by computing the continued fraction for $\sqrt{5}$.

    We know that $\sqrt{5} = 2 + x$ where $x = \sqrt{5} - 2$ and is less than 1. Hence, $\sqrt{5} = (2; \cdots)$. Now, we write $\sqrt{5}$ as $\sqrt{5} = 2 + (\sqrt{5} - 2) = 2 + \dfrac{1}{\dfrac{1}{\sqrt{5} - 2}}$ We must determine the largest integer that does not exceed $\dfrac{1}{\dfrac{1}{\sqrt{5} - 2}}$. Since $\dfrac{1}{\sqrt{5} - 2} = \dfrac{\sqrt{5} + 2}{1}$, we see that this integer is 4.

Hence, we have

$$\sqrt{5} = 2 + \cfrac{1}{4 + (\sqrt{5} - 2)} \qquad \sqrt{5} = (2; 4 \cdots).$$

We now set

$$\sqrt{5} = 2 + \cfrac{1}{4 + \cfrac{1}{\cfrac{1}{\sqrt{5} - 2}}} .$$

But determining the largest integer that is less than $\dfrac{1}{\dfrac{1}{\sqrt{5} - 2}}$ is our old problem. We see that our continued fraction repeats.

$$\sqrt{5} = 2 + \cfrac{1}{4 + \cfrac{1}{4 + \cfrac{1}{4 \cdots}}} , \qquad \sqrt{5} = (2; 4, 4, 4, \cdots).$$

Exercises D:

1. Compute the first few approximating fractions from the continued fraction representation for $\sqrt{5}$ and show that one gets

$$\tfrac{2}{1}, \tfrac{9}{4}, \tfrac{38}{17}, \tfrac{161}{72}, \tfrac{682}{305}, \tfrac{2889}{1292}.$$

2. Square several of the numbers in the sequence of Exercise 1, and compare with $\sqrt{5}$.

3. Show that $\tfrac{2}{1} < \sqrt{5}$, $\tfrac{9}{4} > \sqrt{5}$, $\tfrac{38}{17} < \sqrt{5}$, etc.

4. Find the difference between $\tfrac{38}{17}$ and $\tfrac{161}{72}$. Estimate the error if $\tfrac{161}{72}$ is chosen as an approximation to $\sqrt{5}$.

5. Compute $\tfrac{2889}{1292}$ as a decimal and show that it is larger than $\sqrt{5}$, but only about 0.0000002 larger.

Computing Continued Fractions for Irrational Numbers

There are interesting devices for computing the continued fractions for certain irrational numbers. For example, if we set $\sqrt{2} = 1 + x$ then

$$x^2 + 2x = 1, \qquad x(x + 2) = 1, \qquad x = \frac{1}{2 + x}.$$

Now we replace x, in the denominator above, by $\dfrac{1}{2 + x}$ and have

$$x = \cfrac{1}{2 + \cfrac{1}{2 + x}}. \text{ We replace } x \text{ again and again, getting}$$

$$\sqrt{2} = (1; 2, 2, 2, 2, 2, \cdots).$$

Exercises E:

1. Use the method above to get the continued fraction for $\sqrt{10}$; for $\sqrt{17}$.

2. Compute the repeating continued fraction for $\sqrt{7}$.

Given a repeating continued fraction, we can determine the irrational number it represents. For example, if $x = (1; 1, 1, 1, \cdots)$ where the 1's repeat, we have

$$x = 1 + \cfrac{1}{1 + \cfrac{1}{1 + \cdots}} = 1 + \frac{1}{x}.$$

Hence, $x^2 - x - 1 = 0$, and

$$x = \frac{1 + \sqrt{5}}{2} \qquad \text{or} \qquad x = \frac{1 - \sqrt{5}}{2}.$$

Since obviously x is positive,

$$x = \frac{1 + \sqrt{5}}{2}.$$

Exercises F:

1. Compute the continued fraction for $\dfrac{1 + \sqrt{5}}{2}$ and show that one does get $(1; 1, 1, 1, \cdots)$.

2. Determine the number represented by the continued fraction $(0; 2, 3, 2, 3, 2, 3, \cdots)$ when the block "2, 3" repeats.

3. Use the fact that since $\dfrac{1 + \sqrt{5}}{2}$ is the positive root of the equation $x^2 - x - 1 = 0$, then $\dfrac{1 + \sqrt{5}}{2} = \sqrt{1 + x}$, and we are justified in writing

$$\frac{1 + \sqrt{5}}{2} = \sqrt{1 + \sqrt{1 + \sqrt{1 + \cdots}}},$$

where the square root operation is repeated indefinitely.

ALGEBRA WITH FINITE SETS

Much algebra can be taught with relative ease if the domains of the variables are restricted to finite sets of numbers. We illustrate this by choosing for our domain the set of 11 numbers

$$S = \{0, 1, 2, \cdots, 10\}.$$

The important different roles played by variables can now be described.

1. We may use the sentence "X is even" as an abbreviation for the set of 11 statements, "0 is even, 1 is even, $\cdots$, 10 is even." This is the *open sentence* concept. By this simple sentence we call attention to a *set* of 11 sentences, each of which is either true or false.

2. We may say, "For all X in S, X is even." This is a statement *about* the set of 11 statements above. Obviously we have made a false statement.

3. We may say, "There is a number X in S, such that X is even." This is also a statement about the set of 11 statements above. Obviously this statement is true.

4. We may say, "Consider the set of all numbers X in S, such that X is even." We are asking students to think about a certain subset of S.

Nearly all our uses of variables fall into one or the other of these categories. You should carefully consider what is required to prove or disprove statements like those in (2) and (3) above. Statements like: "For no X in S, $\cdots$"; "there is at least one X in S such that $\cdots$" should be formulated and discussed. A sound intuitive basis for understanding the role of logic in mathematics can be built in this way.

Rather complicated equations, inequalities, and other sentences can be put before students so long as the reference set is finite. The student has the comfortable feeling that he can exhaust all possible cases by trial and error. Problems like the following might be given.

Exercises: Finite Sets

Find all numbers X in S such that

1. $x + 2x = 3x$

2. $x^2 + x$ is even

3. x is odd and greater than 7

4. $2x = x + 3$

5. $x^2 = x$

6. $x = 5 - \dfrac{8}{x + 1}$.

Exercises: Graphs

Students may be asked to graph solutions of equations, inequalities, etc., on a number line consisting of precisely 11 points. Sentences in two variables can be introduced and the solution sets may be graphed on a lattice of 121 points. For example, determine all pairs of numbers (x, y) in S such that

1. $2x - y = 3$

2. $y = x^2$

3. $x^2 + y^2 = 25$

4. $x - 2 = 0$

5. $(x - 3)(y - 4) = 0$

6. $xy = 7$.

Generalizations should occur in such sets of exercises as, for example, the following:

7. $x + y = y + x$

8. $(x + 4)y = xy + 4y$

9. $x(y - 1) = xy - x$.

Note that if negative numbers are not admissible for purposes of calculation, (**9**) is not a generalization.

Students should also be given sentences like $x^2 + y^2 = 7$, whose graphs are the empty set and where (x, y) is in the S described above.

TICKTACKTOE

Almost certainly you have played ticktacktoe, but have you ever taken time to analyze the game carefully? The diagram below indicates that the first player has marked an "✕" in the upper-right-hand corner, and the second player has replied by marking a "○" in the center top space. Now the first player can win by marking in the lower right hand corner.

The second player must mark in the remaining space on the right, and now, marking either in the center or in the lower left corner gives the first player a winning position. (Why?)

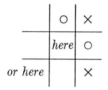

Show that the first player could also have won by marking in the center square at his second move.

One purpose of mathematics is to provide tools for the analysis of games. You may think that at your first move in ticktacktoe you have nine choices. You have only three. For example, whatever corner you might mark, we could rotate the figure so that your marked corner would fall in the upper-left position. The three possible first moves might be described as: center, corner, side.

The Role of Symmetry

If we set ourselves the goal of listing all possible ways that the game of ticktacktoe could be played, we could save much time by noticing the role of symmetry. We shall not analyze the game in this fashion, but if you read this article carefully and give some thought to the ideas, then you should never lose another game of ticktacktoe.

ANALYZING YOUR POSITION. In order to fix the concept of *symmetry* in your mind, answer the following questions.

1. If your opponent marks first in the center, how many plays are at your disposal?
2. If your opponent marks first in a corner, how many replies have you?
3. How many replies have you to an opening side move?
4. If you have marked the center square and your opponent has replied by marking in a corner, how many lines of play are available to you on your second move?
5. If you have marked the center square and your opponent has marked a side square, how many lines of play are now open?
6. Your opponent plays first in a corner; you mark the center; your opponent marks the corner diagonally opposite his occupied corner, so that your three marks are lined up. How many replies have you now?
7. You have marked a corner and your opponent marks an adjacent side square. How many lines of play have you?

The answers to the seven questions posed above are respectively 2, 5, 5, 4, 4, 2, 7.

MAKING YOUR MOVES. The following seven problems refer to the corresponding questions above.

1. Show that at least one of your two possible moves enables your opponent to win.
2. Show that at least four of your five moves enable your opponent to win.
3. Show that two of your five moves enable your opponent to win.
4. Show that none of your four moves enables your opponent to win.
5. Show that three of your four lines of play enable you to win.
6. Show that one of your two possible moves loses and the other draws.
7. Show that three of your possible seven moves enable you to win.

Ticktacktoe Problems

Now, take the quiz (page 255). In each diagram, it is your play. Number the board as shown below analyze the position, and select your move.

NUMBER LINE GAMES

Number line games provide an interesting application of the function concept. We might "move" on the number line according to the rule: Move from any number x to x^2. Beginning at 2 we would move to 4, 16, 256.

1	2	3
4	5	6
7	8	9

Problem 1

		×
	O	
×		O

Play and win.

Problem 2

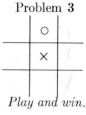

		×
	O	
×		

Play and draw.

Problem 3

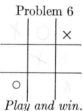

	O	
	×	

Play and win.

Problem 4

	O	O
	×	
×		

Play and win.

Problem 5

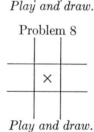

		O
	×	
×		

Play and draw.

Problem 6

		×
O		

Play and win.

Problem 7

		O
×		
	×	

Play and win.

Problem 8

×		

Play and draw.

Problem 9

		×

Play and draw.

Problem 10

	×	
	O	
	×	

Play and win.

Let us use the notation $R: \quad x \to x^2$ to describe the rule for a move on the number line from x to x^2.

As a second example, if $R: \quad x \to \dfrac{x-1}{x}$, beginning at 2, you move to $\frac{1}{2}$, then to -1, then to 2. In three moves we have come back to our starting point, 2. This is not accidental. Try to show that if we start at any point except 1 or 0, upon our third move we return to our starting point. (Why can we not start at 1 or 0?)

We shall limit ourselves to the so-called *linear fractional* functions of the form R: $\quad x \rightarrow \dfrac{ax + b}{cx + d}$ where a, b, c, and d are integers. A few of these functions have the property that after a certain number of moves one returns to the starting point. We shall say that they are *cyclic*.

For example, the rule above given by R: $\quad x \rightarrow \dfrac{x - 1}{x}$ is cyclic. We say that this rule has *order* 3 because we return to our starting point in three moves.

Number line games built around various moves can be restricted to whole numbers or to rational numbers. The most interesting activities require a knowledge of operations with negative numbers. Not only can many thought-provoking problems be formulated but also a great deal of practice in computation is provided.

The following set of problems is built around the simple rule

$$R: \quad x \rightarrow 2x - 4.$$

The diagram (Fig. 1) shows that moving successively according to R one jumps from 3 to 2 to 0 to -4. We write R:$3 \rightarrow 2 \rightarrow 0 \rightarrow -4$.

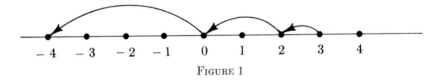

FIGURE 1

Exercises G:

1. Find a point such that in one move you go to 22.
2. Where must you start to reach 20 in two moves?
3. Where must you start in order for one move to carry you 2 units to the right? 10 to the right?
4. Find a point from which two moves take you 30 units to the right; 30 to the left.
5. Describe all points on the line such that two moves from any of these points carry you beyond 100.
6. Show that there is no point with integral coordinates from which we can jump to 30 in two moves.
7. Note that, beginning at 3, the first jump is 1 unit to the left, the second 2 to the left, the third 4 to the left. Does this pattern con-

tinue? How large is the jump that lands on 3? The jump before that? Before that?

8. Some moves are to the right, some to the left. Find a *fixed point* for R, that is, a point from which you move to that point itself. What happens if you start from a point on the left of this point? The right?

The rules R_2, R_3, R_4, and R_6 given below are cyclic and have orders 2, 3, 4, and 6 respectively.

$$R_2: \quad x \to -x, \qquad R_3: \quad x \to \frac{x-1}{x},$$

$$R_4: \quad x \to \frac{x-1}{x+1}, \qquad R_6: \quad x \to \frac{x-1}{x+2}.$$

Beginning at the point 3 and applying each rule we have

$$R_2: \quad 3 \to -3 \to 3$$
$$R_3: \quad 3 \to \tfrac{2}{3} \to -\tfrac{1}{2} \to 3$$
$$R_4: \quad 3 \to \tfrac{1}{2} \to -\tfrac{1}{3} \to -2 \to 3$$
$$R_6: \quad 3 \to \tfrac{2}{5} \to -\tfrac{1}{4} \to -\tfrac{5}{7} \to -\tfrac{4}{3} \to -\tfrac{7}{2} \to 3.$$

Test the cyclic property of each rule using numbers other than 3.

Can you find other cyclic rules? A formula for all cyclic rules of order 2 is given below.

$$R^{(2)}: \quad x \to \frac{ax+b}{cx-a}$$

The formula for all rules of order 3 is more complex.

$$R^{(3)}: \quad x \to \frac{abx+b^2}{-(a^2+ac+c^2)x+bc}$$

One gets R_3 from this formula by setting $a = -1$, $b = 1$, $c = 0$, and multiplying numerator and denominator by -1.

Exercise H:

Find values of a, b, c so that the formula for all rules of order 3 gives the rule

$$R: \quad x \to \frac{x+3}{-7x+4}.$$

There is no rule of order 5 that has integers for coefficients. However, using irrational numbers, one can describe such a rule. Check the one below.

$$R_5: \quad x \to \frac{x-(5+2\sqrt{5})}{x+1}$$

No rules of order higher than 6 can be given using rational coefficients; however, by using combinations of two rules one can construct a series of moves leading from one initial point to either 8 or 12 distinct points. We illustrate this in the problems below.

Problems:

1. Show that $x \rightarrow 2 - x$ and $x \rightarrow -x - 7$ are cyclic moves. Determine all cyclic polynomial moves, that is cyclic moves of the type: $R\colon \quad x \rightarrow ax + b$, where a and b are rational numbers.

2. Show that if $R\colon \quad x \rightarrow 2 - x$, and $S\colon \quad x \rightarrow 3 - x$, then the "double" move, R followed by S, carries you one unit to the right while S followed by R carries you one unit to the left. Let us agree to write $R*S$ and $S*R$ respectively for these composite moves.

3. Determine a and b so that the rule $R\colon \quad x \rightarrow ax + b$ jumps both from 4 to 11 and from 6 to 15.

4. Show that $R\colon \quad x \rightarrow 3x + 6$ can be "factored" as $S*T$ where $S\colon \quad x \rightarrow 3x$, and $T\colon \quad x \rightarrow x + 6$. Show also that $R = U*S$ where $U\colon \quad x \rightarrow x + 2$.

5. Give a geometric interpretation in terms of sliding, rotating, or stretching the line for the rules: $R\colon x \rightarrow x + 1$, $S\colon x \rightarrow -x$, $T\colon \quad x \rightarrow -x + 4$, $U\colon \quad x \rightarrow 2x$.

6. The rule $R\colon \quad x \rightarrow 2x + 4$ has an "inverse" rule S, such that $R*S = S*R$ and for every number x, $(R*S)\colon x \rightarrow x$. That is, $R*S$ is the *identity* function. Determine S.

7. Determine the inverses of

$$S_1\colon \quad x \rightarrow -x \qquad\qquad S_4\colon \quad x \rightarrow \frac{1}{x}$$

$$S_2\colon \quad x \rightarrow 4 - x \qquad\qquad S_5\colon \quad x \rightarrow -\frac{3}{5x}$$

$$S_3\colon \quad x \rightarrow ax + b \qquad\qquad S_6\colon \quad x \rightarrow \frac{x + 1}{x}$$

8. Note that the rule $R\colon \quad x \rightarrow \dfrac{x - 1}{x}$ does not apply to the point 0

From zero there is no place to go! However, observe:

$$R\colon \tfrac{11}{10} \rightarrow \tfrac{1}{11} \rightarrow -10 \rightarrow \tfrac{11}{10}$$
$$R\colon \tfrac{101}{100} \rightarrow \tfrac{1}{101} \rightarrow -100 \rightarrow \tfrac{101}{100}$$
$$R\colon \tfrac{1001}{1000} \rightarrow \tfrac{1}{1001} \rightarrow -1000 \rightarrow \tfrac{1001}{1000}$$
$$R\colon 1 \rightarrow 0 \rightarrow \, ? \rightarrow \, ?$$

Clearly, when we consider a number x *near* 1, but not equal to 1, the first jump lands *near* 0, the second is extremely large, and the third jump returns to x. Let us *invent* a new point P on our line and agree that we jump from 0 to P and from P to 1. This enables us to start at *any* point on our line (even P) and return in three moves. We might call P "the point at infinity."

Consider the rule S: $\quad x \to \dfrac{2x - 1}{4x + 2}$ which has order 4. S is not defined for $x = -\frac{1}{2}$. If we start moving from 0 S: $\quad 0, \to -\frac{1}{2} \to ?$ $\to ? \to ?$ Invent a new point, Q, on the line and show how, using this point, we can return to 0 on the fourth move.

9. If one uses combinations of the rules S and T where

$$S: \quad x \to \frac{x}{x - 1}, \qquad T: \quad x \to \frac{x - 2}{2x - 1}$$

then no more than four points can be reached from any one point. Begin at 3 and move alternately, first according to S, and then according to T. Move in the order T, S, T, S.

10. Using the moves S: $\quad x \to \dfrac{1}{1 - x}$, T: $\quad x \to \dfrac{1}{x}$ at most six different points can be reached from one point. For example beginning with 3 and moving S, T, S, S, T, we have $3 \to -\frac{1}{2} \to -2 \to \frac{1}{3} \to \frac{3}{2} \to \frac{2}{3}$. Now show that neither S nor T produces a new point. Try various orders of moves.

11. Using S: $\quad x \to -x$, T: $\quad x \to \dfrac{x + 1}{1 - x}$ at most eight different points are attainable from one point. For example, begin with 2 and move in the order, S, T, T, T, S, T, T, T. Try to find other orders of moving by which eight points can be reached.

12. The most complicated finite system is generated by the two rules:

$$R: \quad x \to -x - 2 \qquad S: \quad x \to \frac{4}{3x}.$$

These two rules generate a "group" of twelve rules. Begin at 1 and move alternately S, T, S, T, $\cdots$ for 12 moves. You should return to 1.

13. The matrix $\begin{pmatrix} a & b \\ c & d \end{pmatrix}$, with $ad - bc \neq 0$ and a, b, c, d integers, may be used to describe the rule R: $\quad x \to \dfrac{ax + b}{cx + d}$. Show that if $k \neq 0$, the matrices

$$\begin{pmatrix} a & b \\ c & d \end{pmatrix} \quad \text{and} \quad \begin{pmatrix} ka & kb \\ kc & kd \end{pmatrix}$$

describe the same rule.

14. Explain why in (13) it was necessary to require that $ad - bc \neq 0$.

15. Consider a rule R described by the matrix $\begin{pmatrix} a & b \\ c & d \end{pmatrix}$ with $ad - bc \neq 0$.

We call the number $ad - bc$ the *determinant* of the matrix. We call the number $(a + d)$ the *trace* of the matrix. Verify the following statements for certain numerical cases.

    (a) If $(a + d)^2 = 0(ad - bc)$, R has order 2
    (b) If $(a + d)^2 = 1(ad - bc)$, R has order 3
    (c) If $(a + d)^2 = 2(ad - bc)$, R has order 4
    (d) If $(a + d)^2 = 3(ad - bc)$, R has order 6.

16. Try to find rules R and S such that:
    (a) R, S, and $R*S$ have order 2
    (b) R and S have order 2 and $R*S$ has order 3
    (c) R and S have order 2 and $R*S$ has order 4
    (d) R and S have order 2 and $R*S$ has order 6.

17. Rules can be "factored" into the product of "simpler" rules. For example, if

$$R: \quad x \to x + 2, \qquad S: \quad x \to \frac{1}{x}, \qquad T: \quad x \to 3x,$$

and

$$W: \quad x \to \frac{2x + 7}{x + 2},$$

then show that

$$W = R*S*T*R.$$

USING SETS TO STUDY ODD, EVEN, AND PRIME NUMBERS

HAROLD TINNAPPEL

Bowling Green State University
Bowling Green, Ohio

The first part of this article deals with odd and even numbers, some of their properties, and proof of selected theorems. The latter part of the article gives an introduction to prime and composite numbers, and culminates in the proof of the infinitude of primes.

The topics are developed by using the language of sets and set notation. Of course, odd, even, and prime numbers may be studied without using set notation and the language of sets.

Some background in algebra will be needed by students undertaking this unit.

FIRST EXPERIENCES WITH NUMBERS

Perhaps among our first experiences with numbers is our use of them for counting objects. In time, we appreciate the fact that each number after the first in this counting sequence is obtained from the one which immediately precedes it by adding one to that predecessor. We can visualize the pattern in this manner:

$$1, \quad 1 + 1 = 2, \quad 2 + 1 = 3, \quad 3 + 1 = 4, \quad 4 + 1 = 5, \cdots .$$

We soon realize that the numbers which are formed in this manner differ widely one from the other. If we consider two consecutive numbers such as 11 and 12, for example, we see totally different characteristics if they are written as products of numbers in the counting sequence:

$$11 = 1 \cdot 11, \quad 12 = 1 \cdot 12 = 2 \cdot 6 = 3 \cdot 4.$$

NATURAL NUMBERS, ODDS, AND EVENS

The purpose of this section is to examine properties of some subsets of the set of natural numbers, mentioned in the above paragraph. We

261

designate the set of natural numbers by N. Employing a notation widely used we write

$$N = \{1, 2, 3, \cdots \}.$$

We represent elements of the set N by small letters: n, a, b, $\cdots$. We write $n \, \epsilon \, N$ for the statement "n belongs to N," or "n is an element of the set N." Two subsets of N which merit particular attention are the set of odd numbers, designated by $\mathcal{O}$,

$$\mathcal{O} = \{1, 3, 5, 7, \cdots \},$$

and the set of even numbers, designated by $\mathcal{E}$,

$$\mathcal{E} = \{2, 4, 6, 8, \cdots \}.$$

If $n \, \epsilon \, N$, then $(2n - 1) \, \epsilon \, \mathcal{O}$. For example, when

$$n = 1, \qquad 2n - 1 = 1, \qquad \text{and 1 is the first odd number}$$

$$n = 2, \qquad 2n - 1 = 3, \qquad \text{and 3 is the second odd number}$$

$$n = 3, \qquad 2n - 1 = 5, \qquad \text{and 5 is the third odd number.}$$

In fact, we can regard the expression $2n - 1$ as a formula for the nth odd natural number. We will use a notation which is used extensively in mathematics, and which we will have occasion to use later in this chapter. This notation relates the natural number n and the nth odd number, $f(n)$,

$$f(n) = 2n - 1.$$

We then have, as noted above, $f(1) = 2(1) - 1 = 1$, $f(2) = 2(2) - 1 = 4 - 1 = 3$, $f(3) = 2(3) - 1 = 5$, etc. Similarly, a formula for the elements of $\mathcal{E}$ is given by $2n$, $n \, \epsilon \, N$. For example, when

$$n = 1, \qquad 2n = 2 \cdot 1 = 2, \qquad \text{and 2 is the first even number}$$

$$n = 2, \qquad 2n = 2 \cdot 2 = 4, \qquad \text{and 4 is the second even number}$$

$$n = 3, \qquad 2n = 2 \cdot 3 = 6, \qquad \text{and 6 is the third even number.}$$

We shall now turn our attention to three important sets which may be considered in connection with given sets. To explain symbols which are regularly used to designate certain relations among sets, let us designate by A, B, and C the following three sets:

$$A = \{1, 2, 4, 8\}, \qquad B = \{2, 3, 5, 8\}, \quad \text{and} \quad C = \{3, 5, 7\}.$$

The symbol $A \cup B$ (read, "the union of A and B") represents the set of elements which belong to either set A or to set B, or possibly to both.

In this case, $A \cup B = \{1, 2, 3, 4, 5, 8\}$. We could say $1 \,\epsilon\, A \cup B$, $2 \,\epsilon\, A \cup B$, etc. Verify the fact that $A \cup C = \{1, 2, 3, 4, 5, 7, 8\}$ and $B \cup C = \{2, 3, 5, 7, 8\}$. The symbol $A \cap B$ (read "the intersection of A and B") represents the set of elements which are common to A and B, that is, which belong to both A and B. In this case $A \cap B = \{2, 8\}$. Verify the fact that $B \cap C = \{3, 5\}$. In case a set possesses no element (or is said to be empty) we call the set a null set and designate it by the symbol $\varnothing$. Notice the fact that $A \cap C = \varnothing$.

Returning to the set of even and odd numbers, we observe that

$$\vartheta \cup \varepsilon = N \quad \text{and} \quad \vartheta \cap \varepsilon = \varnothing.$$

We now note that it is reasonable to say that there are just as many even numbers as there are natural numbers. In mathematics, we say that one set has as many elements as a second, if we can show a one-to-one correspondence which matches the pairs of elements in the two sets —— one of the pair is in one set and the other is in the second set. One such correspondence is displayed below for the sets N and ε.

$$
\begin{array}{lcccccc}
N: & \mathbf{1}, & 2, & 3, & 4, & 5, & \cdots, & n, & \cdots \\
& \updownarrow & \updownarrow & \updownarrow & \updownarrow & \updownarrow & & \updownarrow \\
\varepsilon: & \mathbf{2}, & 4, & 6, & 8, & 10, & \cdots, & 2n, & \cdots
\end{array}
$$

We match with each element n belonging to N the element of ε which is just twice that number, namely, $2n$. With each element e belonging to ε we pair that number of N which is precisely $\frac{1}{2}e$.

Exercise 1:

Show how to pair the numbers of N with the numbers of ε using a one-to-one correspondence different from the one shown above.

Exercise 2:

(a) Show that there are just as many odd numbers as natural numbers.
(b) Show that there are just as many odd numbers as even numbers.

Counting the elements of a set S is itself a process of setting up a one-to-one correspondence between the elements of a subset of the set N and the elements of set S. For example, we might count the letters of the alphabet by matching these letters with a subset of N as shown below.

$$
\begin{array}{cccccc}
z, & y, & x, & \cdots, & b, & a \\
\updownarrow & \updownarrow & \updownarrow & & \updownarrow & \updownarrow \\
1, & 2, & 3, & \cdots, & 25, & 26
\end{array}
$$

We will now develop a few properties of the set ϑ and the set ε. Before we present the theorems we examine some examples. We have all probably observed that when two even numbers are added, their sum is even. For example:

$$2 + 6 = 8, \quad 10 + 20 = 30, \quad 8 + 154 = 162.$$

THEOREM 1: *If both x and y belong to $\mathcal{E}$, then $x + y$ belongs to $\mathcal{E}$.*
Proof: Since $x \in \mathcal{E}$, we write $x = 2a$, $a \in N$;
and since $y \in \mathcal{E}$, we can write $y = 2b$, $b \in N$.

$$x + y = 2a + 2b = 2(a + b).$$

Now since $a + b = n \in N$ ($a + b$ equals n, and n is an element of N)

$$x + y = 2n \in \mathcal{E}.$$

In a similar way if two even numbers are multiplied the product is even. For example:

$$2 \cdot 6 = 12, \quad 10 \cdot 20 = 200, \quad 8 \cdot 6 = 48.$$

THEOREM 2: *If both x and y belong to $\mathcal{E}$, then xy belongs to $\mathcal{E}$.*
Proof: As in Theorem 1, since $x \in \mathcal{E}$, we can write $x = 2a$, $a \in N$;
since $y \in \mathcal{E}$, we can write $y = 2b$, $b \in N$.

$$xy = (2a)\,(2b) = 2(2a \cdot b).$$

We notice that since 2, a, b are in N, then their product

$$2a \cdot b = n \in N.$$

Hence $$xy = 2n \in \mathcal{E}.$$

If we add two odd numbers, the sum is even. For example:

$$7 + 9 = 16, \quad 11 + 33 = 44, \quad 101 + 13 = 114.$$

THEOREM 3: *If both x and y belong to $\mathcal{O}$, then $x + y$ belongs to $\mathcal{E}$.*
Proof: Since $x \in \mathcal{O}$, we can write $x = 2a - 1$, $a \in N$.
Since $y \in \mathcal{O}$, we can write $y = 2b - 1$, $b \in N$.

$$x + y = 2a - 1 + 2b - 1 = 2(a + b - 1).$$

We notice that $a + b - 1 = n \in N$.
Hence $$x + y = 2n \in \mathcal{E}.$$

If we multiply two odd numbers, the product is also odd. For example:

$$7 \cdot 9 = 63, \quad 11 \cdot 33 = 363, \quad 71 \cdot 9 = 639.$$

THEOREM 4: *If both x and y belong to $\mathcal{O}$, then xy belongs to $\mathcal{O}$.*

Exercise 3:

Proof of Theorem 4 is left to the student.

THEOREM 5: *If $x \in \mathcal{E}$ and $y \in \mathcal{O}$, then $x + y$ belongs to $\mathcal{O}$.*

Exercise 4:

Give at least three examples illustrating Theorem 5. Give the proof for Theorem 5.

THEOREM 6: *If $x \in \mathcal{E}$ and $y \in \mathcal{O}$, then xy belongs to $\mathcal{E}$.*

Exercise 5:

Give at least three examples illustrating Theorem 6. Give the proof for this theorem.

Exercise 6:

Investigate the preceding theorems, if more than two elements are involved. For example, suppose that all three, x, y, z, belong to $\mathcal{O}$. What can be said about the sum $x + y + z$? Can this conclusion be obtained by combining the results of Theorems 3 and 5?

Another way of describing the two subsets $\mathcal{E}$ and $\mathcal{O}$ of N is to say that $\mathcal{E}$ contains all multiples of 2, and $\mathcal{O}$ contains all numbers not divisible by 2. In a similar manner we can consider the sets:

$$T = \{3, 6, 9, \cdots, 3n, \cdots\}$$
$$T' = \{1, 2, 4, 5, 7, 8, \cdots, 3n - 2, 3n - 1, \cdots\}.$$

T contains all multiples of 3, and T' all numbers not divisible by 3.

Exercise 7:

(a) Find $T \cup T'$; $T \cap T'$.
(b) If $x \in T$ and $y \in T$, to what set does $x + y$ belong? xy?
(c) If $x \in T'$ and $y \in T'$, what can be said concerning $x + y$? xy?
The two formulas $3n - 2$ and $3n - 1$, for $n \in N$, together give all the numbers in T'. Find elements of T' using these formulas and convince yourself that this is so.

Exercise 8:

(a) Write two formulas that together give all the integers which are not divisible by 2 or 3.
(b) Can you find formulas that give all integers which are divisible by 2 or 3?

Exercise 9:

(a) Write a formula for the squares of all the even numbers.
(b) Write a formula for the squares of all the odd numbers.
(c) If you know that a number is even and is the square of a natural number n, what can you say about n?
(d) If a square is an odd number, what can be said concerning its square root?

Exercise 10:

(a) Write a formula for the squares of all numbers which are divisible by 3.
(b) Write formulas for the squares of all numbers in N which are not divisible by 3. From the information given by these expressions complete the following statements:
(c) A square which is divisible by 3 is necessarily divisible by _____?
(d) If a square is not divisible by 3, then its square root is _____?

PRIME AND COMPOSITE NUMBERS

A number greater than 1 which is divisible only by 1 and itself is called a *prime* number. Since a prime number is defined to be divisible by 1 and itself, the number 1 is seen to present a special case. As is the general practice, we do not regard 1 as a prime number. A natural number that is greater than 1 and is also not a prime is called a *composite* number. Some examples of such composite numbers together with their prime divisors are as follows:

$$4 = 2^2 \quad \text{(a prime power)}$$

$$6 = 2 \cdot 3$$

$$8 = 2^3 \quad \text{(a prime power)}$$

$$9 = 3^2 \quad \text{(a prime power)}$$

$$10 = 2 \cdot 5$$

$$12 = 2^2 \cdot 3 \quad \text{(a product of two prime powers)}.$$

We note that some of the composites are divisible by only one prime, that is they are prime powers, such as $4 = 2^2$; some are products of two or more different primes, such as $6 = 2 \cdot 3$; and others are products of two or more prime powers, such as $12 = 2^2 \cdot 3$.

Exercise 11:

Suppose we designate the set of primes as

$$P = \{2, 3, 5, 7, 11, \cdots \}$$

and composites as

$$C = \{4, 6, 8, 9, 10, 12, \cdots \}.$$

What can be said concerning: (a) $P \cup C$, (b) $P \cap C$, (c) $C \cap E$, (d) $P \cap \mathcal{E}$, (e) $C \cap \Theta$? Note that, unlike the cases for sets previously discussed, we have not written the formula for the elements of the sets P and C. (f) Can you tell why?

A most important fact is that each composite number is the product of a unique set of prime powers (unique means just one). Any composite number is the product of integers. This set of integers is called a *factorization* of the composite number. A prime factorization (a factorization in which the factors are prime powers) is usually given with the primes stated in order of increasing magnitude. For example: $45 = 3^2 \cdot 5$, $48 = 2^4 \cdot 3$, $56 = 2^3 \cdot 7$, $210 = 2 \cdot 3 \cdot 5 \cdot 7$.

Exercise 12:

Tell which of the following are prime and which are composite, and in the latter case write a prime factorization: 59, 95, 57, 75, 119, 157, 218, 512, 513, 1001.

Exercise 13:

As we noted in Exercise 11, we do not have a formula which gives a prime in the set P, but there are some formulas which do give some primes. As an example, consider $f(n) = n^2 - n + 41$. We observe that $f(1) = 41$, a prime; $f(2) = 43$, a prime; $f(3) = 47$, a prime. Compute $f(4)$, $f(5)$, $\cdots$, $f(42)$, and tell whether these other 39 numbers are prime or composite. For $f(n) = n^2 - 79n + 1601$, compute $f(1)$, $f(2)$, $\cdots$, $f(80)$ and tell whether these 80 numbers are prime or composite.

Exercise 14:

A formula somewhat similar to those given in Exercise 13, which also gives primes for certain integers, is of considerable interest for the work which follows. This formula is illustrated in the following manner.

$$2 \qquad\qquad + 1 = 3, \text{ a prime}$$

$$2 \cdot 3 \qquad\qquad + 1 = 7, \text{ a prime}$$

$$2 \cdot 3 \cdot 5 \qquad\quad + 1 = 31, \text{ a prime}$$

$$2 \cdot 3 \cdot 5 \cdot 7 \qquad + 1 = \cdots\cdots\cdots$$

$$2 \cdot 3 \cdot 5 \cdot 7 \cdot 11 \quad + 1 = \cdots\cdots\cdots$$

$$2 \cdot 3 \cdot 5 \cdot 7 \cdot 11 \cdot 13 + 1 = \cdots\cdots\cdots$$

You will note that the first known primes are multiplied together and this product is increased by 1. Fill in the blanks above and tell whether each number is a prime or is composite.

You will recall that when we previously asked the questions: "How many even numbers are there?" and "How many odd numbers are there?" we were able to establish a one-to-one correspondence between the elements of the set of natural numbers N and the elements of $\mathcal{E}$ or $\mathcal{O}$. We were able to set up this correspondence because we had a simple formula for all even numbers and all odd numbers. Since a formula for the nth prime is not available to us, we cannot use this method to decide "how many" prime numbers there are. For this reason we resort to an indirect method for proving that there is an infinite number of primes.

Suppose that we list all primes in order of magnitude, and that there is a finite number, say k, of them, $1, 2, 3, \cdots$; then P_k is the last prime. The set of all primes, M, would then be the following:

$$M = \{2, 3, 5, 7, \cdots P_k\}.$$

We now form the number $N = 2 \cdot 3 \cdot 5 \cdot 7 \cdot \cdots \cdot P_k + 1$ (see Exercise 14). Since N is a natural number different from 1, it is either prime or composite. N exceeds any prime in the set M, and so is not found among the finite set of primes. Hence N is composite. However, N is not divisible by any of the primes in the set M (why?), and so must be divisible by a prime not in M. *Hence M does not contain every prime.* From this contradiction we conclude that the set of primes is not finite, but is infinite.

HOW FAR CAN YOU SEE?

LAUREN G. WOODBY

U.S. Office of Education
Washington, D.C.

The main question considered in this article is the relationship of the curvature of the earth and sight. The Pythagorean Rule is applied repeatedly to discover the relationship of the earth's curvature and sight distances. Some background in algebra will be helpful in studying this article.

THREE QUESTIONS

How does the curvature of the earth affect sight (or television and radar transmission)? Can you answer these three questions: (a) If you are a life guard on a 20-foot tower, can you see a swimmer 3 miles away? (b) Can a television signal from a 500-foot high transmitter be picked up by a receiver a hundred miles away? (c) How far away could an aircraft, flying at 200-feet altitude, be detected by a ship's radar set whose antenna is on a 50-foot mast?

AN APPLICATION OF THE PYTHAGOREAN RULE

The answers to these questions may be found by means of the Pythagorean Rule, which states a relationship which has been known for more than two thousand years: in terms of the right triangle (Fig. 1), $a^2 + b^2 = c^2$.

Thus, for the answers to our questions consider the section of the earth shown in Figure 2.

BASIC ASSUMPTIONS. An observer at the point P could see an object at A, which is d miles away from P. One assumption we shall use is that light (or other radiation) travels in a straight line. Another assumption is that the radius of the earth is 4,000 miles. Let h be the measure of the height, in feet, of the observer (or antenna) above the earth. All the questions then may be asked as one general question: "Just how does d depend on h?"

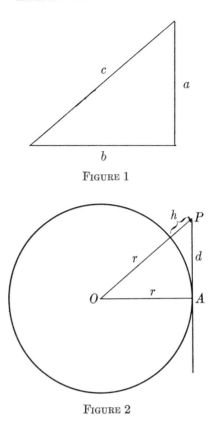

FIGURE 1

FIGURE 2

Application of the Pythagorean Rule gives $r^2 + d^2 = \left(r + \dfrac{h}{5280}\right)^2$.

Note that the hypotenuse, expressed in miles, is $r + \dfrac{h}{5280}$; because h is expressed in feet.

This relation becomes

$$4000^2 + d^2 = 4000^2 + \frac{2(4000)h}{5280} + \left(\frac{h}{5280}\right)^2;$$

which is equivalent to

$$d^2 = 1.52h + \frac{h^2}{5280^2}.$$

ELIMINATING TROUBLESOME NUMBERS. Remember that h is the number of *feet* above the earth. This relation allows us to compute

d^2, if we know the value of h. But the expression is complicated—in particular, the term $\dfrac{h^2}{5280^2}$ is difficult to evaluate. How important is this troublesome term? Let us consider some special cases. For example, if h is 20 feet, $d^2 = 1.52(20) + \dfrac{(20)^2}{5280^2}$, which equals $30.4 + .000014$. Here the troublesome term is insignificantly small compared to 30.4 (which is an approximate number rounded off to the nearest 0.1). The number .000014 would never be missed if we ignored it completely. In this case, then, d is approximately $\sqrt{30.4}$ or 5.51 miles.

A SECOND APPLICATION OF THE RULE

For another example let h be 500 feet. Can we still ignore the term $\dfrac{h^2}{5280^2}$, or is its contribution too large to neglect?

In this case,

$$d^2 = 1.52(500) + \frac{500^2}{5280^2} = 760 + .009 \text{ (approximately)}.$$

Again the number .009 is insignificantly small. When added to the number 760, its contribution is negligible. So $d^2 = 760$ and $d = 27.6$.

Even if h is 100,000 feet, the contribution of the term $\dfrac{h^2}{5280^2}$ will be insignificant when compared with $1.52h$. Do you see why?

Let us therefore simplify our relation between d and h by discarding the term $\dfrac{h^2}{5280^2}$. When we do this we get the much simpler relation,

$$d^2 = 1.52h.$$

This gives an approximation for d,

$$d = 1.23\sqrt{h}.$$

Remember that d is the measure of the distance in miles, and h is the measure of the height in feet. Now can you answer the three questions asked in the beginning?

EXERCISES:

Here are some related questions you might try.

1. Suppose you are on the moon. How far can you see if you are on a 20-foot tower? (Assume the radius of the moon is one-half the radius of the earth.)

2. A surveillance plane flies at 1000 feet. (a) How wide is the strip of the earth's surface that is under observation? (b) How wide is the strip if the plane is 10,000 feet high?

3. A satellite carrying radar and photographic equipment is 200 miles high. (a) How far is the horizon away from the satellite? (b) What is the error in this computation if you use the approximate formula, $d = 1.23\sqrt{h}$? (c) What fraction of the surface of the earth would appear in a photograph taken by the satellite camera? (Assume a single exposure with a sufficiently wide angle lens.)

4. Calculate the distance to the horizon from the third Soviet sputnik at its greatest height of 1880 kilometers.[1] (1 km. = .62 miles.)

[1] This problem was extracted from Karnitskii, *Questions about the Universe in Secondary School Mathematics Problems.* Moscow: Uchpedgiz, 1959.

21

THE ANATOMY OF
A MATHEMATICAL SYSTEM

BRUCE R. VOGELI

Bowling Green State University
Bowling Green, Ohio

This unit shows how a problem of describing the connective characteristics of electrical circuits leads to the development of a mathematical
system. Basic characteristics of a mathematical system are interwoven
with the practical problem and are discussed more fully following the example. It is then shown how the system devised can be applied to another
practical situation, a problem in botany.

Some background in electrical circuits will be helpful but not mandatory.
Only a minimal background in algebra is needed.

MATHEMATICS HAS MANY SYSTEMS

Modern mathematics is comprised of many parts or sub-themes
called mathematical systems. The study of mathematics is essentially
the study of these mathematical systems. Today the number and scope
of these systems is very great. Several important and well known mathematical systems may be familiar to you. The "subjects" of arithmetic,
geometry, and algebra taught in your school are mathematical systems.
Analytic geometry, calculus, and probability and statistics are others of
a more advanced nature. Mathematical research is ever extending existing systems and ever creating new ones. Scientific research, on the other
hand, is ever providing new applications of existing systems and creating
novel situations which require either extension of existing systems or the
creation of new ones for the solution of these novel problem situations.

Students frequently question the possibility of "creating" new mathematical systems. They often interpret "new" to mean mathematical
systems or principles "new" to them but known to others. Is it possible,
after all, to create "mathematics" which is in no textbook and known

to no man? The answer to this question is "yes." If "new" mathematics were not possible, then certainly there would be no mathematics at all. Mathematical concepts, though motivated by nature, are not products of it. They are, in a very real sense, "man-made."

"CREATING" A SYSTEM

Perhaps the best method of learning about mathematical systems is to participate in the creation of one. The mathematical system which we will "create" is a fairly simple one. In fact, it could have been created by a young, inexperienced mathematician, perhaps even a bright junior or senior high school pupil.

The Problem Posed

Electrical engineers in a certain industrial plant wanted a mathematical scheme which would enable them to describe accurately and quickly the connective characteristics of various electrical circuits. These circuits consisted of a power source, a number of relays or switches, and a power outlet or power consuming device such as a motor or light. Two general types of circuit connections among relays were used. Relays were wired in series, that is, one after another as in Figure 1, or in parallel as in Figure 2.

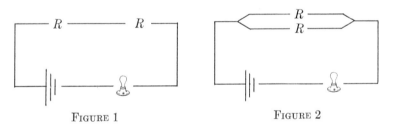

FIGURE 1   FIGURE 2

The relays used in this industry could be either *open* or *closed*. In the open position, no current flowed through the relay. In the closed position, current was permitted to pass through the relay. The engineers posed this question to the plant mathematician: "Is there a mathematical scheme for determining whether or not current will reach the power outlet in a circuit containing perhaps 100 relays in various states and wired in various ways?"

The mathematician noted that the state of a relay could be represented by one of two symbols. Thus, "0" could be used to denote any open relay, while "1" could be used to denote any closed relay. Of course, any other symbols such as p and q or $\triangle$ and $\bigcirc$ could have been employed for this purpose. "0" and "1" were chosen only because they are handy and easy to write and type.

Analyzing the Problem

The mathematician noted that if two relays were "wired in series," as in Figure 1, no current would flow to the power outlet unless *both* were *closed*. That is, two relays in series could be replaced by one closed relay without altering the connective characteristics of the circuit if and only if both relays were closed.

The mathematician expressed this fact on his work sheet by writing:

$$\text{(a) } 1 \otimes 1 = 1$$

where the symbol "$\otimes$" stands for "wired in series." He expressed other series wiring possibilities similarly:

$$\text{(b) } 1 \otimes 0 = 0$$

$$\text{(c) } 0 \otimes 1 = 0$$

$$\text{(d) } 0 \otimes 0 = 0.$$

Statement (b) states that a closed relay "wired in series" with an open one is equivalent to an open relay. Thus, in case (b), no current would reach the power outlet. Statements (c) and (d) may be translated similarly. What are your translations of them? Statements (a), (b), (c), and (d) for two relays wired in series can be summarized in table form as follows.

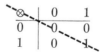

Thus, to determine the state of a single relay equivalent to an open relay (0) "wired in series" with a closed relay (1), one need only look at the entry in the "0" row, "1" column. This entry is "0"; hence, an open relay "wired in series" with a closed one is equivalent to a single *open* relay.

Making Discoveries

The mathematician soon discovered that the situation for parallel relays was quite different. For example, two relays "wired in parallel" permitted current to flow if either or both were closed and prevented the flow of current *only* if *both* relays were open. Thus

$$\text{(a) } 0 \oplus 0 = 0$$

$$\text{(b) } 0 \oplus 1 = 1$$

$$\text{(c) } 1 \oplus 0 = 1$$

$$\text{(d) } 1 \oplus 1 = 1$$

where the symbol "$\oplus$" means "wired in parallel." These four statements can be summarized in tabular form as follows.

$\oplus$	0	1
0	0	1
1	1	1

By this time, the mathematician was very pleased with himself for he had devised a set of elements and two operations with these elements which have many of the properties of the circuits he was studying. He was able to determine with the help of his two operation tables whether or not current would flow through even a complicated circuit. For example, consider the circuit in Figure 3.

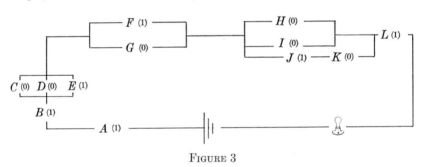

FIGURE 3

The letters denote the relays; the symbols 0 and 1 their states. Remembering that "$\otimes$" means "wired in series" and "$\oplus$" means "wired in parallel," the mathematician represented this circuit by the following expression.

$$A \otimes B \otimes (C \oplus D \oplus E) \otimes (F \oplus G) \otimes [H \oplus I \oplus (J \otimes K)] \otimes L$$

He then replaced the letters representing the relays by the symbols for their states.

$$1 \otimes 1 \otimes (0 \oplus 0 \oplus 1) \otimes (1 \oplus 0) \otimes [0 \oplus 0 \oplus (1 \otimes 0)] \otimes 1$$

Using the tables for "$\otimes$" and "$\oplus$" given above, the mathematician determined that the entire complicated circuit was equivalent to "0," that is, to a *single open* relay. Naturally, no current would reach the lightbulb. Can you design more complicated circuits which can be analyzed by these same means?

DISCOVERING AN ALGEBRAIC SYSTEM

In reality, the mathematician had created a new mathematical system in an effort to answer the question raised by the engineers. He made some discoveries of his own while working with the system. First of all, he found that order in a "series" wiring was unimportant, that is, it makes no difference which relay comes first in a series of two or more. He arrived at this conclusion after noting that the "$\otimes$" table was symmetric with respect to the principal diagonal; that is, if the table were folded along the diagonal (dotted in figure) the entries which would be brought together would be the same in all cases. Upon making this discovery, the mathematician made up several actual circuits and by interchanging series-connected relays, convinced himself that his mathematical discovery had real practical significance. An operation which has this "interchange" property is said to be "commutative." Is "$\oplus$" a *commutative* operation? What practical significance can you attach to your answer?

The mathematician also discovered that if three switches were "wired _ in parallel," he was forced by the limits of his "$\oplus$" table to consider them two at a time. After a number of trials, he became convinced that it made no difference how he "paired" them, since the final results all appeared to be the same. Thus, $(0 \oplus 1) \oplus 1 = 0 \oplus (1 \oplus 1)$. Can you use the "$\oplus$" table to verify the preceding statement as well as the following statements?

$$(0 \oplus 0) \oplus 1 = 0 \oplus (0 \oplus 1)$$

$$(1 \oplus 0) \oplus 1 = 1 \oplus (0 \oplus 1)$$

An operation in which three or more elements can be "associated" in pairs of consecutive elements without affecting the final result is said to be "associative." Is "$\otimes$" also an associative operation? Can you give several "numerical" statements to support your answer?

The mathematician soon discovered a miniature "algebra" which complemented his relay arithmetic. The algebra proved even more useful to the plant engineers than the arithmetic. The mathematician noted that relay A could be in only one of two states; i.e., open or closed, 0 or 1. He decided to indicate the relays' states not by the constants 1 or 0, but by a placeholder or variable a which could be replaced by either 1 or 0 as the situation indicated. The mathematician called the set $D = \{0, 1\}$ the domain of the variable a. Thus, a certain circuit could be depicted as in Figure 4 where the domains of a, b, and c are $D = \{0, 1\}$.

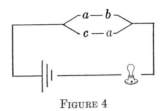

FIGURE 4

Note that *two* relays in the circuit are lettered a, indicating that, although separate, they are controlled — opened or closed — together. The mathematician described this circuit as follows:

$$(a \otimes b) \oplus (c \otimes a)$$

where a, b, and c could be replaced by 0 or 1. The mathematician knew that

$$(a \otimes b) \oplus (c \otimes a) = (a \otimes b) \oplus (a \otimes c)$$

since earlier he had verified that "$\otimes$" was a commutative operation. He also discovered that

$$(a \otimes b) \oplus (a \otimes c) = a \otimes (b \oplus c).$$

Combining these results, he concluded that

$$(a \otimes b) \oplus (c \otimes a) = a \otimes (b \oplus c).$$

Can you verify these statements as the mathematician did by replacing the variables with elements of D?

Practical Applications

The final result, $(a \otimes b) \oplus (c \otimes a) = a \otimes (b \oplus c)$, proved to have very important practical implications since it suggested that the following circuits are equivalent.

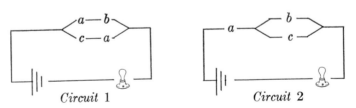

Circuit 1   *Circuit 2*

Can you show that the two circuits actually are electrically equivalent? Note that Circuit 2 requires only 3 switches while Circuit 1 requires 4. Also, Circuit 2 should require less wire than Circuit 1. Both simplifications would permit important savings in the production costs of equipment involving such circuits.

Thus, while the mathematician's relay *arithmetic* answered only questions concerning the electrical characteristics of given circuits, his relay *algebra* actually assisted in the design of the circuits themselves. The mathematician had created a mathematical system, an algebra, which belongs to a general family of systems called "Boolean Algebras" in honor of the English mathematician George Boole.

Components of a System

The mathematician who created the preceding system had a decided advantage over the mathematician of ancient times, for he had numerous examples of mathematical systems to use as patterns. In fact, the anatomy of a mathematical system is well known to the modern mathematician. All mathematical systems appear to be comprised of four ingredients:

1. A basic logic
2. A set of elements
3. Relations (including operations) defined for a set of elements
4. Laws or rules governing these relations.

STEPS IN BUILDING A MATHEMATICAL SYSTEM

When the scientist confronts the mathematician with a novel situation to be described mathematically, probably the mathematician's first step will be to run through his mental (or written) file of existing mathematical systems in an attempt to find a system which accurately depicts the physical phenomena before him. If this is unsuccessful, he may try to adapt or extend an existing system which is similar to the one needed. If this, too, fails he usually sets about the task of creating an entirely new system. The choice of the first ingredient of the new system, its logic, usually is not a difficult one to make. The mathematician frequently chooses the logic, the methods of thought, the "common sense" used in common systems such as geometry, arithmetic, and algebra. In bizarre cases, he may choose a different logic altogether, but such systems are rather rare.

The choice of elements is the next to be made. The mathematician may choose numbers, lines, points, switches, events, or other entities as the elements of his system. Usually he devises a symbolism for representing these elements on paper or on the blackboard. Next he must decide upon the relations and operations which relate elements of his set in a way which is typical of and significant for the physical situation which he is trying to describe mathematically. Finally, he decides upon rules of operation, of procedure, which will cause his mathematical system to agree with the physical phenomena being depicted.

Of course, it is not always possible to devise an abstract mathematical system which accurately depicts a given physical situation on the first try. The mathematician may try again and again to form laws, to define operations in his system, before finding that combination which causes his abstract mathematical system to agree sufficiently with observed phenomena to be useful. Later, his "new" mathematical system may be applied to other very different situations or extended to include more complex phenomena. It frequently happens that "theorems" proved by the mathematician while working in a mathematical system serve as predictions for the scientist working with practical real-world phenomena. Thus, "facts" discovered by the mathematician at the purely mathematical level are "clues" to the scientist. If the mathematical system is indeed a good one, these clues may lead to important "real-world" discoveries.

APPLYING A SYSTEM TO A NEW PROBLEM

In the preceding electrical circuit example, you participated in the creation of a new mathematical system. You studied a given practical situation along with the mathematician. You helped in choosing the elements and in defining the operations ($\oplus$, $\otimes$). You actually performed many of the calculations necessary to establish the *commutative law* for $\otimes$ and the *associative law* for $\oplus$. You observed how discoveries made while working with the abstract mathematical system can be significant at the practical level. You should also realize that the mathematical system created may have other applications or realizations. For example, a botanist may discover that a certain plant disease can be transmitted from plant to plant in two ways: by interbreeding infected specimens and by grafting portions of infected plants. He notes, in the case of interbreeding, that the disease is transmitted to the offspring if and only if both parent plants were infected. On the other hand, in grafting experiments, it is observed that the product plant will be healthy if and only if neither parent portion were infected. Thus, if b stands for the operation of interbreeding, g for grafting, 1 for the state of infection, and 0 for "not infected" or "healthy," we have the following:

b	0	1
0	0	0
1	0	1

g	0	1
0	0	1
1	1	1

Note that the operations b and g are completely analogous to $\otimes$ and $\oplus$ respectively. Hence, the botanical phenomenon described provides a second, significantly different application for the mathematical system created earlier. The generalizations arrived at in connection with relays (commutativity of $\oplus$ and $\otimes$; associativity of $\oplus$ and $\otimes$, and distributivity, i.e., $1 \otimes (0 \oplus 1) = (1 \otimes 0) \oplus (1 \otimes 1)$ and $0 \oplus (1 \otimes 0) = (0 \oplus 1) \otimes (0 \oplus 0)$, should be entirely valid in the case of the plant disease described.

QUESTIONS TO EXPLORE

Perhaps you can think of still other practical, concrete situations which are realizations of these same mathematical systems. Better yet, can you devise other new and different mathematical systems to do certain practical jobs?

Can you devise, for example, a mathematical scheme which will determine on which day of the week a day N days from today will fall for any whole number N?

Can you devise a mathematical system which pictures accurately the possible positions three children playing musical chairs with three chairs might take? With four children and four chairs? If you feel that you can, you will find information concerning the requirements for these systems in the references listed below.

REFERENCES

DENBOW, CARL. "To Teach Modern Algebra." *Mathematics Teacher* 52: 162–70; March 1959.

SCHOOL MATHEMATICS STUDY GROUP. *Mathematics for Junior High School.* Volume I, (part 2), revised edition. New Haven, Conn.: School Mathematics Study Group. pp. 527–77.

SLOAN, ROBERT. "Mathematical Chairs." *Mathematics Student Journal* 8: 4–5; January 1961.

UNIVERSITY OF MARYLAND MATHEMATICS PROJECT. *Mathematics for the Junior High School.* First book. College Park, Md.: College of Education, University of Maryland, 1960. pp. 87–102.

LOGIC:
FOR TEACHER,
FOR PUPIL

ROBERT L. SWAIN†

Rutgers University
New Brunswick, New Jersey

Basic ideas of logic are introduced, such as truth value of statements, "and" and "or" connectives, converses and negations plus combinations of the two, and quantifiers. Also, an analysis of the nature of mathematical proof with several examples and exercises is presented.

Although a few examples and exercises deal with irrational numbers, very little background beyond arithmetic is needed in the reading of this article.

WHAT IS A STATEMENT?

Some sentences convey feelings or attitudes, express commands, or put questions: *What wonderful weather! Thank you. Strike him out, Jimmy! Where are you going?* Such sentences play no direct role in reasoning processes.

Other sentences make definite assertions about facts or relationships.

Bill is 5 feet $4\frac{1}{2}$ inches tall.

Mary sits next to Joan.

Such sentences can be labeled as **true** or as **false**, although sometimes one can be sure about the correct label only after careful investigation of the circumstances. But whether or not the correct label is known, if a

---

† This article is one of many contributions to the improvement of instruction in mathematics made by Robert L. Swain. Tragically it proved to be his last. The Editorial Committee learned with sorrow of his death just after the submission of this article.

sentence is capable of being labeled either true or false, then the sentence is called a **statement**.

These mathematical sentences are statements:

$$2 + 3 = 5. \ (True)$$

$$4 - 1 = 2 + 8. \ (False)$$

$$3 < 2\pi. \ (True)$$

Note that each expresses a relationship involving specific numbers.

The following sentences may or may not be statements.

He is over 21 years of age.

Line AB ⊥ line CD.

$$x + 1 = 3.$$

If the word "he" in the first sentence is understood to designate (name) some specific person, then that sentence is a statement. Likewise, if A, B, C, D name definite points, and if x names a definite number, then the other two sentences are statements. Thus if x is supposed to be another name for the number 2, the sentence $x + 1 = 3$ is a true statement. If x is supposed to be another name for the number 6, then the sentence $x + 1 = 3$ is a false statement. However such usage would be unusual. In most algebraic work, x is a variable rather than the name of a specific number. In that case it is usual to call $x + 1 = 3$ an "open sentence." We will not pursue this topic further, but will confine our discussion to statements.

THE CONNECTIVE "AND"

We determine the **truth value** of a statement by finding whether the label *true* or the label *false* is properly applicable to the statement.

Suppose that we link two statements by the connective **and**. There is a simple relationship between the truth values of the original statements and the truth value of the new combination. If the original statements are both true, so is the new one. If either or both are false, so is the new one. The following examples show all four possibilities:

$$2 + 3 = 5 \quad and \quad 4 = 4. \quad (True)$$

$$2 + 3 = 5 \quad and \quad 4 = 7. \quad (False)$$

$$2 + 3 = 6 \quad and \quad 4 = 4. \quad (False)$$

$$2 + 3 = 6 \quad and \quad 4 = 7. \quad (False)$$

We can make a table, called a *truth table*, to show the relationship of the truth values. For brevity, we write A in place of the first statement, B for the second. The *and* combination then has the form A *and* B.

TRUTH TABLE

A	B	A and B
True	True	True
True	False	False
False	True	False
False	False	False

While pupils will ordinarily accept this assignment of truth values as consistent with everyday usage of *and*, occasions of resistance will arise. When taking true-false tests, pupils often become annoyed at meeting *and* combinations involving one true and one false statement, such as:

Iron is a metal and oxygen is a mineral.

Even though the pupil correctly marks the statement *false*, he may do it under protest, feeling that he ought somehow to be able to label it as "half-right," and that the question is unfair or tricky. (If the test is to be graded by subtracting "wrongs" from "rights," the pupil is justified in feeling aggrieved, since he must get two factual "rights" to earn one test question "right.")

In classroom illustration, the teacher should use both related and unrelated statements to make up combinations, as in the examples below.

Mary is blond and Charles is dark-haired.

Jean's hair is curly and 2 = 3.

Iron is hard and Dick is Ann's brother.

There are other connectives that have the same *logical* significance as *and*, but that have different over-all shades of meaning. "But" suggests contrast, or surprise. Examples:

Mary is blonde but her sister Joan is brunette.

While Nero fiddles, Rome burns.

I was frightened, nevertheless I kept silent.

Since John is over 21, he is voting.

If x names a specific number, then the sentence

$$x > 1 \quad and \quad x < 4$$

asserts that the number in question lies between 1 and 4. In algebra, the sentence might be taken as a direction to find all the numbers that x could name for which the sentence would be true. The "picture," or *graph*, of the sentence would then consist of an entire line segment:

THE CONNECTIVE "OR"

In everyday usage, we employ **or** in two ways: the exclusive, or "either-or" sense (one or the other, but not both); the inclusive, or "and/or" sense (one or the other or both).

When Mother tells young Jimmy, "Stop hitting your brother, or you can't go with us to the movie," she only means to put two choices before him: Stop and go; don't stop and don't go. Jimmy would be outraged if he stopped at once, then was told he could not go. In such a case we do not expect *both* statements to be true; for in this case *or* is used in the exclusive sense.

But Bill may tell Bob, "I heard that song on TV, or perhaps I heard it on the radio." If it turns out later that Bill heard it on both TV and radio, he would not be surprised, nor inclined to withdraw his original statement. Bill would have to retract his statement only if it developed that he had heard it neither on TV nor on radio. Bill's use of *or* was in the inclusive sense.

In logical and mathematical work, we cannot tolerate ambiguity, and must choose one of the two senses. It is the inclusive (and/or) sense that has been chosen, resulting in the following truth table.

TRUTH TABLE

A	*B*	*A or B*
True	True	True
True	False	True
False	True	True
False	False	False

THE CONNECTIVE "IF, THEN"

If Joe catches cold, then he will miss the party.
If that's a real diamond, it will scratch glass.
If Mary is in fifth grade, she must be over eight years old.
If the New York Times printed it, then it must be so.

When we make **if A, then B** statements like those above, we commonly focus our attention upon the case in which the first statement, *A*, is true. We then think of its truth as compelling the truth of the second statement, B. Following along this line of thought, we may put down a partial truth table as follows:

PARTIAL-TRUTH TABLE

A	B	If A, then B.
True	True	True
True	False	False

What if the first statement, A, is false? On the everyday level, we usually just forget the "if, then" statement, discarding it as no longer applicable. Take the statement, "If Joe catches cold, then he will miss the party." Joe's mother may be concerned about what truth value to attach to this statement. She and Joe may even disagree about the value to be attached. But suppose Joe does not catch cold. Neither he nor his mother is likely to have any further interest in the statement, much less in its truth value.

In logic and in mathematics we cannot neglect the two cases in which the "if" clause is false. Here are instances of them.

$$If\ 2\ =\ 3,\quad then\quad 10\ =\ 10$$

$$If\ 2\ =\ 3,\quad then\quad 10\ =\ 100$$

Both can be proved true by accepted mathematical principles. If $2 = 3$, then according to the definition of equality, we also have $3 = 2$. Addition gives:

$$\begin{array}{r} 2 = 3 \\ 3 = 2 \\ \hline 5 = 5 \end{array}$$

Multiplying by 2 on both sides now gives $10 = 10$. In the other case we may pass from $2 = 3$ to $0 = 1$ by subtracting 2 on both sides, then to $0 = 90$ by multiplication, then to $10 = 100$ by addition.

It would appear from the above that an "if, then" statement must be regarded as true whenever the "if" clause is false. In fact it is possible, by making use of the principle of contradiction (mentioned later), to show that by assuming true a statement which is false, one can prove true any statement whatsoever! Possibly Dave, age ten, would be simply confused by the discussion above, but he shows his implicit grasp of the situation when he exclaims, "If Jerry makes the team, I am a monkey's uncle!" Despite the fact that the false "if" clause cases give rise to interesting examples, a formal treatment is probably best avoided at the elementary level, even for the top pupils.

Note on Terminology

An "and" combination is called a *conjunction*, an "or" combination a *disjunction*. An "if, then" form is a *conditional statement* or an *implication*. Some youngsters love to use such technical words. Others resist.

CONVERSE

In making "if, then" statements we may put the "if" clause first or second, without affecting the meaning:

> *If George sinks his putt, he will win.*
>
> *George will win, if he sinks his putt.*

But switching the "if" from one statement to the other alters the meaning, and results in a new combination called the **converse** of the original:

> STATEMENT: *If A, then B* (or: *B, if A*).
>
> CONVERSE STATEMENT: *If B, then A* (or: *A, if B*).

Knowing that an "if, then" statement is true or false will not usually help you to find out whether its converse is true or false. Suppose John says, "If that catcher's mitt is under \$5, I will buy it." In so avowing, he is not also promising the converse: "If I buy the mitt, it will have been under \$5." (Note that alteration of verb tenses is common, and does not affect the logical meaning.) If the price turns out to be \$6.50, John is free to buy or not buy, and may do either.

The converse is often formed by replacing the word "if" by the phrase "only if." "John will pass only if he studies harder" has the same logical meaning as, "If John passed, he studied harder." Here again a formal treatment is probably best avoided, involving subtleties too profound for elementary school pupils.

In high school mathematics, when a statement "*A if B*" and its converse "*A only if B*" (i.e., "*B if A*") are *both* true, this relationship may be expressed in a single statement of equivalence: "*A if and only if B*." Examples:

> *A natural number is divisible by 6 if and only if it is divisible by both 2 and 3.*
>
> *A triangle is equilateral if and only if it is equiangular.*

Mathematicians use the "if and only if" phrase so much that they often abbreviate it as "iff" — then face the risk that a typist or printer will correct their spelling!

NEGATION

The **negation** of a statement is a statement with opposite truth values: false when the original statement is true; true when the original statement is false. Observe that negating a negation restores the original truth values: *Not (Not A)* is equivalent to *A* itself.

TRUTH TABLE

A	*Not A*
True	False
False	True

To obtain the negation, prefix the original statement with the phrase "It is not true that." It will usually be possible, with care, to simplify the result.

To negate "Mary is beautiful," you may write:

It is not true that Mary is beautiful.

You may rewrite this:

Mary is not beautiful.

In the simplifying process, the word "not" can sometimes be eliminated. For human beings, "not male" and "female" are synonymous. Hence the negation of "Jackie Bryant is male" can be put in the form, "Jackie Bryant is female." Likewise, the negation of "8 is an odd number" is "8 is an even number."

A word of caution: Negation is a matter of merely denying the truth of a given statement rather than of asserting its extreme opposite. In asserting the negation of "Mary is beautiful," we do not claim that "Mary is ugly." "Beautiful" and "not beautiful" embrace all the possibilities, while "beautiful" and "ugly" do not.

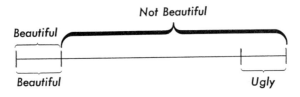

NEGATING COMBINATIONS

The negation of an "and" combination is an "or" combination (and vice versa) of the negations of the separate statements:

Not (A and B)   means   *(Not A) or (Not B).*

Not (A or B)   means   *(Not A) and (Not B).*

Thus to say that it is not true that

Mary is blond and Jim is tall

is equivalent to saying

Mary is not blond or Jim is not tall.

Similarly, the negation of "Jackie is male or Dot is not blond" is "Jackie is female and Dot is blond."

The above facts are easily proven by showing the forms to have the same truth values. Thus when A is true and B is false, $(A$ *and* $B)$ is false, so *Not* $(A$ *and* $B)$ is true; but also *Not* A is false and *Not* B is true, so that $(Not$ $A)$ *or* $(Not$ $B)$ is likewise true. This is easily done for all four cases at once, in truth table form. We abbreviate T for "true", F for "false". Noting that the last two columns are identical completes the proof.

A	*B*	*Not A*	*Not B*	*A and B*	*Not (A and B)*	*(Not A) or (Not B)*
T	*T*	*F*	*F*	*T*	*F*	*F*
T	*F*	*F*	*T*	*F*	*T*	*T*
F	*T*	*T*	*F*	*F*	*T*	*T*
F	*F*	*T*	*T*	*F*	*T*	*T*

The truth table technique above can be used to find the negation of an "if, then" statement. It turns out that

Not (If A, then B)   means   *A and (Not B).*

To say that it is not true that

If Joe catches cold, he will miss the party

is equivalent to saying:

Joe catches cold and he does not miss the party.

But even at this initial plunge, we find ourselves in water that is both rough and deep, and if we have brought any elementary pupils with us, they are likely floundering. In other words, it may be best just not to bring up the topic of negating "if, then" statements in the elementary classroom.

QUANTIFICATION

Many everyday and most mathematical statements are about sets or groups of objects rather than about single objects. Phrases that tell whether **all** or **some** of a group are involved in a statement are called **quantifiers**.

As synonyms for *all*, we may use *each*, *every*, *any*, *an* or *a*. The following quantified statements are all equivalent:

> *All fractions are numbers.*
>
> *Every fraction is a number.*
>
> *Each fraction is a number.*
>
> *Any fraction is a number.*
>
> *A fraction is a number.*

A statement like "No cat is a dog" can be rewritten in "all" form, and hence seen to be quantified: "Anything that is a cat is not a dog", or "Any cat is a non-dog."

Some means one *or more*. The following are all equivalent:

> *Some number has a square equal to 4.*
>
> *There is a number whose square is 4.*
>
> *There is at least one number whose square is 4.*
>
> *There exists a number whose square is 4.*

Since an "all" statement is actually an "and" combination of separate statements, its negation is a "some" statement — a denial of some one of the statements making up the combination. Suppose a teacher says to three pupils, "All of you are tall." Let us say that the names of the pupils are, for brevity, x, y, and z. The teacher is therefore asserting:

> *(x is tall) and (y is tall) and (z is tall).*

The negation is

> *(x is not tall) or (y is not tall) or (z is not tall).*

This can be rewritten as:

> *At least one of you is not tall.*

Since *Not (Not A)* is equivalent to *A*, the negation of "At least one of you is not tall" must be the statement with which we began, "All of you are tall." Apparently the negation of a "some" statement is an "all" statement. The negation of "Some cat is a dog" is "All cats are non-dogs," or "No cat is a dog." The negation of "All numbers are rational" (i.e., expressible in fractional form) is "There exists an irrational number."

NATURE OF PROOF

The principal activity of man is the making of decisions. Each of us makes thousands daily, most of them trivial, such as "There—that pile in the spoon should be just about the right amount of instant coffee to use," but some decisions are important, such as "I shall take the new job." We may look upon decision–making as essentially a matter of accepting or rejecting statements; either because we know definitely whether they are true or false, or because we judge or wish them true or false.

The process of showing that a statement is definitely true is called *deductive proof*. A proof is usually structured in the form of a "theorem": *Hypothesis → Conclusion*. We try to show that the truth of the statement comprising the hypothesis (in conjunction with whatever statements are assumed as axioms) guarantees the truth of the statement comprising the conclusion. For an adequate discussion of the nature of proof, we would have to take up the topics of axioms and of rules of inference. Instead we shall partly bypass these matters and partly take them for granted — as is indeed often done, even in college-level instruction.

It might seem that we ought at the very least to take up the idea of the "structure" of a proof — the way a proof is put together. But this too is a highly technical topic, as we shall now attempt to illustrate.

Do you visualize a proof as a chain of statements leading from hypothesis to conclusion, the truth of each statement guaranteeing the truth of the next?

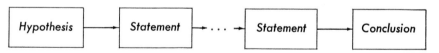

Unfortunately, such a simple pattern is rarely found either in everyday life or in mathematical work.

Suppose you wish to show that to add 102 to any number, you may add 100, then add 2 to the result. In practice you may just write

$$N + 102 = N + (100 + 2) = (N + 100) + 2,$$

taking for granted that everyone will agree that the first equality is justified by simple number facts from early grades as well as by an obvious substitution principle, while the second may be justified as an instance of the associative property of addition. Your efficient use of a "continued equality" conceals the actual proof structure involved, which looks something like Figure 1.

Actually, Figure 1 is incomplete. Had we put in the particular statements from the top box needed to justify the steps, there would have been more boxes and arrows.

The study of proof structure would appear to be a topic which is beyond the capabilities even of most college students to handle. Yet the idea of proof is basic to mathematics and to science, and must enter the school curriculum at the earliest possible time.

Actually, there is no real dilemma here. Youngsters can learn about proof, what proofs are like, and how to do them, much as they learn their own language. We learn to use language by example, precept, and practice; and we use it fairly correctly, long before we discuss its technical mechanism in school. (Nor does school grammar bear much resemblance

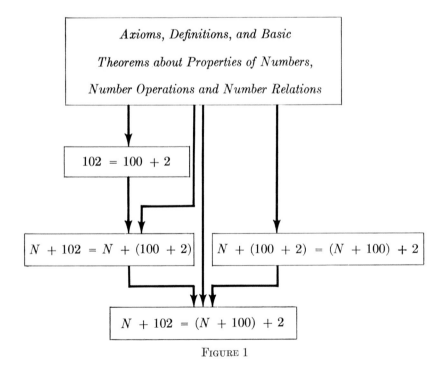

FIGURE 1

to the structural analysis of linguists.) Proof processes are much a part of everyday life and of school work in all areas. From early grades, teachers can assist children to recognize proof processes and to understand them as best they can.

Suppose a child is asked to "prove" that

$$2 + 3 = 3 + 2.$$

This means that he is supposed to derive this statement from certain other statements about numbers whose truth he may assume. A troublesome matter is that of deciding what those certain statements should be which are to be used as axioms, theorems, etc. One child may be unable or unwilling to make a choice. Another may make a choice without being aware of its arbitrary nature, and proceed merrily with a proof. He may think in terms of number sums and counting, and may count on successive fingers 1–2–1–2–3, then recount 1–2–3–4–5, verifying in this way that $2 + 3 = 5$. Then he may count 1–2–3–1–2, and recount 1–2–3–4–5, verifying that $3 + 2 = 5$. Having found that "$2 + 3$" and "$3 + 2$" are both names for the same number, 5, he knows he can write: $2 + 3 = 3 + 2$.

Another child may associate the sums $2 + 3$ and $3 + 2$ with sets of blocks, basing his proof upon the idea illustrated in Figure 2.

A child with an abstract bent may think of 2 and 3 as

$$2 = 1 + 1 \quad \text{and} \quad 3 = 1 + 1 + 1,$$

then conceive that by regrouping, $(1 + 1) + (1 + 1 + 1)$ may be rewritten in the form $(1 + 1 + 1) + (1 + 1)$.

Possibly another child may object to the whole procedure, saying to the teacher, "You told us that numbers obey the commutative law of addition, and $2 + 3 = 3 + 2$ is just a case of that law, so there is nothing to prove." The teacher might point out to this child that the recognition

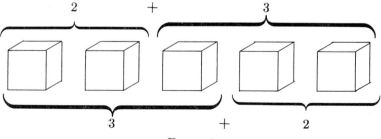

FIGURE 2

that a statement is an instance of an accepted law is itself a proof procedure.

Of course, all the children are "right." Ideally, whenever a proof of any sort is demanded, the assumptions to be used should in some way be specified. In practice, complete specification is usually impractical, and one does the best one can, leaving much up the pupil's judgment. This is what we shall also do in the ensuing discussion. (Complete specification of assumptions is possible only in formal axiomatic treatments. In some modern high school courses now being developed, rough approximations of such treatments are attempted both in algebra and in geometry.)

PROOF PROCEDURES

Perhaps the very simplest form of proof is that in which a general law (which may itself be an axiom, definition, or theorem) is applied to a specific instance or to a restricted set of instances. We do this constantly and routinely in all mathematical work. But unfamiliar applications appear from time to time, forcing us to stop and think about what law is to be used.

Applying a General Law

A child may discover, for example, that instead of dividing by 25, it is easier to multiply by 4, then divide by 100. Asked to "prove it," he may be able to show that the rule is nearly an immediate consequence of the basic law of fractional equality $\left(\dfrac{a}{b} = \dfrac{na}{nb}\right)$:

$$N \div 25 = \frac{N}{25} = \frac{4 \times N}{4 \times 25} = \frac{4 \times N}{100} = (4 \times N) \div 100.$$

Likewise, a child may have added a column of 20 different figures representing daily wages paid 20 workers. Asked to find the weekly total, $5\frac{1}{2}$-day basis, he may, we hope, multiply his previous total by $5\frac{1}{2}$ instead of multiplying each of the 20 figures by $5\frac{1}{2}$ and adding. If he adopts the more efficient procedure, he is making an application of the extended distributive law:

$$a(b + c + d + \cdots) = ab + ac + ad + \cdots.$$

For a less trivial illustration of a simple direct proof, we suppose that a child has been told that an average of two numbers is found by taking half their sum. We also suppose that the child has had some practice in using letters or "frames" as variables. We ask him to prove that the average of two numbers a, b lies halfway between them. Proof, of course, consists in showing that the following two differences are equal ($b \geqq a$):

$$d_1 = b - \frac{1}{2}(a + b) = \frac{2b}{2} - \frac{a + b}{2} = \cdots = \frac{b - a}{2}$$

$$d_2 = \frac{1}{2}(a + b) - a = \frac{a + b}{2} - \frac{2a}{2} = \cdots = \frac{b - a}{2}$$

$$\therefore\ d_1 = d_2.$$

Proof by Cases

Proof by cases is a common procedure. We illustrate by a purely arithmetic proof that *the square of any odd number is an odd number.*

A given odd number must end in one of the five digits 1, 3, 5, 7, or 9. We take the following to be our five "cases":

Case I.   The given odd number ends in 1.

Case II.   The given odd number ends in 3.

Cases III, IV, and V.   The given odd number ends in 5, 7, or 9.

Now the last digit in a product of two whole numbers is the last digit in the product of their last digits. This is easily verified by inspecting the pattern of a long multiplication, as is suggested by the example at the right. In the case of a square, this last digit will be the last digit in the square of the number. But the squares of 1, 3, 5, 7, 9 are

$$243$$
$$\times 37$$
$$\textcircled{1}$$
$$\textcircled{1}$$

$$1, 9, 25, 49, 81$$

and have as their last digits only 1, 9, 5, 9, 1, respectively, as you see above. In other words:

Case I.   The square of a number that ends in 1 is a number that ends in 1, hence is odd.

Case II.   The square of a number that ends in 3 is a number that ends in 9, hence is odd.

Cases III, IV, and V.   The squares of numbers ending in 5, 7, or 9 end in 25, 49, or 81, respectively, hence are odd.

We may now argue that given *any* odd number, the statement in some one of the five cases applies to it, so its square must be odd.

Proof by Elimination

Another common proof scheme is that of *elimination*. Here again we have a breakdown into cases, or alternatives. We must be able to show that some case holds (though we need not show which particular one); then, we must show that all but a particular one may be ruled out. This is a favorite scheme of detectives, at least in fiction. If there happen to be exactly seven people, let us say, who could possibly have committed a crime, and if six of them can somehow be eliminated as suspects, then the one person who remains must be the criminal.

Suppose Dorothy wants to send an air-mail letter to Haiti, the postage being 10¢. She has a supply of 3¢ and 4¢ stamps, and does not want to waste postage. What are her choices?

TABLE I

Number of 3¢ stamps

		0	1	2	3
Number	0	0	3	6	9
of 4¢	1	4	7	10	13
stamps	2	8	11	14	17

Table I shows the values of all the combinations she may try. Not only does the combination of one 4¢ stamp and two 3¢ stamps give the required 10¢ postage, but this is proved to be the only solution to the problem because all other alternatives are eliminated. (Note that there were actually infinitely many possibilities, but Dorothy had at once ruled out those involving more than two 4's or three 3's, which obviously result in values exceeding 10¢.)

INDIRECT PROOF

Proofs fall into two main classes, **direct** and **indirect.** It is hard to say which type appears more frequently in mathematics and in everyday life. An indirect proof is one in which a **logical contradiction** is obtained at some stage of the work, such as an assertion that some particular statement is both true and false.

To prove a statement by the indirect method, one usually begins by supposing that the *negation* of the statement is true.

Johnny may tell his mother, "Billy wasn't at the party," then may add by way of explanation, "I'd have seen him if he'd been there."

Johnny is employing indirect argument in a commonplace way, with some hidden assumptions and unvoiced steps of reasoning. The statement to be proved is "Billy was not at the party." To prove it, Johnny supposes the negation to be true: "Billy was at the party." But he also assumes true the statement: "If Billy was at the party, I would have seen him." This, together with the negation statement, leads to the conclusion that Johnny did see Billy at the party. But this stands in contradiction to the statement that Johnny left unvoiced, that he did not see Billy.

In other words, if the negation statement "Billy was at the party" is true, then so is the statement "Johnny saw Billy and Johnny did not see Billy." But this is impossible, since this latter statement is of form A *and (Not A)*, hence is always false. Therefore, the negation statement, "Billy was at the party," is false, and the original statement that was to be proved, "Billy was not at the party," is true.

Let us apply the indirect method to show that the fractions $\frac{2}{3}$ and $\frac{7}{10}$ are unequal. Suppose that the negation is true: $\frac{2}{3}$ and $\frac{7}{10}$ are equal. Form the difference: $\frac{7}{10} - \frac{2}{3}$. Since the fractions are equal, this difference is zero. But we also find

$$\frac{7}{10} - \frac{2}{3} = \frac{21}{30} - \frac{20}{30} = \frac{1}{30} \neq 0.$$

Hence the "and" combination of the last two statements is true:

The difference is zero and the difference is not zero.

As before, this contradiction means that the negation is false, hence that the original statement is true.

"If, Then" Statements

Most mathematical statements that we wish to prove are of the "if, then" type. To prove an *If A, then B* statement, we begin by supposing true the statement A *and (Not B)*, then show that a contradiction can be obtained. We conclude from this that the statement A *and (Not B)* is false. This means that whenever A is true, *(Not B)* is false, so that B is true; that is—*If A, then B*—as was to be proved.

As an illustration, we prove:

If $\sqrt{2}$ is irrational (i.e., not rational), so is $3\sqrt{2}$.

What we must do is to rule out the possibility A *and (Not B)*:

$\sqrt{2}$ *is irrational and $3\sqrt{2}$ is rational.*

Assuming this possibility true, we can express $3\sqrt{2}$ as a quotient of two integers:

$$3\sqrt{2} = \frac{\text{An Integer}}{\text{An Integer}}.$$

But then division on both sides by 3 gives

$$\sqrt{2} = \frac{\text{An Integer}}{3 \times (\text{An Integer})} = \frac{\text{An Integer}}{\text{An Integer}} = \text{A Rational Number.}$$

We have obtained the desired contradiction: "$\sqrt{2}$ is rational, and $\sqrt{2}$ is not rational." The conclusion is that we have proved the original statement.

Quantified Statements

As final illustration of indirect proof, we prove a theorem whose statement is in quantified form.

THEOREM: *If the square of a (natural) number is even, so is the number.*

Proof: First rewrite the statement so that the quantifier appears explicitly:

For each number, if its square is even, it is even.

To prove this, we try to show the following impossible:

For some number, its square is even and it is not even.

But that a number cannot be odd (i.e., not even) and fail to have an odd square was proven in the section on "direct proof procedures."

Now suppose that we consider any number whose square is even. Since we have shown the "some" statement above to be false, we know that the number under consideration cannot be not even; hence must be even. And this is what we set out to prove.

The general pattern seen in the above proof occurs very frequently in mathematical work. There is a set of objects of some sort (e.g. natural numbers). We wish to prove that whenever statement A is true for *any* one of the objects, so is B. We do this by proving false the assertion that there is *some* one of the objects for which statement A is true *and* statement B is *not* true. (What we are actually considering here is the negation of the quantified "if, then" statement, a topic which we are bypassing.)

COUNTEREXAMPLES

Suppose a person claims:

Every number is greater than 24.

When you object, the person points out that he can cite innumerable instances of the truth of his claim, such as "63 is greater than 24," "272 is greater than 24," and so on.

You have a right to turn a deaf ear to his recital of instances, since you may counter by citing just a single example:

20 is a number which is not greater than 24.

The number 20 furnishes a **counterexample** that suffices to disprove the original statement.

When a mathematician is faced with an apparently significant statement whose truth status he does not know, he will usually alternate between trying to prove it and seeking a counterexample to disprove it. Whichever alternative works out, the result may be important. A whole school of Greek thought received a shattering blow when it was found that $\sqrt{2}$ furnished a counterexample for the proposition that all numbers could be expressed in fractional form.

To illustrate the idea of a counterexample, a teacher may say to the class, "Everyone in this room is male." This is a cue for Mary to rise and say, "Your statement is false. *I* am a counterexample." We leave it to the teacher to deal properly with the pupil who claims that "the exception proves the rule!"

Teachers frequently use counterexamples to show pupils that wrong methods are being used. John may try to add two fractions by adding the numerators and the denominators. To show John that his rule of procedure is incorrect, the teacher may cite a counterexample:

$$\frac{1}{1} + \frac{1}{1} \neq \frac{2}{2}.$$

Logically, only a single counterexample is required. Pedagogically, several may be helpful. A child's (or adult's) faith in a wrong method dies hard. He may hopefully regard the unwelcome counterexample as an accident, best overlooked, just as he might overlook a single wicked act of a friend. Also, to be sure, the teacher must continue with the mathematical therapy by making a strong effort to show the pupil the *why* of the correct method, so that the pupil can begin to build a new faith to replace that which was lost.

SAMPLE PROOF-PROBLEMS

Use direct proofs in problems 1 and 2 below. Problem 3 involves a proof by cases and a proof by elimination. Problems 4, 5, 6, 7, 13, 15 may be treated as indirect proofs. The rest mainly call for the discovery of counterexamples.

1. Prove that to multiply a number by 5, you may multiply it by 10, then divide the result by 2.

2. Prove that doubling just one factor of a product doubles the product.

3. Room plans for two houses of four and six rooms, respectively, with doorways from each room into each adjoining room are

shown in Figures 3 and 4 below. Prove that it is impossible for a person to enter House I at A, go through each room without ever returning to an earlier visited room, and leave at B. In the case of House II, prove that there is exactly one such path from A to B.

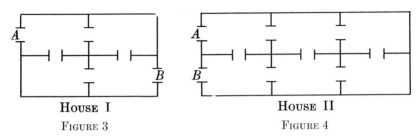

HOUSE I | HOUSE II

FIGURE 3 | FIGURE 4

4. Without multiplying out, how can you quickly convince a friend that his result, shown below, is wrong?

$$242.34 \times 67.62 = 1687.0308$$

5. Prove that if the product of a certain pair of positive numbers is 100, and the first number is less than 10, then the second number is greater than 10.

6. Prove that the number

$$(2 \times 3 \times 4 \times 5 \times 6 \times 7 \times 8 \times 9) + 1$$

is not divisible by any natural number from 2 to 9 inclusive.

7. Prove that if 20 boys have 105 marbles among them, there is at least one boy with more than 5 marbles.

8. Prove or disprove: The reciprocal of a sum is the sum of the reciprocals.

9. Prove or disprove: The value of a fraction is unaltered if numerator and denominator are increased by the same amount.

10. Prove or disprove: A price marked down by a certain percent can be restored by a markup of the same percent.

11. Prove or disprove: If a circle is circumscribed about a triangle, the center of the circle lies within the triangle.

12. Prove or disprove: To find the area of a quadrilateral, average the lengths of each pair of opposite sides, then multiply these two averages.

13. In this and in problems 14, 15, we will use the following terminology. A fraction is any number that can be named by a form

$$\frac{\text{Some Natural Number}}{\text{Some Natural Number}}$$

the natural numbers being 1, 2, 3, etc. Thus, $\frac{2}{3}$, $\frac{18}{4}$, $\frac{6}{3}$, $\frac{2}{1}$ are examples of fractions. Note that $\frac{2}{1}$ and $\frac{6}{3}$ also equal natural numbers. We now define a "true fraction" as a fraction that is not equal to any natural number. Thus, $\frac{2}{3}$ and $\frac{18}{4}$ are true fractions, while $\frac{6}{3}$ and $\frac{2}{1}$ are not. Also, we assume that sums, products, and differences (larger minus smaller) of natural numbers are natural numbers.

Prove that the sum of a true fraction and a natural number is a true fraction.

14. Prove or disprove: (a) The sum of two true fractions is a true fraction; (b) The product of a true fraction and a natural number is a true fraction; (c) The quotient of a true fraction by a natural number is a true fraction.

15. Prove that if the square root of a true fraction is a fraction, then it is a true fraction.

SUGGESTED READINGS

The books listed below contain chapters on logic which many teachers will be able to read and understand. For those who wish more complete and technical treatments, we mention the names of Irving M. Copi, Robert M. Exner and Myron F. Rosskopf, Robert R. Stoll, and Patrick C. Suppes as authors of such works. There are also high school geometry texts and first-year college texts that contain material on the logic of statements and on mathematical proof procedures, as well as on the logic of classes—a topic which we did not discuss.

ALLENDOERFER, C. B., and OAKLEY, C. O. *Principles of Mathematics.* New York: McGraw-Hill Book Co., 1955 (Chapter 1).

CHRISTIAN, ROBERT R. *Introduction to Logic and Sets.* Boston: Ginn and Co., 1958.

EVES, HOWARD, and NEWSOM, CARROLL V. *An Introduction to the Foundations and Fundamental Concepts of Mathematics.* New York: Rinehart, 1958 (Chapters 1, 6, 8, 9).

KEMENY, J. G.; SNELL, J. L.; and THOMPSON, G. L. *Introduction to Finite Mathematics.* Englewood Cliffs, N. J.: Prentice-Hall, 1957 (Chapter 1).

LIEBER, L. R. *Mits, Wits and Logic.* New York: W. W. Norton and Co., 1947.

MAY, KENNETH O. *Elements of Modern Mathematics.* Reading, Mass.: Addison-Wesley, 1959 (Chapters 1, 2, 3).

NATIONAL COUNCIL OF TEACHERS OF MATHEMATICS, a department of the National Education Association. *23rd and 24th Yearbooks,* Washington, D. C.: The Council, 1958, 1959 (Chapters on deductive methods and proof).

STABLER, E. R. *An Introduction to Mathematical Thought.* Reading, Mass.: Addison-Wesley, 1953 (Chapters 3, 4).

UNIVERSITY OF MARYLAND MATHEMATICS PROJECT. *Mathematics for the Junior High School,* 1960 ed. College Park, Md.: College of Education, University of Maryland, 1960. (First Book, pp. 157–70; Second Book, pp. 47–67)

GEOMETRY AND TRANSFORMATIONS

DANIEL E. SENSIBA

Kimberly High School
Kimberly, Wisconsin

This study of transformations as they relate to geometric figures relies heavily on intuition and diagrams. Familiarity with the newer notation and ideas in geometry will be helpful although not absolutely necessary. New terms and notation are carefully defined. If a unit on geometry is taught to the entire class, this topic could easily accompany or follow it for the more interested students.

The purpose of this article is to consider some simple transformations of certain geometric figures. The material presented is by no means exhaustive. On the contrary, it does little more than "scratch the surface" of geometry. Perhaps it will whet your appetite for further investigation and study of these and other topics. You may find the world of geometry intriguing and exciting.

EXAMPLE OF A TRANSFORMATION

Consider some points in a plane. Represent them by dots on a paper and call them A, A', A'', etc. Think of these points as very small marbles resting in tiny depressions on a table top, so there is one and only one marble in each depression. Shaking the table causes the marbles to leave the depressions and roll to new positions in such a way that every depression has exactly one marble in it. Now some marbles may be in the same position while others may have "moved."

In Figure 1, point A is now in the position formerly occupied by A', and A' has "moved" somewhere else, etc. As point A "moved" from A to A', it occupied many intermediate positions which make up a *path* from A to A'. In this way, each point of the plane is paired with one and

FIGURE 1

only one point, the point it replaces. This is denoted by $A \rightarrow A'$, $A' \rightarrow A''$, etc. and is called a *transformation* of the plane onto itself. In the pairing $A \rightarrow A'$, A' is called the *image* of A, and A is called the *pre-image* of A' under the transformation. If $A_1 \rightarrow A_1$, then A_1 is called a *fixed point* of the transformation. If each point A is assigned one and only one image point A'; and also, each point A' has one and only one pre-image A, then the transformation is called a *one-to-one mapping*. In terms of the marble and table top example above, this means that when we jar the table, no marble can go to two different depressions, and after the marbles have come to rest, no marble could have come from two different positions.[1]

Exercise 1:

Consider a set of two points, A and B. Under a given transformation, points A and B exchange positions.
(a) What is the image of A?
(b) What is the image of B?
(c) What is the pre-image of B?
(d) Is the transformation a one-to-one mapping?
(e) Is there a limit to the number of paths from A to B?

Keep in mind that points are ideas and not physical objects like the marbles used above, and that to "move" a point is simply to think of that point as paired with a second point in the plane. Here, we are concerned with the relative positions of points before and after a transformation. For example, consider an automobile with two fixed headlights. Driving the auto two miles down the highway does not change the position of the headlights relative to each other (barring accidents of course), although their position relative to the highway is changed considerably.

---

[1] The author is indebted to Charles Brumfiel for suggesting the above example and presentation.

THREE TYPES OF TRANSFORMATIONS

In this article, three types of transformations will be studied. These are called *parallel translations, rotations,* and *reflections.* The first two of these leave essentially all relationships between points unchanged. The latter changes the plane in a very special way.

Parallel Translations

To begin the investigation of parallel translations pick any direction in the plane (use a sheet of paper to represent the plane and dots on the paper to represent points), and move every point the same distance in that direction. If, as in Figure 2 below, A' and B' are the images of A and B respectively, it is quite clear that the length of $\overline{A'B'}$ ($\overline{A'B'}$ is the symbol for the straight line segment beginning at A' and ending at B') is the same as that of $\overline{AB}$. The images of all the points of the line segment $\overline{AB}$ form the segment $\overline{A'B'}$. That is, if C is between A and B, then C' is between A' and B'.

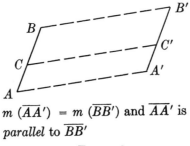

$m\ (\overline{AA'})\ =\ m\ (\overline{BB'})$ and $\overline{AA'}$ is
parallel to $\overline{BB'}$

FIGURE 2

The term "measure of" will be used when length, area, angle size, etc. are to be considered, and will be denoted by "m———." Example: "$m\angle\ BAC$" is read "the measure of angle BAC." Assuming points A and B are one unit apart ($m\overline{AB}\ =\ 1$), what is $m\overline{A'B'}$ in Figure 2? We see that the measure of the distance between any two images of A and B under such a motion is equal to $m\overline{AB}$. This illustrates that *distance between two points is* invariant *(left unchanged) under a parallel translation.* We also know that *a parallel translation carries the points of one line into the points of another line.*

Consider a triangle ABC (denoted $\triangle ABC$): in a plane move all the points of the plane two units by means of the parallel translation described below in Figure 3.

What properties of triangle ABC are left invariant under the parallel translation of Figure 3? Because the distance between any two of its points is left unchanged, it is possible to show that the angle measures are unchanged, the lengths of the altitudes are unchanged, etc. This leads to the important concept of *congruence*. Two geometric figures, say R and T, are congruent if and only if there is a one-to-one correspondence between the points of R and the points of T which leaves distance invariant. This is denoted $R \cong T$ and is read "R is congruent to T." Thus, in Figure 3, $\triangle ABC \cong \triangle A'B'C'$.

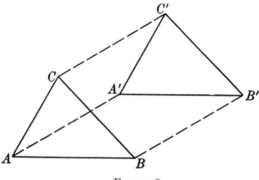

FIGURE 3

Before leaving the discussion of parallel translations, try to extend these ideas to figures other than those discussed here, such as rectangles, circles, hexagons (closed figures of six sides), etc.

Rotations

To introduce a second type of transformation, visualize a circular disk rotating about its center O, with a point A somewhere on the disk other than at the center.

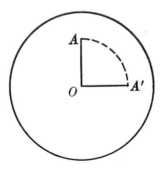

FIGURE 4

Assume the point is in position A before rotation and in the position A' after a clockwise rotation of the disk about point O.

Exercise 2:

    (a) Is the measure of $\overline{AO}$ left invariant under the rotation?

    (b) Should A' be called the image of A under the rotation?

    (c) Draw in the line segments $\overline{OA}$ and $\overline{OA'}$. Is $\overline{OA'}$ the image of $\overline{OA}$ under the rotation? (Keep in mind that a rotation is applied to every point of the plane and that the center is the only fixed point of the transformation.)

Take the disk, with point A as before, and add a point B, as in Figure 5. Perform the rotation as before so A' is the image of A, and B' is the image of B.

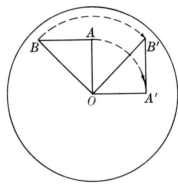

FIGURE 5

Exercise 3:

    In Figure 5, $\overline{OA'}$ is the image of $\overline{OA}$; and $\overline{OB'}$ is the image of $\overline{OB}$.

    (a) Is $m\overline{AB} = m\overline{A'B'}$?

    (b) Is $m\overline{OB'} = m\overline{OB}$?

    (c) If the answer to these questions is "yes," what might be said about distance under a rotation transformation, i.e., is distance invariant under a rotation transformation?

ANGLES AND RAYS. Two ideas necessary for proper consideration of rotation (included previously in the assumption of invariants under a parallel translation) are those of *angle* and *ray*. Roughly speaking, a ray is a point on a line together with all points of the line on one side of the given point. We define a *ray* as the set of points on a line L consisting of two distinct points A and B and all points on B's side of A on the line L. The darkened portion of the line is called a ray.

An *angle* is a set of points consisting of two rays with the same end point.

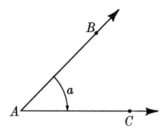

   To name this angle we shall use the symbol $\angle a$ or $\angle BAC$. A different angle might be called $\angle b$ or $\angle c$, etc. Thus, $\angle a$ may also be called $\angle BAC$ or $\angle CAB$. (It is incorrect to say $\angle ABC$ or $\angle ACB$. Why?)

   In Figure 5, it is seen that $m \angle AOB = m \angle A'OB'$. Is $m \angle BAO = m \angle B'A'O$? Is $m \angle ABO = m \angle A'B'O$? Are some angles left invariant under a rotation transformation? Could we get the image $\overline{A'B'}$ of the segment $\overline{AB}$ by means of parallel translation? To answer this, observe that under a parallel translation, the image of a line segment is always parallel to the segment itself.

   Given a triangle ABC in a plane, and given a fixed point O as in Figure 6, a rotation of $\triangle ABC$ through an angle, $\angle a$, gives us the image $\triangle A'B'C'$. Each of the segments $\overline{AB}$, $\overline{BC}$, and $\overline{CB}$, is rotated through an angle with $m \angle a$ giving segments $\overline{A'B'}$, $\overline{B'C'}$ and $\overline{C'B'}$.

   Is $m \angle CAB = m \angle C'A'B'$? What properties of $\triangle ABC$ are invariant under the rotation? Are the two triangles congruent?

   As was the case with parallel translations, there may be repeated applications of rotations, using either the same or different centers of rotation.

Exercise 4:

   In Figure 6, rotate $\triangle A'B'C'$ about fixed point A' getting $\triangle A''B''C''$ so that $\overline{AC}$ and $\overline{A''C''}$ (A' and A'' are the same point) are parallel,

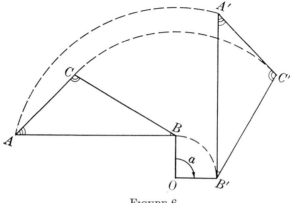

FIGURE 6

and so that $\overline{AB}$ and $\overline{A''B''}$ are parallel. Sketch the figure and show a parallel translation on $\triangle ABC$ which gives $\triangle A''B''C''$.

Exercise 5:

Look at Figure 7. The image $\overline{A''B''}$ of $\overline{AB}$ has been obtained by two rotations, the first through $\angle a$ about O as the fixed point resulting in $\overline{A'B'}$ and the second through $\angle b$ about O', as the fixed point. $\overline{A''B''}$ is parallel to $\overline{AB}$.

(a) Show that the transformation of $\overline{AB}$ into $\overline{A''B''}$ could also be accomplished by a parallel translation.

(b) Could you accomplish the above transformation using only a single rotation?

(c) Using three rotations? (d) Four?

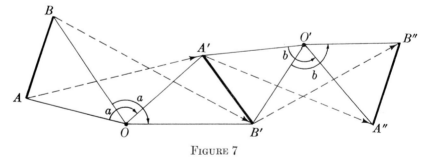

FIGURE 7

Exercise 6:

In Figure 8, knowing that some transformation has acted on $\triangle ABC$ giving the image $\triangle A'B'C'$, carefully study the figure and answer the following questions.

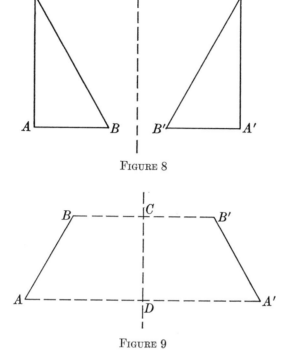

FIGURE 8

FIGURE 9

(a) Could the transformation be a set of parallel translations?
(b) Could the transformation be a set of rotations?
(c) Is there any combination of rotations and parallel translations on $\triangle ABC$ which gives $\triangle A'B'C'$ as an image?

If the transformation is neither of those studied previously, what type is it? To answer this, consider the following: A boy stands before a mirror to comb his hair. Let the boy and his mirror image be A and A' respectively. If A parts his hair on the left side of his head, on what side does A' part his hair? On the left also? If the answer is "yes," look again, since plane mirror images are reversed in direction.

SYMMETRY. In figure 9, $m\overline{BC} = m\overline{B'C}$ and $m\overline{AD} = m\overline{A'D}$. The line CD divides the figure $ABB'A'$ into congruent parts, and we say that the figure is *symmetric* with respect to line CD. The line L in Figure 8 and line CD in Figure 9 are called *axes of symmetry* for the respective figures. Notice that line CD contains the midpoints of AA' and BB' and is perpendicular to both. Two lines are *perpendicular* if they intersect to form equal adjacent angles with the same measure. (Examples: In Figure 9, $m \angle DCB = m \angle DCB'$ and, therefore, $\overline{CD}$ is perpendicular to $\overline{BB'}$. A horizontal and a vertical line through the same point are perpendicular.) Line CD is perpendicular to BB' and AA', and since CD bisects BB' and AA', CD is the *perpendicular bisector* of both segments. A figure is symmetric with respect to a line L if for each point R on the figure (except points of L) there is a point R' on the figure such that L is perpendicular bisector of $\overline{RR'}$.

In Figure 8, are the line segments $\overline{AC}$ and $\overline{A'C'}$ symmetric with respect to line L? Are $\overline{BC}$ and $\overline{B'C'}$ symmetric with respect to L? Are $\overline{AB}$ and $\overline{A'B'}$ symmetric with respect to L? Given $\triangle ABC$ and line L, (Fig. 8) the transformation which gives the image $\triangle A'B'C'$ is called a *reflection transformation*. Does a reflection transformation leave distances, angles, and areas invariant? Try to detect a property of $\triangle ABC$ which is changed by the reflection (one other than position).

ORIENTATION. *Orientation of a closed figure* shall mean the clockwise order of the points of that figure. In Figure 10, the orientation of quadrilateral $ABCD$ is A–B–C–D, while the orientation of the image is A'–D'–C'–B'.

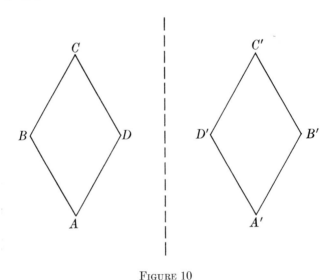

FIGURE 10

Does a reflection give opposite (counterclockwise) orientation? From Figures 8 and 10, it is obvious that orientation is not invariant under a reflection transformation.

Consider Figure 11 with congruent line segments $\overline{AB}$ and $\overline{CD}$.

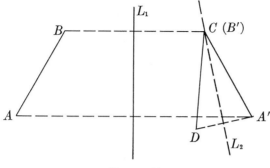

Is there a set of reflections which will have $\overline{CD}$ as the image of $\overline{AB}$? Find the perpendicular bisector of $\overline{BC}$, and then perform a reflection transformation L_1, to get $\overline{CA'}$ as the image of $\overline{AB}$. Now find the perpendicular bisector of $\overline{DA'}$ and perform reflection L_2 on $\overline{CA'}$ to get $\overline{CD}$ as the image of $\overline{CA'}$. Then $\overline{CD}$ is the image of $\overline{AB}$ under the set of reflections about L_1 and L_2. Is $\overline{AB} \cong \overline{CD}$?

Exercise 7:

Given $\triangle ABC \cong \triangle DEF$, where the triangles lie in the same plane. Describe a set of reflections on $\triangle ABC$ such that $\triangle DEF$ is the resulting image.

You may wish to continue the investigation started here with respect to the following:

1. Different sets of transformations acting on the same figure
2. A given transformation acting on different figures
3. Symmetries about a point or about a plane rather than a line
4. Transformations in three dimensions rather than two.

For further work in geometry see: "Geometry: Right and/or Left" by Charles Brumfiel, Chapter 8, in the Twenty-Eighth Yearbook.

REFERENCES

Jones, Burton W. "Reflections and Rotations." *Mathematics Teacher* 54: 406–10; October 1961.

Kutuzov, B. V. *Studies in Mathematics* (Volume IV, *Geometry*). Translation distributed by the School Mathematics Study Group, University of Chicago Press. Chicago, 1960.

School Mathematics Study Group. *Mathematics for High School, Geometry,* Part 1. New Haven, Conn.: Yale University Press, 1960.

ANSWERS TO PROBLEMS

Section II: The Junior High School Years

1. **The Bookworm:** $\frac{1}{2}$ inch   2. **Making a Chain:** $1.40
   3. **Socks:** Three
4. **Age and Month:** Multiplying by 10 and again by 10 places the age in the hundreds' and thousands' places. Adding 5 and subtracting 50 have no effect — this is merely for confusion. Adding the number of the month places this number in the tens' and units' places so there is no overlap in the age.
5. **The Bear Truth:** White (North Pole); penguin (South Pole); many points.
6. **Long Division:** $A = 6, B = 3, D = 7, E = 1, H = 2, J = 5, K = 0$.
7. **Double or Nothing:** $21,474,836.47 for the month of March— so take the job.   8. **Spending:** $8.75
9. **Help Needed:** $S = 9, E = 5, N = 6, D = 7, M = 1, O = 0, R = 8, Y = 2$. $106.52
10. **Plugging a Hole:**

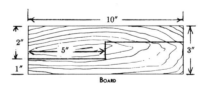

11. **Disappearing Square:**

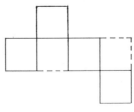

312

12. Another Disappearing Square:

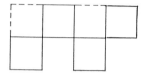

13. The Double Cross:

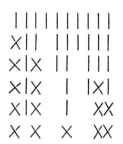

14. Twenty-one: To win, get the total number of toothpicks removed up to 4, 8, 12, 16, 20 in turn.

15. Nim: To win the game from an uninitiated opponent, the player writes all the numbers in the binary system, adds the coefficients of each power of two so obtained, and at each play reduces one of the given numbers so that the sum of the coefficients of each power of two becomes an even number.

A brief description of the binary system of numeration is given in Chapter 1 of this yearbook. For a more detailed account, as well as a discussion of the game of Nim, see: Johnson and Glenn *Understanding Numeration Systems*, pp. 32–44, in the "Exploring Mathematics On Your Own" series of Webster Publishing Company.

A classical and complete solution of the game of Nim was given by C. L. Bouton in *Annals of Mathematics*, Series 2, Vol. 3, pp. 35–39 (October 1901).

16. Huntsmen and Noblemen: Two noblemen and one huntsman. The first man had to say, "I am a nobleman," whether he was or not. The second man told the truth and is a nobleman. If the first man lied, the third told the truth; if the first man told the truth, the third lied.

17. A Fair Problem:

Take the goose and leave it on the far side.

Then take the corn and return with the goose, leaving the corn
on the far side.

Leave the goose and take the fox across and leave it with the corn.

Return for the goose.

18. **The Cannibals and the Missionaries:**

The cannibals must cooperate.

A missionary and a cannibal cross and the missionary returns.

Two cannibals then cross and one returns with the boat.

Two missionaries cross and a missionary and a cannibal return.

Two missionaries cross and the cannibal takes the boat back.

Two cannibals cross and one of them returns.

The two remaining cannibals take the boat across.

19. **Three Jealous Men and Their Wives:**

Mrs. B and Mrs. C cross, and Mrs. B returns with the boat.

Mrs. A and Mrs. B cross, and Mrs. A returns with the boat.

Mr. B and Mr. C cross, and Mr. and Mrs. B return.

Mr. A and Mr. B cross, and Mrs. C returns with the boat.

Mrs. B and Mrs. C cross, and Mr. A returns to pick up his wife.

20. **A Refreshing Problem:**

Fill the 5-pint container; from the 5-pint container fill the 3-pint
container.

Empty the 3-pint container into the 8.

Empty the 5 into the 3.

Fill the 5 from the 8.

Fill the 3 from the 5.

Pour the contents from the 3 into the 8, and it is done.

21. **An Odd Problem:** There is no known proof for this conjecture;
however, it seems to be true.

22. **A Square to Complete:**

(a) 136

(b) 34

(c) They should be.

(d) Yes.

(e) Yes.

(f) By dividing or subtracting.

16	3	2	13
5	10	11	8
9	6	7	12
4	15	14	1

23. **Magic Squares with Letters and Numerals:**

(a) 45

(b) 15

(c) Hints:

(2) 15, 15, 15, 45

(3) 45

(4) $e = 5$.

CHAPTER 15——UNIT FRACTIONS

1. (a) $\frac{1}{2} = \frac{1}{3} + \frac{1}{6}$; $\frac{1}{3} = \frac{1}{4} + \frac{1}{12}$; $\frac{1}{4} = \frac{1}{6} + \frac{1}{12} = \frac{1}{5} + \frac{1}{20}$.

2. (a) $\frac{1}{27} + \frac{1}{54} = \frac{1}{18}$;   (b) $\frac{1}{33} + \frac{1}{66} = \frac{1}{22}$; $\frac{1}{60} + \frac{1}{120} = \frac{1}{40}$;

$\frac{1}{75} + \frac{1}{150} = \frac{1}{50}$.

3. $\frac{7}{8} = \frac{1}{2} + \frac{1}{4} + \frac{1}{8}$.

4. Subtract successively $\frac{1}{2}$, $\frac{1}{4}$, $\cdots$ until the remainder has a denominator in which the power-of-two factor exceeds the numerator, then apply the technique given in Exercise 3.

$\frac{3}{5} = \frac{1}{2} + ? = \frac{1}{2} + \frac{1}{10}$

$\frac{9}{10} = \frac{1}{2} + ? = \frac{1}{2} + \frac{2}{5} = \frac{1}{2} + \frac{1}{4} + ?$

$\qquad = \frac{1}{2} + \frac{1}{4} + \frac{3}{20} = \frac{1}{2} + \frac{1}{4} + \frac{1}{5}(\frac{3}{4})$

$\qquad = \frac{1}{2} + \frac{1}{4} + \frac{1}{5}(\frac{1}{2} + \frac{1}{4}) = \frac{1}{2} + \frac{1}{4} + \frac{1}{10} + \frac{1}{20}$

$\frac{4}{15} = \frac{1}{4} + ? = \frac{1}{4} + \frac{1}{60}$

$\frac{7}{15} = \frac{1}{4} + \frac{1}{8} + \frac{1}{16} + \frac{1}{60} + \frac{1}{120} + \frac{1}{240}$

CHAPTER 16——ON DIVISIBILITY RULES

1. 1, 3, 5, 7, 9.   **2.** 1592, 1816, 1968, 2856.

3. (a) Yes. (b) No. (c) No, since 14 is not divisible by 4.

4. The number formed by the last three digits must be divisible by 8.

5. (a) Yes. (b) No.

6. The sum of the digits of n is the same as the sum of the digits of n'.

7. The difference between the sums of the digits is zero, which is divisible by 9.

$a - a' = 10^3(M - U) + 10^2(H - T) + 10(T - H) + (U - M)$

The test for divisibility by 9 does not require that the coefficients $M - U$, $H - T$, $T - H$, $U - M$ be digits; the test works just as well when they are integers. The conclusion follows immediately.

8. 5; 8; 4; 6. Yes, in each case given, but in 3152——7 the missing digit could be zero or nine.

9. 8;   9;   3;   9.

10. 830456:

83045	8303	829	81	
−12	−6	−14	−10	*No*
83033	8297	815	71	

91362:

9136	913	90	
−4	−4	−18	*No*
9132	909	72	

5873:	587	58		
	-6	-2	Yes	
	581	56		

31029:	3102	308	30	
	-18	-8	-0	No
	3084	300	30	

69146:	6914	690	68	
	-12	-4	-12	Yes
	6902	686	56	

142857:	14285	1427	142	13	
	-14	-2	-10	-4	No
	14271	1425	132	9	

11. $m = \dfrac{n - U}{10} - 9U;$ $10m = n - U - 90n = n - 91n;$ so

$n = 10m + 91n.$

If m is divisible by 7, $n = 7\left(10 \cdot \dfrac{m}{7} + 13n\right)$, and n is divisible

by 7 since it has 7 as a factor.

By changing the coefficient of U, divisibility by many numbers can be tested in this manner.

12. (a) $\frac{3}{4}$ (b) $\frac{1}{55}$ (c) $\frac{97}{907}$

    (d) $\frac{149}{1009}$ (e) $\frac{1}{91}$ (f) $\frac{1052}{3333}$

    (g) $\frac{71}{10101}$ (h) $\frac{1234}{4321}$

CHAPTER 17——NUMERATION SYSTEMS

Exercises A:

1. 68_{ten} **2.** $14_{\text{ten}};$ 20_{seven} **3.** $121_{\text{nine}};$ 400_{five}

4. $473_{\text{eight}} = 4 \cdot 8^2 + 7 \cdot 8 + 3 \cdot 1;$ $265_{\text{twelve}} = 2 \cdot 12^2 + 6 \cdot 12 + 5 \cdot 1$

5. (a) 34_{five} (b) 62_{seven} (c) 49_{twelve} (d) 1221_{seven}

    (e) 11223_{seven}

6. (a) 25_{six} (b) 114_{six} (c) 216_{seven} (d) 257_{eight}

7. (a) 42_{seven} (b) 24_{eight} (c) 63_{ten} (d) 71_{nine} (e) 52_{six}

    (f) 201_{seven}

Exercises B:

1. Twelve.

2. (a) 71_{ten} (b) 34_{ten} (c) 537_{ten} (d) 7266_{ten} (e) 35_{ten}

    (f) 1174_{ten} (g) 33_{ten} (h) 241_{ten} (i) 35_{ten} (j) 100_{ten}

3.

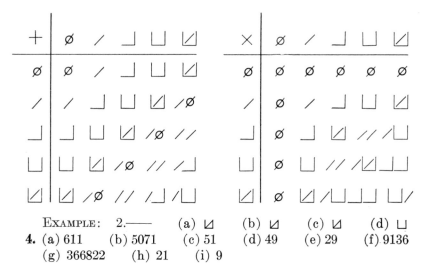

EXAMPLE: 2.——— (a) ◿ (b) ◿ (c) ◿ (d) ⊔

4. (a) 611 (b) 5071 (c) 51 (d) 49 (e) 29 (f) 9136
(g) 366822 (h) 21 (i) 9

Exercises C:

1. Yes. 2. Yes. 3. Yes; Yes. 4. Yes; No. 5. Yes; No.
6. Yes.
7. If the last digit of the number is divisible by 2, the number is also divisible by 2.

Exercises D:

1. Yes. 2. No. 3. Yes. 4. No. 5. Yes.
6. When the first digit is divisible by 2.
7. Yes; Yes; Yes; Yes. 8. No; Yes; No; No.
9. Yes; No; No; No. 10. No; No; Yes; Yes.
11. Yes; Yes; Yes; Yes. 12. No; Yes; No; No.
13. Yes; No; No; No. 14. No; No; Yes; Yes.
15. Yes; Yes; No.
16. When the sum of the digits is divisible by 2.

Exercises E:

1. 54 2. 144 3. 96 4. 162 5. 450 6. 75,600
7. 50 8. 350 9. 7 10. 17 11. 34 12. 870

Exercises F:

1. $2 \cdot 5 \cdot 5$ 2. $2 \cdot 2 \cdot 5 \cdot 5$ 3. $5 \cdot 5 \cdot 5$ 4. $2 \cdot 2 \cdot 2 \cdot 2 \cdot 5 \cdot 5$
5. $2 \cdot 2 \cdot 2 \cdot 5 \cdot 5 \cdot 5$ 6. $2 \cdot 2 \cdot 5 \cdot 13$ 7. $2 \cdot 5 \cdot 5 \cdot 7$ 8. $3 \cdot 3 \cdot 3 \cdot 5$

9. $2 \cdot 2 \cdot 2 \cdot 2 \cdot 5 \cdot 5 \cdot 5$ **10.** $2 \cdot 3 \cdot 3 \cdot 5 \cdot 7$ **11.** 4240_p **12.** 457_p

13. 2005424_p **14.** 4000_p **15.** 200000_p **16.** 5000000000

17. 201_p **18.** 202_p **19.** 300_p **20.** 204_p **21.** 303_p

22. 100102_p **23.** 1201_p **24.** 130_p **25.** 304_p **26.** 1121_p

27.

$1 = 0$	$38 = 10000001$
$2 = 1$	$39 = 100010$
$3 = 10$	$40 = 103$
$4 = 2$	$41 = 1000000000000$
$5 = 100$	$42 = 1011$
$6 = 11$	$43 = 10000000000000$
$7 = 1000$	$44 = 10002$
$8 = 3$	$45 = 120$
$9 = 20$	$46 = 100000001$
$10 = 101$	$47 = 1000 \cdots 000$ [14 0's]
$11 = 10000$	$48 = 14$
$12 = 12$	$49 = 2000$
$13 = 100000$	$50 = 201$
$14 = 1001$	$51 = 1000010$
$15 = 110$	$52 = 100002$
$16 = 4$	$53 = 1000 \cdots 000$ [15 0's]
$17 = 1000000$	$54 = 31$
$18 = 21$	$55 = 10100$
$19 = 10000000$	$56 = 1003$
$20 = 102$	$57 = 10000010$
$21 = 1010$	$58 = 1000000001$
$22 = 10001$	$59 = 1000 \cdots 000$ [16 0's]
$23 = 100000000$	$60 = 112$
$24 = 13$	$61 = 1000 \cdots 000$ [17 0's]
$25 = 200$	$62 = 10000000001$
$26 = 100001$	$63 = 1020$
$27 = 30$	$64 = 6$
$28 = 1002$	$65 = 100100$
$29 = 1000000000$	$66 = 10011$
$30 = 111$	$67 = 1000 \cdots 000$ [18 0's]
$31 = 10000000000$	$68 = 1000002$
$32 = 5$	$69 = 100000010$
$33 = 10010$	$70 = 1101$
$34 = 1000002$	$71 = 1000 \cdots 000$ [19 0's]
$35 = 1100$	$72 = 23$
$36 = 22$	$73 = 1000 \cdots 000$ [20 0's]
$37 = 100000000000$	$74 = 100000000001$

27. *Continued*

$75 = 210$	$88 = 10003$
$76 = 10000002$	$89 = 1000\cdots000$ [23 0's]
$77 = 11000$	$90 = 121$
$78 = 100011$	$91 = 101000$
$79 = 1000\cdots000$ [21 0's]	$92 = 100000002$
$80 = 104$	$93 = 10000000010$
$81 = 40$	$94 = 100000000000001$
$82 = 1000000000001$	$95 = 10000100$
$83 = 1000\cdots000$ [22 0's]	$96 = 15$
$84 = 1012$	$97 = 1000\cdots000$ [24 0's]
$85 = 1000100$	$98 = 2001$
$86 = 10000000000001$	$99 = 10020$
$87 = 1000000010$	$100 = 202$

Exercises G:

1. 939_p **2.** $7[14]_p$ **3.** $[17]6_p$ **4.** $[11]8_p$ **5.** $988[14]8_p$
6. $[14]9_p$ **7.** 332_p **8.** 39411_p **9.** 1111111_p **10.** $3[34]4_p$
11. By 2 when last digit is not zero
     By 3 when 3's place digit is not zero
     By 4 when last digit is 2 or greater
     By 5 when 5's place digit is not zero
     By 6 when neither of last two digits is zero
     By 7 when 7's place digit is not zero
     By 8 when last digit is 3 or greater
     By 9 when 3's place digit is 2 or greater

CHAPTER 18——NUMBERS AND GAMES

Exercises A:

1. (a) $0.66\overline{6}\cdots$ (b) $0.99\overline{9}\cdots$ (c) $1.99\overline{9}\cdots$
2. (a) $0.33\overline{3}\cdots$ (b) $0.33\overline{3}\cdots$ (c) $0.00\overline{0}\cdots$
3. (a) $0.99\overline{9}\cdots$ (b) $1.99\overline{9}\cdots$ (c) $1.99\overline{9}\cdots$
     (d) $0.99\overline{9}\cdots$ (e) $0.99\overline{9}\cdots$ (f) $0.99999\overline{9}\cdots$
     (g) $0.33\overline{3}\cdots$ (h) $0.11\overline{1}\cdots$
4. (a) $\frac{1}{9}$ (b) $\frac{1}{99}$ (c) $\frac{2}{9}$ (d) $\frac{2}{99}$ (e) $\frac{14}{99}$ (f) $\frac{1}{999}$
     (g) $\frac{5}{999}$ (h) $\frac{214}{999}$ (i) $1\frac{25}{99}$ (j) $\frac{1}{90}$

Exercises B:

1. $(1; 2, 4, 3) = 1 + \cfrac{1}{2 + \cfrac{1}{4 + \frac{1}{3}}} = 1 + \cfrac{1}{2 + \cfrac{1}{\frac{13}{3}}} = 1 + \cfrac{1}{2 + \frac{3}{13}}$

$$= 1 + \cfrac{1}{\frac{29}{13}} = 1 + \frac{13}{29} = \frac{42}{29}$$

$(1; 3, 3, 3) = 1 + \cfrac{1}{3 + \cfrac{1}{3 + \frac{1}{3}}} = 1 + \cfrac{1}{3 + \cfrac{1}{\frac{10}{3}}} = 1 + \cfrac{1}{3 + \frac{3}{10}}$

$$= 1 + \cfrac{1}{\frac{33}{10}} = 1 + \frac{10}{33} = \frac{43}{33}$$

2. $\frac{11}{4} = 2\frac{3}{4} = 2 + \cfrac{1}{\frac{4}{3}} = 2 + \cfrac{1}{1 + \frac{1}{3}} = (2; 1, 3)$

3. $(0; 2, 2, 1, 4, 2) = \cfrac{1}{2 + \cfrac{1}{2 + \cfrac{1}{1 + \cfrac{1}{4 + \frac{1}{2}}}}} = \cfrac{1}{2 + \cfrac{1}{2 + \cfrac{1}{1 + \frac{1}{\frac{9}{2}}}}}$

$$= \cfrac{1}{2 + \cfrac{1}{2 + \cfrac{1}{2 + \frac{1}{\frac{11}{9}}}}} = \cfrac{1}{2 + \cfrac{1}{2 + \frac{31}{11}}} = \cfrac{1}{2 + \frac{1}{\frac{73}{31}}} = \frac{31}{73}$$

Exercises C:

1. $(1; 2, 2, 2, 2, 2, 2)$ $r_0 = \frac{1}{1}$ $r_1 = \frac{3}{2}$

$$r_2 = \frac{1 + 2 \cdot 3}{1 + 2 \cdot 2} = \frac{7}{5} \qquad r_3 = \frac{3 + 2 \cdot 7}{2 + 2 \cdot 5} = \frac{17}{12} \qquad r_4 = \frac{7 + 2 \cdot 17}{5 + 2 \cdot 12} = \frac{41}{29}$$

$$r_5 = \frac{17 + 2 \cdot 41}{12 + 2 \cdot 29} = \frac{99}{70} \qquad r_6 = \frac{41 + 2 \cdot 99}{29 + 2 \cdot 70} = \frac{239}{169}$$

2. $r_0^2 = 1$ $= 2 - 1$

$$r_1^2 = \frac{9}{4} \qquad = 2 + \frac{1}{4}$$

$$r_2^2 = \frac{49}{25} \qquad = 2 - \frac{1}{25}$$

$$r_3^2 = \frac{289}{144} \qquad = 2 + \frac{1}{144}$$

$$r_4^2 = \frac{1681}{841} = 2 - \frac{1}{841}$$

$$r_5^2 = \frac{9801}{4900} = 2 + \frac{1}{4900}$$

$$r_6^2 = \frac{57121}{28561} = 2 - \frac{1}{28561}$$

3. $(3; 7, 16, \cdots)$

$r_0 = \frac{3}{1}, \qquad r_1 = \frac{22}{7}, \qquad\qquad r_2 = \frac{355}{113}.$

$r_0 = 3, \qquad r_1 = 3.142856 \cdots, \qquad r_2 = 3.1415929 \cdots.$

Exercises D:

1. $r_0 = \dfrac{2}{1} \qquad r_1 = \dfrac{9}{4} \qquad r_2 = \dfrac{2 + 4 \cdot 9}{1 + 4 \cdot 4} = \dfrac{38}{17} \qquad r_3 = \dfrac{9 + 4 \cdot 38}{4 + 4 \cdot 17} = \dfrac{161}{72}$

2. $\left(\frac{2}{1}\right)^2 = 4; \qquad \left(\frac{9}{4}\right)^2 = 5.0625; \qquad \left(\frac{38}{17}\right)^2 = 4.996 \cdots;$

$\left(\frac{161}{27}\right)^2 = 5.0001 \cdots; \qquad \left(\frac{682}{305}\right)^2 = 4.9998 \cdots.$

3. $\sqrt{5} = 2.2360679 \cdots; \qquad \frac{9}{4} = 2.25; \qquad \frac{38}{17} = 2.235 \cdots;$

$\frac{161}{72} = 2.2361 \cdots; \qquad \frac{682}{305} = 2.236065 \cdots.$

4. $\frac{161}{72} - \frac{38}{17} = \frac{1}{1224} = .000817 \cdots. \qquad$ Error about .00005.

5. $\dfrac{2889}{1292} = 2.2360681 \cdots$

$$\begin{array}{r} 2.2360681 \cdots \\ -2.2360679 \cdots \\ \hline .0000002 \cdots \end{array}$$

Exercises E:

1. $\sqrt{10} = 3 + x; \qquad 10 = 9 + 6x + x^2; \qquad 1 = 6x + x^2 = x(6 + x);$

$x = \dfrac{1}{6 + x}; \qquad 3 + x = 3 + \dfrac{1}{6 + \dfrac{1}{6 + x}} = (3; 6, 6, \bar{6}, \cdots);$

$\sqrt{17} = 4 + x; \qquad 17 = 16 + 8x + x^2; \qquad 1 = 8x + x^2 = x(8 + x);$

$x = \dfrac{1}{8 + x}; \qquad 4 + x = 4 + \dfrac{1}{8 + x} = 4 + \dfrac{1}{8 + \dfrac{1}{8 + x}}$

$= (4; 8, 8, \bar{8}, \cdots).$

2. $\sqrt{7} = 2 + x;$ $\quad 7 = 4 + 4x + x^2;$ $\quad 3 = 4x + x^2 = x(4 + x);$

$$x = \frac{3}{4 + x}; \quad x = \frac{1}{\dfrac{4 + x}{3}} = \frac{1}{\dfrac{4}{3} + \dfrac{x}{3}} = \frac{1}{\dfrac{4}{3} + \dfrac{1}{4 + x}};$$

$$2 + x = 2 + \cfrac{1}{\dfrac{4}{3} + \cfrac{1}{4 + \cfrac{1}{\dfrac{4}{3} + \cfrac{1}{4 + x}}}} = (2; \tfrac{4}{3}, 4, \overline{\tfrac{4}{3}, 4}, \cdots).$$

Exercises F:

1. $\dfrac{1 + \sqrt{5}}{2} = 1 + x;$ $\quad 1 + \sqrt{5} = 2 + 2x;$ $\quad \sqrt{5} = 1 + 2x;$

$5 = 1 + 4x + 4x^2;$ $\quad 4 = 4x + 4x^2;$

$1 = x + x^2 = x(1 + x);$ $\quad x = \dfrac{1}{1 + x};$

$$1 + x = 1 + \cfrac{1}{1 + x} = 1 + \cfrac{1}{1 + \cfrac{1}{1 + x}} = (1; 1, 1, \overline{1}, \cdots).$$

2. $(0; 2, 3, \overline{2, 3}, \cdots) = \cfrac{1}{2 + \cfrac{1}{3 + \cfrac{1}{2 + \cfrac{1}{3 + \cdots}}}};$

$$x = \cfrac{1}{2 + \cfrac{1}{3 + x}} = \cfrac{1}{\dfrac{7 + 2x}{3 + x}} = \frac{3 + x}{7 + 2x};$$

$7x + 2x^2 = 3 + x;$ $\quad 2x^2 + 6x - 3 = 0;$ $\quad x = \dfrac{\sqrt{15} - 3}{2}.$

3. $x = \dfrac{1 + \sqrt{5}}{2};$ $\quad 1 + x = \dfrac{3 + \sqrt{5}}{2};$

$$\sqrt{1 + x} = \sqrt{\frac{3 + \sqrt{5}}{2}} = \sqrt{\frac{6 + 2\sqrt{5}}{4}}$$
$$= \frac{\sqrt{5 + 2\sqrt{5} + 1}}{\sqrt{4}} = \frac{\sqrt{5} + 1}{2}.$$

Then $x = \sqrt{1+x} = \sqrt{1+\sqrt{1+x}} = \sqrt{1+\sqrt{1+\sqrt{1+\cdots}}}$

and $\dfrac{1+\sqrt{5}}{2} = \sqrt{1+\sqrt{1+\sqrt{1+\cdots}}}$.

Exercises: Finite Sets

1. $\{0, 1, 2, 3, 4, 5, 6, 7, 8, 9, 10\} = S$
2. $\{0, 1, 2, 3, 4, 5, 6, 7, 8, 9, 10\} = S$
3. $\{9\}$ 4. $\{3\}$ 5. $\{0, 1\}$ 6. $\{1, 3\}$

Exercises: Graphs

1. $\{(2, 1), (3, 3), (4, 5), (5, 7), (6, 9)\}$
2. $\{(0, 0), (1, 1), (2, 4), (3, 9)\}$
3. $\{(0, 5), (3, 4), (4, 3), (5, 0)\}$
4. $\{(2, y)\}$ where y can be any element of set S.
5. $\{(3, y), (x, 4)\}$ where x, y can be any elements of S.
6. $\{(1, 7), (7, 1)\}$ 7. $\{(x, y)\}$ Any pair will work.
8. $\{(x, y)\}$ 9. $\{(x, y)\}$

Ticktacktoe Problems

1. 1 wins, all others lose.   2. 2, 4, 6, 8, draw.   3. Every move except 8 wins.   4. 1 wins, all others lose.   5. 1 and 9 draw, the rest lose.   6. 1 and 9 win, 2 and 6 lose, all others draw.
7. 1 and 9 win, 2 and 6 draw, 5 and 7 lose.   8. A corner play draws, a side play loses.   9. Center move draws, all others lose.
10. Every move wins (by symmetry only two moves available).

Exercises G:

1. 13   2. 8   3. 6; 14   4. 14; -6.
5. $\{x \mid x > 28\}$ 6. First move $x \to 2x - 4$.
Second move $2x - 4 \to 2(2x - 4) - 4 = 4x - 12$.
There is no integer x so that $4x - 12 = 30$.
7. Yes; $\frac{1}{2}$; $\frac{1}{4}$; $\frac{1}{8}$.   8. 4; move left; move right.

Exercise H:

Either $a = 1, b = 3$, and $c = 4$, or $a = -1, b = -3$, and $c = -4$.

Problems

1. $x \to 2 - x; 2 - x \to 2 - (2 - x) \qquad = x$
$x \to -x - 7; -x - 7 \to -(-x - 7) - 7 = x$

To be cyclic R_2, $a^2 = 1$ and $b(a + 1) = 0$.
To be cyclic R_3, $a^3 = 1$ and $b(a^2 + a + 1) = 0$.
To be cyclic R_4, $a^4 = 1$ and $b(a^3 + a^2 + a + 1) = 0$.
To to cyclic R_n, $a^n = 1$ and $b(a^{n-1} + a^{n-2} + \cdots + a + 1) = 0$.

2. By R, $x \to 2 - x$; and by S, $2 - x \to 3 - (2 - x) = x + 1$, and $x + 1$ is 1 to the right of x.

3. $x \to 2x + 3$

4. $R: x \to 3x + 6$; $S: x \to 3x$ and $S*T: 3x \to 3x + 6$
　　$R: x \to 3x + 6$; $U: x \to x + 2$ and
　　　$U*S: x + 2 \to 3(x + 2) = 3x + 6$.

5. R slides line 1 unit to right so each number falls upon its successor.
　　S rotates the line through 180 degrees.
　　T rotates through 180 degrees, then slides 4 units to right.
　　U stretches line to twice its original length.

6. $S: x \to \frac{1}{2}x - 2$

7. Inverse of $S_1 : x \to -x$
　　Inverse of $S_2 : x \to 4 - x$
　　Inverse of $S_3 : x \to \dfrac{1}{a}x - \dfrac{b}{a}$
　　Inverse of $S_4 : x \to \dfrac{1}{x}$
　　Inverse of $S_5 : x \to \dfrac{-3}{5x}$
　　Inverse of $S_6 : x \to \dfrac{1}{x - 1}$

8. By $S: x \to \dfrac{2x - 1}{4x + 2}$, $0 \to \dfrac{-1}{2} \to \dfrac{-2}{0}$ or $Q \to \dfrac{2Q - 1}{4Q + 2}$

$$\to \dfrac{\dfrac{4Q - 2}{4Q + 2} - 1}{\dfrac{8Q - 4}{4Q + 2} + 2} = \dfrac{\dfrac{-4}{4Q + 2}}{\dfrac{16Q}{4Q + 2}} = \dfrac{-4}{16Q} = \dfrac{-4}{16} \cdot \dfrac{1}{Q}$$

$$= \dfrac{-4}{16} \cdot \dfrac{1}{\left(\dfrac{-2}{0}\right)} = \dfrac{-4}{16} \cdot \dfrac{0}{-2} = 0$$

　　so $0 \to \dfrac{-1}{2} \to Q \to \dfrac{2Q - 1}{4Q + 2} \to 0$.

9. $3 \xrightarrow{S} \dfrac{3}{2} \xrightarrow{T} \dfrac{-\frac{1}{2}}{2} = -\dfrac{1}{4} \xrightarrow{S} \dfrac{-\frac{1}{4}}{-\frac{5}{4}} = \dfrac{1}{5} \xrightarrow{T} \dfrac{-\frac{9}{5}}{-\frac{3}{5}} = 3$

　　$3 \xrightarrow{T} \dfrac{1}{5} \xrightarrow{S} \dfrac{\frac{1}{5}}{\frac{1}{5} - 1} = -\dfrac{1}{4} \xrightarrow{T} \dfrac{-\frac{9}{4}}{-\frac{3}{2}} = \dfrac{3}{2} \xrightarrow{S} \dfrac{\frac{3}{2}}{\frac{1}{2}} = 3$

10. $\frac{2}{3} \xrightarrow{S} \frac{1}{\frac{1}{3}} = 3$ and $\frac{2}{3} \xrightarrow{T} \frac{1}{\frac{2}{3}} = \frac{3}{2}$.

11. $2 \xrightarrow{S} -2 \xrightarrow{T} \frac{-1}{3} \xrightarrow{T} \frac{1}{2} \xrightarrow{T} 3 \xrightarrow{S} -3 \xrightarrow{T} -\frac{1}{2} \xrightarrow{T} \frac{1}{3} \xrightarrow{T} 2$

12. $1 \xrightarrow{S} \frac{4}{3} \xrightarrow{T} -\frac{10}{3} \xrightarrow{S} -\frac{2}{5} \xrightarrow{T} -\frac{8}{5} \xrightarrow{S} -\frac{5}{6} \xrightarrow{T} -\frac{7}{6} \xrightarrow{S}$

$-\frac{8}{7} \xrightarrow{T} -\frac{6}{7} \xrightarrow{S} -\frac{14}{9} \xrightarrow{T} -\frac{4}{9} \xrightarrow{S} -3 \xrightarrow{T} 1$

13. $\begin{pmatrix} a & b \\ c & d \end{pmatrix}$ means $x \to \dfrac{ax + b}{cx + d}$ $\begin{pmatrix} ka & kb \\ kc & kd \end{pmatrix}$ means $x \to \dfrac{kax + kb}{kcx + kd}$

$= \dfrac{k(ax + b)}{k(cx + d)} = \dfrac{ax + b}{cx + d}.$

14. It is desirable that $ad - be \neq 0$, so that the game remains interesting. If $ad - bc = 0$, all points for which $\begin{pmatrix} a & b \\ c & d \end{pmatrix}$ is defined map to the same point.

15. (a) $\begin{pmatrix} 2 & 1 \\ 0 & -2 \end{pmatrix}$ has order 2;   (b) $\begin{pmatrix} 2 & -7 \\ 1 & 1 \end{pmatrix}$ has order 3;

(c) $\begin{pmatrix} 3 & 9 \\ -1 & 3 \end{pmatrix}$ has order 4;   (d) $\begin{pmatrix} 2 & -1 \\ 1 & 1 \end{pmatrix}$ has order 6.

16. (a) $R = \begin{pmatrix} 1 & 1 \\ 0 & -1 \end{pmatrix}$, $S = \begin{pmatrix} 3 & 2 \\ -6 & -3 \end{pmatrix}$, and $R*S = \begin{pmatrix} 3 & 1 \\ -6 & -3 \end{pmatrix}$

(b) $R = \begin{pmatrix} 0 & 1 \\ 1 & 0 \end{pmatrix}$, $S = \begin{pmatrix} 1 & 0 \\ 1 & -1 \end{pmatrix}$, and $R*S = \begin{pmatrix} 0 & 1 \\ -1 & 1 \end{pmatrix}$

(c) $R = \begin{pmatrix} 0 & 1 \\ 1 & 0 \end{pmatrix}$, $S = \begin{pmatrix} 1 & 1 \\ 1 & -1 \end{pmatrix}$, and $R*S = \begin{pmatrix} 1 & 1 \\ -1 & 1 \end{pmatrix}$

(d) $R = \begin{pmatrix} 1 & -1 \\ -2 & -1 \end{pmatrix}$, $S = \begin{pmatrix} 1 & 0 \\ -1 & -1 \end{pmatrix}$, and $R*S = \begin{pmatrix} 1 & -1 \\ 1 & 2 \end{pmatrix}.$

17. $x \xrightarrow{R} x + 2 \xrightarrow{S} \dfrac{1}{x + 2} \xrightarrow{T} \dfrac{3}{x + 2} \xrightarrow{R} \dfrac{3}{x + 2} + 2 = \dfrac{2x + 7}{x + 2}$

CHAPTER 19——USING SETS TO STUDY ODD, EVEN, AND PRIME NUMBERS

1. Separate N into odds and evens and consider separately:

$$1, 3, 5, 7, \cdots, 2n - 1, \cdots ; 2, 4, 6, 8, \cdots, \quad 2n, \cdots$$
$$\updownarrow \updownarrow \updownarrow \updownarrow \qquad\quad \updownarrow \qquad\qquad \updownarrow \updownarrow \updownarrow \updownarrow \qquad\qquad \updownarrow$$
$$4, 8, 12, 16, \cdots, \quad 4n, \cdots ; \quad 2, 6, 10, 14, \cdots, 4n - 2, \cdots.$$

2. (a) N: $1, 2, 3, 4, 5, 6, 7, \cdots, \quad n, \cdots$
$$\updownarrow \updownarrow \updownarrow \updownarrow \updownarrow \updownarrow \updownarrow \qquad\qquad \updownarrow$$
O: $1, 3, 5, 7, 9, 11, 13, \cdots, 2n - 1, \cdots$

(b) E: $2, 4, 6, 8, 10, 12, \cdots, \quad 2n, \cdots$
$$\updownarrow \updownarrow \updownarrow \updownarrow \updownarrow \updownarrow \qquad\qquad \updownarrow$$
O: $1, 3, 5, 7, 9, 11, \cdots, 2n - 1, \cdots$

3. $x \in \mathrm{O}$, so $x = 2k - 1$; $y \in \mathrm{O}$, so $y = 2m - 1$
$xy = (2k - 1)(2m - 1) = 4km - 2k - 2m + 1 = 4km - 2k - 2m + 2 - 1 = 2(km - k - m + 1) - 1$ and, since $(km - k - m + 1) \in N$, call it n and $xy = 2n - 1 \in \mathrm{O}$.

4. $x = 4 \in \mathrm{E}$, $y = 7 \in \mathrm{O}$, $x + y = 11 \in \mathrm{O}$
$x = 8 \in \mathrm{E}$, $y = 1 \in \mathrm{O}$, $x + y = 9 \in \mathrm{O}$
$x = 106 \in \mathrm{E}$, $y = 171 \in \mathrm{O}$, $x + y = 277 \in \mathrm{O}$.
Proof: $x \in \mathrm{E}$, so $x = 2k$; $y \in \mathrm{O}$, so $y = 2m - 1$,
$x + y = (2k) + (2m - 1) = (2k + 2m) - 1 = 2(k + m) - 1$;
but $(k + m) \in N$, so call $(k + m) = n$ and $x + y = 2n - 1 \in \mathrm{O}$.

5. $x = 6 \in \mathrm{E}$, $y = 5 \in \mathrm{O}$, $xy = 30 \in \mathrm{E}$
$x = 2 \in \mathrm{E}$, $y = 17 \in \mathrm{O}$, $xy = 34 \in \mathrm{E}$
$x = 18 \in \mathrm{E}$, $y = 11 \in \mathrm{O}$, $xy = 198 \in \mathrm{E}$
Proof: $x \in \mathrm{E}$, so $x = 2k$; $y \in \mathrm{O}$, so $y = 2m - 1$,
$xy = (2k)(2m - 1) = 4km - 2k = 2(2km - k)$;
but $(2km - k) \in N$, so call $(2km - k) = n$ and $xy = 2n \in \mathrm{E}$.

6. If $x, y, z \in \mathrm{O}$, $x + y + z = (x + y) + z$. But $(x + y) \in \mathrm{E}$ (by Thm. 3) and $(x + y) + z \in \mathrm{O}$ (by Thm. 5), so $x + y + z \in \mathrm{O}$.

7. (a) $T \cup T' =$
$\{1, 2, 3, 4, 5, 6, 7, 8, 9, \cdots, 3n - 2, 3n - 1, 3n, \cdots\} = N.$
$T \cap T' = \{\ \} = \varnothing.$
(b) $x + y \in T$ and $xy \in T$.
(c) $xy \in T'$, but nothing can be said about $x + y$.

8. (a) $(6n - 1)$ and $(6n + 1)$ give all numbers not divisible by 2 or 3.
(b) $(6n)$, $(6n + 2)$, $(6n + 3)$, and $(6n + 4)$ give all numbers divisible by either 2 or 3.

9. (a) Since $2n \in \mathrm{E}$, $(2n)^2 = 4n^2$ is a formula for squares of even numbers.
(b) Since $2n - 1 \in \mathrm{O}$, $(2n - 1)^2 = 4n^2 - 4n + 1$ is a formula for squares of odd numbers.
(c) n is even.
(d) n is odd.

10. (a) $9n^2$ (b) $(3n - 1)^2 = 9n^2 - 6n + 1$ and $(3n - 2)^2 = 9n^2 - 12n + 4$ (c) 9   (d) not divisible by 3

11. (a) $P \cup C = N$, with 1 deleted.   (b) $P \cap C = \varnothing$
(c) $C \cap \mathrm{E} = \{4, 6, 8, \cdots, 2n, \cdots\}$ (Note: 2 is not in this set.)
(d) $P \cap \mathrm{E} = \{2\}$ (e) $C \cap \mathrm{O} = \{9, 15, 21, 25, 27, 33, \cdots\}$
(f) No such formula is known.

12. Prime: 59, 157.

Composite: $95 = 5 \cdot 19$; $57 = 3 \cdot 19$; $75 = 3 \cdot 5 \cdot 5$; $119 = 7 \cdot 17$;
$218 = 2 \cdot 109$; $512 = 2 \cdot 2 \cdot 2 \cdot 2 \cdot 2 \cdot 2 \cdot 2 \cdot 2 \cdot 2 = 2^9$;
$513 = 3 \cdot 3 \cdot 3 \cdot 19$; $1001 = 7 \cdot 11 \cdot 13$.

13. $n^2 - n + 41$ does not give a prime for $n = 41$ or for $n = 42$.
$n^2 - 79n + 1601$ does not give a prime for $n = 80$.

14. 211; 2311; 30031; all are prime.

CHAPTER 20——HOW FAR CAN YOU SEE?

1. $2000^2 + d^2 = \left(2000 + \dfrac{20}{5280}\right)^2 = 2000^2 + \dfrac{80000}{5280} + \left(\dfrac{20}{5280}\right)^2$;

so $d^2 = \dfrac{80000}{5280}$ approximately; so $d^2 = 15.15$ and $d = 3.9$ miles.

2. (a) Strip about 80 miles wide (distance 40 miles)   (b) Strip about 250 miles wide

3. (a) 1281 miles   (b) Formula gives 1027 miles, an error of 254 miles.   (c) About $\frac{1}{40}$ of the earth's surface

4. 3272 miles, or 5265 kilometers

CHAPTER 22——LOGIC: FOR TEACHER, FOR PUPIL

Sample Proof-Problems

1. $5x = \left(\dfrac{10}{2}\right)x = \left(\dfrac{10}{2}\right)\left(\dfrac{x}{1}\right) = \dfrac{10x}{2}$

2. Double a; then $(2a)b = 2(ab)$—twice the product of a and b.

3. *House I*: Enter at A. Must go either to room 2 or to room 4.

House I

If go to room 2, must go to room 3. From 3 must go to 4, but then cannot get out B without re-entering room 3. If, from room 1, go to room 4, must go to room 3 and then to room 2. Again cannot get out B without re-entering either room 1 or room 3. It cannot be done.

House II: A possible path is from A through rooms 1, 2, 3, 4, 5, 6

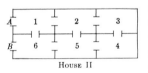
House II

and out door B. If enter room 1 at A, cannot go to 6 because, to get out B would have to return to 6 after entering the other rooms. Must, then, go from 1 into 2. From 2 can go either to
3 or to 5. If go to 5, must go either to 6 or to 4. By going to 6, cannot get to 3 or 4 without re-entering 1 or 5; by going to 4 must then go to 3 and must re-enter 2 or 4. Since cannot go to 5 from 2, must go to room 3. From 3 must go to 4; from 4 to 5; from 5, since

cannot re-enter 2, to 6; from 6 out door B. Therefore, the path given above is the only one available.

4. If both factors are reduced, the product will be reduced. But $200 \times 60 = 12000$; the factors are respectively smaller while the product is greater. This cannot be.

5. We assume $ab = 100$, $a < 10$, and $b \leq 10$; and try to show that this cannot be true. If $a < 10$, then $a = 10 - x$ where x is a positive number. If $b \leq 10$, then $b = 10 - y$ where y is positive or zero. Then $ab = 100 = (10 - x)(10 - y) = 100 + xy - 10x - 10y$. If this is true, $xy - 10x - 10y = 0$ and $xy = 10x + 10y = 10(x + y)$. Since $x \neq 0$ we divide and get $y = 10(1 + y/x)$. But $(1 + y/x)$ is a number at least as large as 1. This means y must be at least 10×1 or 10. If y is 10, then b is zero. This, however, is impossible since b had to be a positive number. Here is the contradiction that shows our assumptions must be impossible.

6. When $(2 \times 3 \times 4 \times 5 \times 6 \times 7 \times 8 \times 9) + 1$ is divided by any integer, 2 through 9, a remainder of 1 is left. This number, then, is not divisible by any of these integers.

7. If each of twenty boys has at most five marbles, they have a total of at most one hundred marbles. Since the "then" part of the statement is false, the "if" part must also be false and at least one of the boys must have more than five.

8. False; since it does not work for 2 and 3: $\dfrac{1}{2+3} \neq \dfrac{1}{2} + \dfrac{1}{3}$.

9. False; since $\dfrac{2}{3} \neq \dfrac{2+1}{3+1}$.

10. False, $100 marked down 10% ($10) is $90, but $90 marked up 10% ($9) is not $100.

11. False.

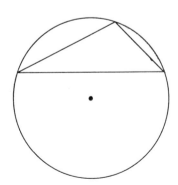

12. False; since for a trapezoid with bases 10, 16, and altitude 4, and arms of length 5, we would have

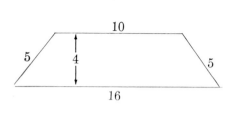

$$\text{area} = \frac{h(b_1 + b_2)}{2}$$

$$= \frac{4(10 + 16)}{2} = 52,$$

but the formula would give

$$\left(\frac{10 + 16}{2}\right) \times \left(\frac{5 + 5}{2}\right) = 65.$$

13. If the sum of a true fraction f and a natural number n were a natural number m, we should have $f = m - n$, which contradicts $m - n$ is a natural number.

14. (a) False, since $\frac{1}{2} + \frac{1}{2} = \frac{2}{2}$ which is not a true fraction.

(b) False, since $\frac{2}{3} \times 3 = \frac{6}{3}$.

(c) If $\dfrac{a}{bc} = n$, then $\dfrac{a}{b} = nc$, a contradiction.

15. If $\sqrt{\dfrac{a}{b}} = n$, then $\dfrac{a}{b} = n^2$, a contradiction.

CHAPTER 23——GEOMETRY AND TRANSFORMATIONS

Exercise 1:

(a) B (b) A (c) A (d) Yes   (e) No

Exercise 2:

(a) Yes   (b) Yes   (c) Yes

Exercise 3:

(a) Yes   (b) Yes   (c) Yes, distance is invariant under a rotation transformation.

Exercise 4:

Parallel translation on $\triangle ABC$ which gives $\triangle A''B''C''$ (Fig. 1, p. 330).

Exercise 5:

(a) Figure 2 (b) No, not with a single rotation. (c) Yes, with three. (d) Yes, with four.

Exercise 6:

(a) No   (b) No   (c) No

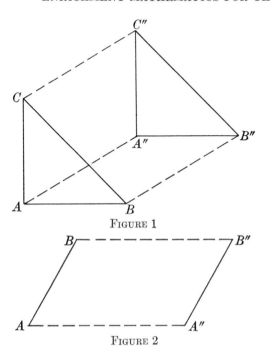

FIGURE 1

FIGURE 2

Exercise 7:

The set of reflections is shown in Figure 3: l_1 is the perpendicular bisector of $\overline{CF}$. Reflect $\triangle ABC$ in l_1 ; the resulting image is $\triangle A'B'F$. l_2 bisects $\angle DFA'$. Reflect $\triangle A'B'F$ in l_2 ; the resulting image is $\triangle DEF$.

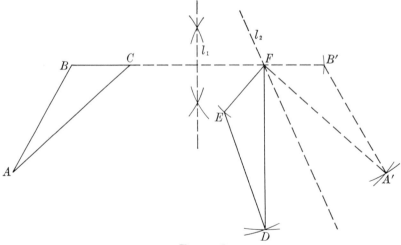

FIGURE 3

MATHEMATICS BOOKLIST
FOR SCHOOL LIBRARIES

The publication of books and paperbacks of interest to students and teachers has been accelerating at such a rate that a bibliography is out-of-date by the time it is published. The following selection of books will undoubtedly share that fate. However this list can be used both as a source for further reading as well as a guide for the building of individual libraries. Among the series that are being published now that hold promise of additions to this list might be mentioned the following: Dover Publications, the New Mathematical Library, the Blaisdell Scientific Paperbacks, and the publications of W. H. Freeman Company. Many of the books listed below are now available in paperback.

This bibliography has been based on and adapted from the *Mathematical Book List for High School Libraries* by permission of the publishers, Mu Alpha Theta, National High School and Junior College Mathematics Club. Appreciation for that permission is hereby expressed to Richard Vernon Andrée.

J. H. H.

ABBOTT, EDWIN. *Flatland.* New York: Dover Publications, 1953. 103 pp. A science-fiction classic of life in a two-dimensional world. Also a first-rate introduction to such aspects of modern science as relativity and hyper–space.

ADLER, IRVING. *The Magic House of Numbers.* New York: New American Library of World Literature, 1961.
Explains the basic whys and hows of our number system in a clear, entertaining way. Packed with curiosities, riddles, tricks, games.

ADLER, IRVING. *The New Mathematics.* New York: John Day Publishing Co., 1958. 122 pp.
With elementary algebra and plane geometry as tools, Adler skillfully builds up many interesting concepts of modern mathematics.

ADLER, IRVING. *Thinking Machines.* New York: John Day Publishing Co., 1961. 189 pp.
An explanation of the theory of electronic computers, made intelligible to anyone who has studied high school mathematics.

ALLENDOERFER, CARL BARNETT, and OAKLEY, C. O. *Fundamentals of Freshman Math.* New York: McGraw-Hill Book Co., 1959. 475 pp.
An excellent college freshman text which introduces calculus in a framework of college algebra and analytic geometry. This is another text which is currently being used by high schools for their senior-level course with considerable success.

ANDERSON, RAYMOND W. *Romping Through Mathematics.* New York: Alfred A. Knopf, 1961.
The story of how the necessary mental tools for counting and measuring have been collected and arranged over a period of four thousand years.

*ANDRÉE, RICHARD VERNON. *Programming the IBM 650 Magnetic Drum*

---

* Believed to be of interest to junior high school teachers and students.

Computer and Data Processing Machine. New York: Henry Holt and Co., 1958. 109 pp.

Basic programming for modern computers. This clearly written book has been used in both high school and college level courses.

ANDRÉE, RICHARD VERNON. *Selections from Modern Abstract Algebra.* New York: Henry Holt and Co., 1958. 212 pp.

A college-level book which includes congruences (modular algebra), Boolean Algebra (set theory), groups, fields, matrices, etc. Has been used successfully as a text for advanced high school students, as well as in dozens of summer institutes for teachers.

BALL, WALTER WILLIAM ROUSE, and COXETER, H. S. M. *Mathematical Recreations and Essays.* New York: Macmillan Co., 1960. 418 pp.

A classic book on recreational mathematics which should be on every library shelf. Very enjoyable.

BARNETT, ISAAC ALBERT. *Some Ideas about Number Theory.* Washington, D.C.: National Council of Teachers of Mathematics, a department of the National Education Association, 1961. 71 pp.

An informal account of some of the more elementary results of number theory. Provides many direct applications for the classroom.

BECKENBACH, EDWIN FORD, and BELLMAN, RICHARD E. *An Introduction to Inequalities.* New York: Random House, 1961. 133 pp.

This book provides an introduction to the fascinating world of quantities that can be ordered according to their size.

BEGLE, EDWARD GRIFFITH. *Introductory Calculus with Analytic Geometry.* New York: Henry Holt and Co., 1954. 304 pp.

Covers topics usually treated in a first course in calculus; includes analytic geometry through the conics. Carefully written and authoritative.

BELL, ERIC TEMPLE. *Men of Mathematics.* New York: Simon and Schuster, 1961. 590 pp. (Also in paperback from Essandess Paperbacks of New York.)

A book which brings the student to a realization of the great men in mathematics and gives him insight into their lives.

BOEHM, GEORGE A. W. *New World of Mathematics.* New York: The Dial Press, 1959. 128 pp.

An excellent, inexpensive book derived from three articles in *Fortune* magazine. Highly recommended.

BOWERS, HENRY and JOAN E. *Arithmetical Excursions.* New York: Dover Publications, 1961. 320 pp.

Lighthearted collection of facts and entertainments for anyone who enjoys manipulating numbers or solving arithmetical puzzles; 529 numbered problems and diversions, all with answers.

BRIXEY, JOHN CLARK, and ANDRÉE, RICHARD VERNON. *Fundamentals of College Mathematics.* New York: Holt, Rinehart and Winston, 1961. 750 pp.

This lively book carefully integrates introductory calculus and statistical inference with college algebra, trigonometry, and analytic geometry. Currently being used in many high schools for the senior mathematics course.

Standard Mathematical Tables. Cleveland: Chemical Rubber Co.

Long established as a time-saving reference aid, this contains a wealth of comprehensive, up-to-date mathematical data for use in algebra, geometry, calculus, trigonometry, statistics, differential equations, finance and investment, statistical analysis, and in virtually any problematic area.

*CARROLL, LEWIS. *Pillow Problems and a Tangled Tale.* New York: Dover Publications, 1958. 152 pp.

Pillow Problems (1893) contains 72 mathematical puzzles, all typically ingenious. The problems in *A Tangled Tale* (1895) are in story form. Carroll not only gives the solutions but uses answers, sent in by readers, to discuss wrong approaches and misleading paths, and grades these readers for insight.

COURANT, RICHARD, and ROBBINS, HERBERT. *What Is Mathematics?* New York: Oxford University Press, 1941. 521 pp.

Not an easy book, but well worth the effort. Contains excellent work on basic mathematical analysis. Portions can be read and enjoyed by all, but some parts require mature cogitation.

COURT, NATHAN ALTSHILLER. *College Geometry.* New York: Barnes and Noble, 1952. 313 pp.

Probably the best known of all advanced plane geometry texts. Has even been translated into Chinese. If your students like plane geometry, they will certainly wish to use this as a reference. Requires no background beyond the usual high school course.

*COURT, NATHAN ALTSHILLER. *Mathematics in Fun and Earnest.* New York: The Dial Press, 1958. 250 pp.

An entertaining book which illustrates Dr. Court's thesis that mathematics in earnest should be fun; mathematics in fun may be in earnest. Requires little background, but will exercise your reasoning power.

COXETER, H. S. M. *Introduction to Geometry.* New York: John Wiley & Sons, 1961. 443 pp.

A survey of all important areas of modern geometry, with relationships to modern algebra stressed.

*CUNDY, HENRY MARTYN, and ROLLETT, A. P. *Mathematical Models.* New York: Oxford University Press, 1961. 286 pp.

A fine discussion of the construction of various mathematical models. You will always have students who prefer to do something rather than study or think. This book will encourage them to do both.

*DADOURIAN, HAROUTUNE M. *How to Study; How to Solve.* Reading, Mass.: Addison-Wesley Publishing Co., 1958. (Out of print.)

This book is designed to help students who are having trouble with mathematics. An excellent volume, and well worth the modest price.

DANTZIG, TOBIAS. *Number: The Language of Science.* New York: Macmillan Co., 1954. 340 pp.

Subtitled, "A Critical Survey Written for the Cultured Non-Mathematician," this is a historical treatment of the number concept and its importance in modern life. Well written.

DAVIS, PHILIP J. *The Lore of Large Numbers.* New York: Random House, 1961. 165 pp.

Arithmetic and the uses of large numbers are explained to introduce the reader to the horizons of modern mathematics.

DEGRAZIA, JOSEPH. *Math Is Fun.* New York: Emerson Books, 1961.

A book of wit-piquing and brain-teasing problems and puzzles—from trifles to cryptograms, and from clock and speed puzzles to problems of arrangement.

For the more faint in spirit—readers who need answers—the author has included answers at the back of the book.

DUBISCH, ROY. *Nature of Number; An Approach to Basic Ideas of Modern Mathematics*. New York: Ronald Press Co., 1952. 159 pp.
A direct and understandable way to gain an over-all view of what mathematics is about, and an insight into the nature of its theory.

*DUDENEY, HENRY ERNEST. *Amusements in Mathematics*. New York: Dover Publications, 1958.
A selection of over 400 mathematical puzzles. Everyone will enjoy this book.

EVES, HOWARD, and NEWSOM, CARROLL V. *An Introduction to the Foundations and Fundamental Concepts of Mathematics*. New York: Rinehart and Co., 1958. 363 pp.
A sound book which good high school students will enjoy reading.

FADIMAN, CLIFTON. *Fantasia Mathematica*. New York: Simon and Schuster. (Also in paperback from Essandess Paperbacks, New York.)
A fantastic collection of pseudo-mathematical materials. No hard math here, but a lot of fun.

*FREEMAN, MAE BLACKER, and IRA, MAXIMILIAN. *Fun with Figures*. New York: Random House, 1946. 60 pp.
Following the "do-it-yourself" idea, the authors show in this book how to have fun with geometric figures. Not much math, but worth the price for fun.

FUJII, JOHN N. *An Introduction to the Elements of Mathematics*. New York: John Wiley and Sons, 1961. 312 pp.
Presents the fundamental ideas underlying mathematics for the reader with the non-technical background. It also discusses history of mathematics and gives an elementary treatment of analysis and problem solving.

*GAINES, HELEN FOUCHÉ. *Cryptanalysis: A Study of Ciphers and Their Solution*. New York: Dover Publications, 1956. 237 pp.
This introductory intermediate-level text is the best book in print on cryptograms and their solutions.

*GAMOW, GEORGE. *One, Two, Three . . . Infinity*. New York: Mentor Press, 1947. 340 pp.
Problems of mathematics, physics, and astronomy clarified for the layman. This paperback should be available at your corner store. Also available in hardback edition from Viking Press.

*GAMOW, GEORGE, and STERN, MARVIN. *Puzzle-Math*. New York: Viking Press, 1958. 119 pp.
Interesting brain-twisters and puzzles based on everyday situations that can be untangled by mathematical thinking.

*GARDNER, MARTIN. *Mathematical Puzzles of Sam Loyd*. Vols. 1 and 2. New York: Dover Publications, 1960.
A delightful collection of puzzles by one of the greatest puzzle makers of all time. A classic.

*GARDNER, MARTIN. *Mathematics, Magic and Mystery*. New York: Dover Publications, 1956. 176 pp.
Another interesting, inexpensive paperback by the mathematical editor of *Scientific American*. Well worth reading.

GARDNER, MARTIN. *The Scientific American Book of Mathematical Puzzles and Diversions.* New York: Simon and Schuster, 1959. 178 pp.

Paradoxes and paperfolding, Moebius variations and mnemonics, fallacies, brain-teasers, magic squares, topological curiosities, probability and parlor tricks, and a variety of ancient and new games and problems from polyominoes, Nim, and the Tower of Hanoi to four-dimensional ticktacktoe—all with mathematical commentaries by Gardner.

GARDNER, MARTIN. *The 2nd Scientific American Book of Mathematical Puzzles and Diversions.* New York: Simon and Schuster, 1961. 253 pp.

A new selection: from origami to recreational logic, from digital roots and Dudeney Puzzles to the diabolic square, from the golden ratio to the generalized ham sandwich theorem—all with mathematical commentaries by Gardner.

GUILBAUD, GEORGES T. *What Is Cybernetics:* New York: Criterion Books, 1959. 126 pp.

A clear exposition of the new field—updated, inclusive, and intended for informed laymen (economists, psychologists, and sociologists as well as all scientists) who find their work area influenced by cybernetics.

HEAFFORD, PHILIP ERNEST. *The Math Entertainer.* New York: Emerson Books, 1959. 176 pp.

Mathematical teasers, ticklers, and twisters designed to puzzle and delight the reader. Answers and complete explanations are given for all problems.

HILBERT, DAVID, and COHN-VOSSEN, STEPHAN. *Geometry and the Imagination.* New York: Chelsea Publishing Co., 1952. 357 pp.

One of the world's outstanding mathematicians shows how to gain insight and intuitive understanding and how to use this understanding to obtain mathematical results.

*HUFF, DARRELL, and GEIS, IRVING. *How to Lie With Statistics.* New York: W. W. Norton and Co., 1954. 142 pp.

Humorous, but penetrating and authoritative explanation of the basic conceptions and misconceptions of statistics. Vivid illustrations in cartoon style fully capture and even extend the content.

*HUFF, DARRELL, AND GEIS, IRVING. *How to Take a Chance.* New York: W. W. Norton and Co., 1959. 173 pp.

Entertaining but soundly exact discussions of various aspects of chance, probability, and error—especially as applied to everyday life.

*HUNTER, JAMES ALSTON HOPE. *Figurets: More Fun with Figures.* New York: Oxford University Press, 1958. 116 pp.

More fascinating mathematical puzzles, these problems for the adult layman are cast in the form of entertaining anecdotes.

*JOHNSON, DONOVAN A., and GLENN, WILLIAM H. *Exploring Mathematics on Your Own.* New York: Webster Publishing Co., 1960. 56 pp.

A set of inexpensive booklets on current mathematical topics. Well worth reading.

*JOHNSON, DONOVAN A. *Paper Folding for the Mathematics Class.* Washington, D.C.: National Council of Teachers of Mathematics, a department of the National Education Association, 1957. 32 pp.

Illustrated directions for folding the basic constructions. Presents geometric concepts, circle relationships, products and factors, polygons, knots, polyhedrons, symmetry, conic sections, and recreations. A fine project for geometry students. Order a supply for class use.

*JONES, BURTON WADSWORTH. *Elementary Concepts of Mathematics*. New York: Macmillan Co., 1963. 294 pp.

Designed for the non-math students, this text clarifies such concepts of everyday importance as compound interest, averages, probability, games of chance, and graphs.

*KASNER, EDWARD, and NEWMAN, JAMES. *Mathematics and the Imagination*. New York: Simon and Schuster, 1940. 380 pp.

An outstanding book which can be read and enjoyed by all. Not merely a "puzzle book"; this volume contains some excellent mathematical ideas.

*KEMENY, JOHN GEORGE, and OTHERS. *Introduction to Finite Mathematics*. Englewood Cliffs, N. J.: Prentice-Hall, 1957. 372 pp.

One of the most modern of the new college freshman texts. An excellent book for good mathematics students. Well written and sound.

KHINCHIN, ALEKSANDR YAKOVLEVICH. *Three Pearls of Number Theory*. New York: Graylock Press, 1952. 64 pp.

Brief history and complete elementary proofs of three famous theorems of additive number theory. Delightful and profitable reading for all who love number theory—from amateur to expert.

*KLINE, MORRIS. *Mathematics in Western Culture*. New York: Oxford University Press, 1953. 484 pp.

This book gives a remarkably fine account of the influence mathematics has exerted on the development of philosophy, the physical sciences, religion, and the arts in Western life.

KOJIMA, T. *The Japanese Abacus*. Rutland, Vt.: Charles E. Tuttle Co., 1961.

The only book available in English which describes the Japanese abacus and its use and gives appropriate problems. The abacus is in standard use in the Orient and is amazing for the speed and accuracy it affords in computation.

*KRAITCHIK, MAURICE. *Mathematical Recreations*. New York: Dover Publications, 1953. 330 pp.

One of the most thorough compilations of recreational mathematical problems. Highly recommended.

KRAMER, EDNA EARNESTINE. *The Mainstream of Mathematics*. New York: Oxford University Press, 1951. 321 pp.

A historical treatment of mathematical thought from primitive number to relativity.

LEVINSON, HORACE CLIFFORD. *The Science of Chance: From Probability to Statistics*. New York: Rinehart, 1950. 348 pp. (Out of print.)

Compact, highly readable survey of chance and statistics, covering many forms of speculation and risk in business, as well as the odds in games of chance.

*LIEBER, LILLIAN ROSANOFF and HUGH G. *The Education of T. C. Mits*. New York: W. W. Norton and Co., 1944. 230 pp.

A delightful, easy-to-read book which contains some interesting philosophy as well as mathematics.

LIEBER, LILLIAN ROSANOFF. *Non-Euclidean Geometry* (Drawings by Hugh G. Lieber). Brooklyn: Galois Institute of Mathematics and Art, 1940. 40 pp.

Various geometries needed for different surfaces, treated postulationally, making Euclidean and Non-Euclidean geometries easier and interesting. Very readable.

LOGSDON, MAYNE IRWIN. *A Mathematician Explains*. Chicago: University of Chicago Press, 1961. 189 pp.
This book tells informally what a student or interested layman should know about mathematics.

MAXWELL, EDWIN ARTHUR. *Fallacies in Mathematics*. New York: Cambridge University Press, 1959. 95 pp.
Some mathematical fallacies traced in depth, often unexpectedly, to the source of error, with a will to please as well as to instruct.

McCRACKEN, DANIEL D. *Digital Computer Programming*. New York: John Wiley and Sons, 1957. 253 pp.
An excellent book on advanced computer programming using a mythical machine TYDAC.

MENGER, K. *You Will Like Geometry*. Chicago: Museum of Science and Industry, 1961.
An especially interesting booklet which discusses unusual curves and amusements of geometry. Believe it or not, you will like geometry if you catch the infectious spirit of the guide book which was published as an adjunct to the Illinois Institute of Technology's permanent geometry exhibit at the Museum of Science and Industry.

MESERVE, BRUCE E. *Fundamental Concepts of Geometry*. Reading, Mass.: Addison-Wesley Publishing Co., 1955. 321 pp.
A book giving the main concepts behind much of classical geometry. Of interest to both students and teachers.

MEYER, JEROME SYDNEY. *Fun with Mathematics*. New York: McClelland, 1952. 172 pp. (Also in paperback from Fawcett Co., Greenwich, Conn.)
A collection of puzzles, problems, number facts, and curiosities which will interest anyone for whom mathematics has an appeal.

MOSTELLER, FREDERICK, and OTHERS. *Probability: A First Course*. Reading, Mass.: Addison-Wesley Publishing Co., 1961. 319 pp.
An introduction to probability and statistics for a one-semester course; requires only two years of high school algebra.

*MOTT-SMITH, GEOFFREY. *Mathematical Puzzles*. New York: Dover Publications, 1954. 248 pp. (Out of print.)
A collection of 188 interesting mathematical puzzles in an inexpensive edition.

*NATIONAL COUNCIL OF TEACHERS OF MATHEMATICS. *The Growth of Mathematical Ideas Grades K-12*. Twenty-Fourth Yearbook. Washington, D.C.: the Council, a department of the National Education Association, 1959. 507 pp.
An excellent book not only for teachers but also for students interested in modern concepts.

NEWMAN, JAMES R. *World of Mathematics*. New York: Essandess Paperbacks, 1961.
A history of the main advances in the science of mathematics.

NIVEN, IVAN MORTON, and ZUCKERMAN, HERBERT S. *An Introduction to the Theory of Numbers*. New York: John Wiley and Sons, 1960. 250 pp.
A substantially complete introduction to the theory of numbers, using an

analytical (not historical) approach. The basic concepts are covered in the first part of the book, followed by more specialized material in the final three chapters.

NIVEN, IVAN MORTON. *Numbers: Rational and Irrational*. New York: Random House, 1961. 136 pp.
This book deals with the number system, one of the basic structures in mathematics. It is concerned especially with ways of classifying numbers into various categories.

*ORE, OYSTEIN. *Number Theory and Its History*. New York: McGraw-Hill Book Co., 1948. 370 pp.
One of the most readable books on elementary number theory with many interesting historical references.

PECK, LYMAN C. *Secret Codes, Remainder Arithmetic, and Matrices*. Washington, D.C.: National Council of Teachers of Mathematics, a department of the National Education Association, 1961. 54 pp.
A sprightly enrichment pamphlet using fun with secret codes to introduce ideas from modern mathematics. Many problems, with answers.

PÓLYA, GEÖRGY. *Mathematical Discovery*. New York: John Wiley and Sons, 1962. 216 pp.
Presents ways and means which lead to the discovery of the solution of problems.
This is a continuation of Polya's two earlier works, *How to Solve It* and *Mathematics and Plausible Reasoning*.

PÓLYA, GEÖRGY. *Mathematics and Plausible Reasoning*. Vols. 1 and 2. Princeton, N. J.: Princeton University Press, 1954.
This is a guide to the practical art of plausible reasoning. The first volume deals with induction and analogy in mathematics, the second with patterns of plausible inference. These are among the most suitable books your panel could find on these important topics.

RADEMACHER, HANS, and TOEPLITZ, OTTO. *The Enjoyment of Mathematics*. Princeton, N. J.: Princeton University Press, 1957. 204 pp.
Probably the most outstanding "popular" mathematics book. Each chapter starts with simple observations easily within the grasp of all, and smoothly catapults the reader into the heart of a genuine mathematical research-type problem. Highly recommended for good students.

*RINGENBERG, LAWRENCE A. *A Portrait of 2*. Washington, D.C.: National Council of Teachers of Mathematics, a department of the National Education Association, 1956. 42 pp.
Written to enlarge the reader's concept of number, this pamphlet discusses the number 2 as an integer, a rational number, real number, and a complex number. Lays an excellent foundation for modern mathematical ideas.

SALKIND, CHARLES T. *The Contest Problem Book*. New York: Random House, 1961. 154 pp.
Composed of problems from the annual high school contests of the Mathematical Association of America, and co-sponsored by the Society of Actuaries. The interest on the part of these organizations in developing and administering the high school contests is based on the firm belief that one way of learning mathematics is through solving of selected problems.

SAWYER, W. W. *A Concrete Approach to Abstract Algebra.* San Francisco: W. H. Freeman and Co., 1961.
Gives excellent introductory material for preparation in modern abstract algebra.

SAWYER, W. W. *Math Patterns in Science.* Middletown, Conn.: American Education Publications, 1960.
Shows how math and science are interrelated and employs a series of simple illustrated exercises which tie algebra patterns into everyday general science applications.

*SAWYER, W. W. *Mathematician's Delight.* Baltimore, Md.: Penguin Books, 1943.
A well written, popular volume; dispels the fear that surrounds mathematics.

*SAWYER, W. W. *Prelude to Mathematics.* Baltimore, Md.: Penguin Books, 1955.
A delightful account of some stimulating and surprising branches of mathematics. Highly recommended.

SAWYER, W. W. *What Is Calculus About?* New York: Random House, 1961. 118 pp.
Calculus is described in simple nontechnical language, understandable to any interested reader.

SCHAAF, WILLIAM L. *Basic Concepts of Elementary Mathematics.* New York: John Wiley and Sons, 1960. 386 pp.
Discusses the nature of number and enumeration, the logical structure of arithmetic, the number system of arithmetic and algebra, informal and formal geometry, computation measurement, functional relations, and certain concepts of probability.

*STEINHAUS, HUGO. *Mathematics Snapshots.* New York: Oxford University Press, 1960. 328 pp.
Pictures help visualize mathematics; the simple text and clear illustrations enable this book to be enjoyed by anyone with a knowledge of algebra. Many interesting suggestions for models and projects are presented.

TURNBULL, HERBERT WESTERN. *The Great Mathematicians.* New York: New York University Press, 1961. 141 pp.
A readable and fascinating biographical history of mathematics from the early Egyptians to the great men of the Twentieth Century. Presents the ideas and lives of the men who have dedicated themselves to the first and most exacting of man's skills.

*VAN DER WAERDEN, B. L. *Science Awakening: Egyptian, Babylonian and Greek Mathematics.* New York: Oxford University Press, 1961.
Presents a history of mathematics for the general reader—from the Egyptians as the "inventors" of geometry through the decline and final decay of Greek mathematics.

*WHITEHEAD, ALFRED NORTH. *An Introduction to Mathematics.* New York: Oxford University Press, 1958. 191 pp.
A highly recommended book written by a well known English mathematician.

WILDER, RAYMOND LOUIS. *Foundations of Math.* New York: John Wiley & Sons, 1952. 305 pp.

A fine book which stresses set theory and logic. Not always easy to read, but well worth the effort.

WILLIAMS, JOHN DAVIS. *The Compleat Strategyst*. New York: McGraw-Hill Book Co., 1954. 234 pp.

An excellent book on the theory of game strategy. Although written in a light vein, it provides a sound introduction to this fast growing field of modern mathematics.

*WYLIE, CLARENCE RAYMOND. *101 Puzzles in Thought and Logic*. New York: Dover Publications, 1957.

New problems you need no special knowledge to solve. Introduction with simplified explanation of general scientific method and puzzle solving. A fine book to stimulate logical thinking.

ANNOTATED BIBLIOGRAPHY

OF ARTICLES IN PERIODICALS

The articles in this yearbook represent the merest sampling of the materials which can be enjoyed by talented students and by their teachers. Some additional materials are represented by the following list of articles, carefully selected from the most easily accessible periodicals and found in issues current during the last ten years. Articles are organized in broad categories, and each selection is annotated with a brief description of the contents and an estimate of the background required for understanding and appreciating the article.

A good use of this bibliography for teacher or student would consist of browsing through it and then hunting down an article, if the title or description sounds exciting. Looking up an article may open the way to further exploration, for most articles give other references to the topic discussed.

This bibliography has been prepared by the members of the Subcommittee for the High School Years, and thanks are hereby expressed to them. Special acknowledgment is due to Karl H. West, Jr., Needham Public Schools, Needham, Mass., who is responsible for the thorough and exhaustive selections from the *Mathematics Teacher*.

It will be seen, even from a cursory examination of the following bibliography, what a rich source of enrichment materials our periodicals are. Every high school library ought to subscribe to at least some of the periodicals listed below.

PERIODICALS

American Mathematical Monthly (10 times a year), Mathematical Association of America, University of Buffalo, Buffalo 14, N. Y.

* *Arithmetic Teacher* (8 times during school year), National Council of Teachers of Mathematics, 1201 Sixteenth Street, N.W., Washington 6, D.C.

Mathematical Gazette (4 times during school year), G. Bell & Sons, Ltd., York House, Portugal Street, London, WC 2.

Mathematical Log (journal of the National High School and Junior College Mathematics Club, Mu Alpha Theta), Box 504, University of Oklahoma, Norman, Okla.

Mathematics Magazine (5 times during school year), R. E. Horton, Los Angeles City College, 855 N. Vermont, Los Angeles 29, Calif.

** *Mathematics Student Journal* (4 times during school year), National Council of Teachers of Mathematics, 1201 Sixteenth Street, N.W., Washington 6, D.C.

** *Mathematics Teacher* (8 times during school year), National Council of Teachers of Mathematics, 1201 Sixteenth Street, N.W., Washington 6, D. C.

Pi Mu Epsilon Journal (biennial), St. Louis University, St. Louis, Mo.

** *Scientific American* (monthly), 415 Madison Avenue, New York 17, N. Y.

---

* Of interest to elementary schools.
** Believed to be of interest to junior high schools.

Recreational Mathematics Magazine (bimonthly), Box 35, Kent, Ohio.

School Science and Mathematics (9 times during school year), P.O. Box 408, Oak Park, Ill.

Scripta Mathematica (quarterly), 186th Street and Amsterdam Avenue, New York 33, N. Y.

** *University of Oklahoma Mathematics Letter* (quarterly), Norman, Okla.

ARITHMETIC

CARNAHAN, WALTER H. "Methods for Systematically Seeking Factors of Numbers." *School Science and Mathematics* 52: 429–35; June 1952.

Presents several methods for factoring numbers, especially suitable methods for factors not in the neighborhood of $\sqrt{N}$.

A student of elementary algebra or even of arithmetic should be able to gain from reading this.

DIAMOND, LOUIS E. "Some Elementary Properties of the Relation, Congruence, Modulo *M*." *Mathematics Magazine* 28: 213–20; March–April 1955.

This article is more advanced than the one presented in Vol. 31; no. 4 of same magazine. The article starts with the terms ". . . sequence by a cyclic permutation . . ." but explains. Discusses "a set of integers which are congruent, modulo *M*, in ordered number pairs;" and also the properties of reflexive, symmetric, and transitive—equivalence relation. Also discusses nonequivalence relation by the use of nonzero integers with the relations: (a) is the negative of, (b) is divisible by. The main body of the article is modular addition and multiplication, and shows the properties of closure, associativity, commutation, and distribution, in a fine and clear, precise fashion.

Could be used in 11th and 12th grades. Some familiarity with modern algebra would be helpful.

DUBISCH, ROY. "Applications of Finite Arithmetic, I." *Mathematics Teacher* 53: 322–24; May 1960.

Shows the advantages of using finite arithmetics in problems involving factorization of polynomials.

Worth considering for some second-year algebra classes.

DUBISCH, ROY. "Applications of Finite Arithmetic, II." *Mathematics Teacher* 53: 430–32; October 1960.

A follow-up to an article in the May 1960 issue on finite arithmetic in factoring; this one covers finite arithmetic in coding and decoding. Matrices are required for the study of the work or it can be considered as an application of matrices.

Suitable for an advanced math class studying matrices.

EVES, HOWARD W., editor. "Prime Numbers." *Mathematics Teacher* 51: 201–03; March 1958.

A fine article discussing some of Euclid's theorems on prime numbers, and the method of proof. The sieve of Eratosthenes is also discussed for small primes, and generalities are mentioned for the large primes.

For any algebra class.

FLEMING, ROBERT E. "A Family of Numbers." *School Science and Mathematics* 54: 410–13; May 1954.

A discussion of the set of number pairs which have the property that the product of the two numbers is equal to their sum.

Best usable by students who have had a second course in algebra and in trigonometry.

GRAVES, LAWRENCE M. "The Postulates of Algebra and Non-Archimedean Number Systems." *Mathematics Teacher* 52: 72–77; February 1959.

This article is a bit deep in spots, but is a valuable reprint presenting the structures of algebra being stressed in high school today. This would be useful for eleventh grade SMSG[1] teachers when referring to the Archimedean postulates.

Good reading for teachers.

HAHN, HANS. "Is There an Infinity?" *Scientific American* 187: 76–84; November 1952.

Condensed "popular" lecture; discusses Cantor's set theoretical approach to infinite sets and infinite numbers, discusses also an introduction to the question of infinite space, leaves it with two questions: (a) How are we to go about fitting observed physical events into a mathematical space? (b) Is a finite or an infinite mathematical space better adapted for arrangement and interpretation of physical events?

HAWKINS, DAVID. "Mathematical Sieves." *Scientific American* 199: 105–12; December 1958.

Contains some simple number theory which very quickly becomes profound. Prime series, random series, and "lucky numbers" are investigated with detailed tables, and the methods of sieving are well described.

JONES, BURTON W. "Miniature Number Systems." *Mathematics Teacher* 51: 226–31; April 1958.

Starts out with the traditional algebraic system, and then by the rules of groups, rings, and fields continues to discuss number systems of the base seven and twelve.

For better students in algebra.

JONES, PHILLIP S., editor. "More About Big Numbers." *Mathematics Teacher* 46: 265–69 and 273; April 1953.

This is a collection of correspondence from various persons to the editor; discusses big numbers, giving references to their whereabouts and their names. Includes the names, from Henkle, of big numbers; large Pythagorean numbers; Mersenne numbers; and factorials.

Good reference for teachers.

JONES, PHILLIP S. "Irrationals or Incommensurables; Their Discovery, and a 'Logical Scandal'." *Mathematics Teacher* 49: 123–27; February 1956.

Discusses irrationals and incommensurables, including the history of their discovery and the earliest proof of an incommensurable. Has some interesting facts about Pythagorean triples and the pentagram. This is the first of a series of articles.

Good reading for algebra or geometry classes.

LAW, CAROL. "Arithmetical Congruence with Practical Application." *Mathematics Magazine* 31: 221–27; March–April 1958.

A good, clear, and elementary introduction to congruence. Mentions Carl F. Gauss in the introduction. The main features of the article are: (a) What is meant by two numbers being congruent. (b) The possibility to solve for unknowns in linear, quadratic, and cubic congruences. (c) How we might ap-

---

[1] School Mathematics Study Group.

ply congruences practically. Contains good diagrams using circles, etc. Works calendar problems at the end. Gives a good list of references, e.g.: *Elementary Number Theory* by Uspensky and Heaslet; *What Is Math?* by Courant and Robbins; *The Elements of the Theory of Algebraic Numbers* by Reid; *Elementary Theory of Numbers* by Griffin.
Could be used at any level from 9–12.

McELWAIN, FERD W. "Digital Computer—Nonelectronic." *Mathematics Teacher* 54: 224–28; April 1961.
Presents some interesting computations on one of our oldest digital computers, the fingers. This has a drawback in that to make the program a success one must spend quite a bit of time learning what each finger represents.
This might be fun for a math club.

MIKSA, FRANCIS L. "A Table of Integral Solutions of $a^2 + b^2 + c^2 = r^2$." *Mathematics Teacher* 48: 251–55; April 1955.
This is a chart of all the odd values of r from 3 to 207 that yield integral solutions to the equation.
Could be incorporated into the studies of any class.

REID, CONSTANCE. "Perfect Numbers." *Scientific American* 188: 84–86; March 1953.
A short article reporting that some work with a computer has led to the discovery of five more perfect numbers, bringing the total known to seventeen. In the preceding 2000 years, only twelve had been found. Euclid knew that a number was 'perfect' when it could be factored into $2^{n-1} (2^n - 1)$, where the second factor is a prime number.

ROSENTHAL, EVELYN B. "A Pascal Pyramid for Trinomial Coefficients." *Mathematics Teacher* 54: 336–38; May 1961.
This is a step beyond the famous Pascal triangle. Shows the method of forming the Pascal pyramid for the coefficients of a trinomial raised to a power.
For any second-year algebra class.

SWALLOW, KENNETH P. "The Factorgram." *Mathematics Teacher* 48: 13–17; January 1955.
The factorgram, a device similar to the sieve of Eratosthenes, is used to find the prime numbers. This article includes a sample chart, or factorgram, and the reasons why it works.
Interesting for an algebra class.

VON BARAVALLE, H. "The Number π." *Mathematics Teacher* 45: 340–48; May 1952.
An excellent article showing uses of π, and series and other methods of computing π. Contains a geometric method of finding π; and a method of squaring the circle, based on an approximation type construction by Kochansky.
Helpful for students making a study of series.

ALGEBRA

BAUMGART, JOHN K. "Axioms in Algebra—Where Did They Come From?" *Mathematics Teacher* 54: 155–60; March 1961.
Gives an historical background to the origin of axioms and some of the reasons why they exist. Includes the Rhind papyrus, Euclid, groups, fields, and quaternions.
Good background material for the algebra teacher.

BEBERMAN, MAX, and MESERVE, BRUCE E. "Graphing in Elementary Algebra." *Mathematics Teacher* 49: 260–66; April 1956.
An argument to broaden the scope of graphing in the algebra class. Includes some new ideas of graphing on the number line and graphing inequalities.
Good for use in first-year algebra classes.

BERGDAL, ED. "Complex Graphs." *Mathematics Magazine* 24: 195–202; March–April 1951.
An elementary article on complex and real numbers, illustrating equations with complex roots. Good explanation and illustration of some curves that come in pairs of one real and one complex, real ellipse, and hyperbola with corresponding complex curves. The complex hyperbola with the form of a real ellipse and also some all-complex functions are discussed and illustrated. Could be used at any level from 9–12.

BROTHER U. ALFRED, FSC. "Problems Involving Overlapping Finite Sets." *Mathematics Teacher* 53: 524–33; November 1960.
A nice article illustrating the use of a Venn diagram or Euler's circle to solve problems. Shows a chart as a method of collecting data and concludes with many problems to solve by this method.
Good for use in algebra classes or possibly, due to some of the notations, in advanced math classes.

BRUMFIEL, CHARLES. "The Foundations of Algebra." *Mathematics Teacher* 50: 488–92; November 1957.
Introduces a "modern" idea or method of developing definitions, proofs, and logical structure in algebra. Proofs of some theorems are presented, based on nine basic axioms.
Good for use in first-year algebra classes.

CONKWRIGHT, N. B. "Practical Determination of the Rank of a Matrix." *Mathematics Teacher* 49: 344–46; May 1956.
This work of determining the rank of a matrix usually is not part of the work of the high school curriculum; however, with the talk of matrices coming into the high school (SMSG) this article should be more than interesting.
Worthwhile for advanced senior students.

DALTON, LEROY C. "Complex Numbers and Loci." *Mathematics Teacher* 54: 229–33; April 1961.
An excellent article on the graphing of equations, or loci, in the complex plane. Incorporates modern ideas of absolute values and their graphs. This should be an interesting unit to teach.
Suitable for second-year algebra students.

GROSS, FRANCIS. "Graphing Pictures." *Mathematics Teacher* 52: 295–96; April 1960.
A good article to file away for the last days of the year, or possibly as an assignment near the Christmas vacation. Most students enjoy making pictures by plotting graphs.
Suitable for any class at certain times in the year.

GRUENBERGER, FRED. "Imaginaries That Are Not Imaginary." *Mathematics Teacher* 47: 11–12; January 1954.
This article coincides with the practice the reviewer uses in class to show a practical application of the non-imaginary complex number.
Good for use in algebra two (second-year) classes.

HOHFELD, JOSEPH F. "An Analysis of the Quadratic." *Mathematics Teacher* 54: 138–41; March 1961.
Discusses the parabola only, showing how changes in the constants affect changes in the shape and position of the curve. Calculus is used, but the discussion could be handled without this advanced topic.
A good study for second-year algebra students.

JEFFREY, JAY M. "Analysis of a Spider Nomograph." *Mathematics Teacher* 54: 239–40; April 1961.
A discussion of a subject, nomographs, which should be but is not covered fully enough in the mathematics of science teaching. Presents the solution of reciprocal sum equations by a ruler and a graph.
Worthwhile for algebra classes and all teachers.

JOHNSON, PAUL B. "Mathematical Induction, $\sum i^p$ and Factorial Powers." *Mathematics Teacher* 53: 332–34; May 1960.
Mathematical induction, needed by all students these days, is the subject of this article. The simple problems which are worked out may give the teacher ideas for more.
For advanced or senior mathematics classes.

KLIMCZAK, W. J. "The Solution of Linear Equations and Inequalities." *Mathematics Teacher* 47: 460–63; November 1955.
This is an interesting method of finding the solutions of linear equations and inequalities by the use of graphs and rotations of axes.
Good material for analytic geometry classes.

LEVITT, H. "An Extension of Pascal's Triangle." *Mathematical Gazette* 45: 99–107; May 1961.
The definition of the binomial coefficients is extended, and various identities are derived.
For upper-grade level.

MAZKEWITSCH, D. "Coefficients in Expansions of Polynomials." *School Science and Mathematics* 63: 703–09; December 1958.
An investigation of the interesting patterns that occur in polynomials in which the exponents of the literal factor form an arithmetic progression, expansions of squares of polynomials with coefficients 1, and others.
Subject matter is at the level of a second course in algebra.

PAGE, ROBERT L. "First-Degree Equations in More Than Three Variables." *Mathematics Teacher* 54: 170–71; March 1961.
This article brings out the warning of the pitfalls of attempting to solve all groups of equations by the elimination method. With four equations, it is noted, there are six combinations and three incorrect ways of choosing the pairs of equations.
Worth discussing with any algebra class.

PERISHO, CLARENCE R. "A Non-Commutative Algebra." *School Science and Mathematics* 43: 727–30; December 1958.
A simple example of a counterexample—an algebra in which the commutative law does not hold. The group of symmetries of a square discussed in a simple manner.
A second course in algebra and some mathematical maturity required.

RANSOM, WILLIAM R. "Elementary Calculations of Logarithms." *Mathematics Teacher* 47: 115–16; February 1954.

A scheme to introduce logarithms so as to remove the mystery about possible origins of tables. Emphasizes the fact that common logarithms are exponents of ten.

Good for use in second-year algebra or trigonometry.

RIO, S. T., and SANDERS, W. J. "Interval Graphing." *Mathematics Teacher* 54: 194–200; April 1961.

Describes interval graphing as a method of finding solution sets for a number of varied problems. It shows a good application of the set terminology, including: union, subset, closed or open set. Contains many illustrations.

Suitable for any algebra or advanced mathematics class.

ROSENBERG, HERMAN. "Modern Applications of Exponential and Logarithmic Functions." *School Science and Mathematics* 60: 131–38; February 1960.

Interesting treatment, as can be judged from the subtitles: Population and the Exponential Function; Love and the Logarithmic Function; Love and Population Go Together Just Like the Exponential Function and the Logarithmic Function.

Requires some acquaintance with the concepts in a second course in algebra.

SEEBECK, C. L., JR., and HUMMEL, P. M. "Logarithmic and Exponential Functions—A Direct Approach." *Mathematics Teacher* 52: 439–43; October 1959.

An argument for the direct approach, by means of properties of logarithms, to the common logarithms, and later to the general base and change-in-base procedures. From the initial properties, the authors build up several theorems for logarithms.

Worth considering for a second-year algebra class.

SHUSTER, CARL N. "Graphic Solution of a Quadratic Equation." *Mathematics Teacher* 54: 142–44; March 1961.

Covers only the graph and roots of a parabola. Calculus is used to a small degree. The idea is to show that imaginary roots of a quadratic equation can be found graphically.

Possibly for second-year algebra or senior math classes.

SMITH, G. S. "A New Method of Calculating Roots of Algebraic Equations." *Mathematical Gazette* 44: 241–45; December 1960.

A method of calculating real roots by a simple division process when an integral approximation is known.

For middle-grade level.

STEWART, LURLINE. "The Binomial Theorem." *Mathematics Teacher* 53: 344–48; May 1960.

Gives a description of the historical background of the binomial theorem, then goes on to describe methods of calculating the coefficients, and describes some applications pertaining to probability.

Very good material for a second-year algebra class.

TWEEDIE, M. C. K. "A Graphical Method of Solving Tartaglian." *Mathematical Gazette* 23: 278–82; July 1939.

Measuring puzzles are presented.

For lower-grade level.

WENDT, ARNOLD. "A Simple Example of a Noncommutative Algebra." *Mathematics Teacher* 52: 534–40; November 1959.

Shows an algebra in which the multiplication is noncommutative, thus impressing on the students the importance of these laws. The article is based on simple matrices and their applications.

The ideas expressed could be used in any algebra class, but the whole unit might have to wait until later—for a senior mathematics class.

WENDT, ARNOLD. "Solving Simultaneous Linear Equations." *Mathematics Teacher* 53: 12–17; January 1960.

Presents a useful method of solving simultaneous linear equations, especially if there are more than three unknowns. The method is taught in the SMSG ninth-grade course.

Suitable for all traditional first-year algebra courses.

WHEELER, NORMAN E. "Determinants Whose Values Are Zero." *Mathematics Teacher* 52: 443–44; October 1960.

A timely as well as interesting article on determinants, in view of all the talk today about modern math and matrices. Describes a method of inspection to tell when the value of a determinant is zero.

Suitable and worthwhile for a second-year algebra class.

GEOMETRY

ARTIN, EMIL. "The Theory of Braids." *Mathematics Teacher* 52: 328–33; May 1959.

Presents an interesting topic from topology that could be used to liven up any geometry class. The article on braids is complete with diagrams.

Excellent for a better class in geometry.

BARDIS, PANOS D. "Evolution of *Pi*: An Essay in Mathematical Progress from the Great Pyramid to Eniac." *School Science and Mathematics* 60: 73–78; January 1960.

An interesting historical study of evaluations of *pi*, covering those made in Japan, China, India, Babylonia, Greece, and modern nations.

Only prerequisite for understanding and appreciating this article is an interest in *pi*.

BOTTS, TRUMAN. "Finite Planes and Latin Squares." *Mathematics Teacher* 54: 300–06; May 1961.

Describes a finite geometry with its limited number of points and lines. A related topic: Latin squares—which for order n have n rows and n columns, with each row and each column containing the same symbols exactly once— are discussed. (A 1961 issue of *Scientific American* also has a good article on Latin squares.)

Excellent for a good class in geometry.

BOYER, CARL B. "The Invention of Analytic Geometry." *Scientific American* 180: 40–45; January 1949.

A fascinating article which presents the historical development (as opposed to the "invention" suggested by the title) of analytic geometry from the original Delian problem: solve $x^3 = 2$. Descartes is herein given his due and no more.

BRADFORD, OWEN. "Polyhedra of Any Dimension." *School Science and Mathematics* 60: 589–92; November 1960.

An extension of Euler's theorem on numbers of vertices, edges, faces, and solids to higher dimensions by a high school student.

For plane and solid geometry classes.

BRISTOL, JAMES D. "Construction and Evaluation of Trigonometric Functions of Some Special Angles." *Mathematics Teacher* 54: 4–7; January 1961.

A background in trigonometry is recommended to appreciate the whole article, but it may be of value to the geometry student because of the construction of angles of 18°, 36°, 54°, 72°, and 108°.

Recommended for geometry or trigonometry classes.

BRUECKEL, FRANK J. "Parallelograms with Integer Sides and Diagonals." *School Science and Mathematics* 56: 687–96; December 1956.

The article grew out of a problem which is an extension of the problem of finding integral right triangles, or Pythagorean triples. There is a general solution and a method for obtaining any number of such parallelograms.

Plane geometry, some second course algebra, and a little trigonometry needed as background.

CHENEY, WILLIAM FITCH, JR. "Can We Outdo Mascheroni?" *Mathematics Teacher* 46: 152–56; March 1953.

This article extends the challenge of the Mascheroni constructions by changing from the modern compass, which retains its setting, to the classical compass which returns to a zero setting after each use.

Good material for a geometry class.

CHRISTIAN, ROBERT R. "A New Role for High School Geometry." *Mathematics Teacher* 53: 433–36; October 1960.

The author brings out the fact that in geometry the student can meet some elementary examples of important abstract ideas. Several examples of this statement are illustrated in the article.

Good reading for plane geometry classes.

COURT, NATHAN ALTSHILLER. "Imaginary Elements in Pure Geometry—What They Are and What They Are Not." *Scripta Mathematica* 17: 55–64; March, June 1951; 190–205; September, December 1952.

In these twenty-odd pages the author discusses the "imaginary" pair of real numbers obtained in an algebraic solution of a geometry problem. He goes into the various geometric interpretations of these "imaginary solutions."

After two years of algebra and a year of geometry the high school student has seen the relationship of the "reals" and the point in geometry. He questions the place of the "imaginary" in this. The article can give an interpretation of this concept, though it is rather difficult reading.

COURT, NATHAN ALTSHILLER. "Mascheroni Constructions." *Mathematics Teacher* 51: 370–72; May 1958.

This background article on Mascheroni contains a bibliography of his works, and gives credit to some of the people who have kept his constructions alive through the centuries.

Should be read by all geometry teachers who discuss Mascheroni constructions in class.

COXETER, H. S. M. "Map Coloring Problems." *Scripta Mathematica* 23: 11–25; Memorial Issue 1957.

Begins with "The Four Color Problem," then goes on to discussion of other "coloring" problems. Finishes on a note of challenge, in that problems still waiting to be solved are listed for interested readers.

Students need at least a year of geometry before they can gain much from this article. For a student with imagination and curiosity this could be a starting point for research into this topic.

COXETER, H. S. M. "The Four Color Map Problem, 1840–1890." *Mathematics Teacher* 52: 283–89; April 1959.

This is a very fine article on topology, dealing with the four color problem. The article deals with two- and three-dimensional maps and is complete with several illustrations.

Excellent for a geometry class or club.

CUNDY, H. MARTYN. "25-Point Geometry." *Mathematical Gazette* 36: 158–66; September 1952.

A discussion of structure and theorems in a finite geometry of 25 points.

For upper-grade level.

DAWSON, T. R. " 'Match-Stick' Geometry." *Mathematical Gazette* 23: 161–68; May 1939.

An account of constructions which are possible, given an unlimited supply of match-sticks all of one length, L, and subject to certain allowable rules.

For middle-grade level.

EULER, LEONHARD. "The Koenigsberg Bridges." *Scientific American* 189: 66–70; July 1953.

A reprint of Euler's original memoir, setting forth his solution of the problem of the Koenigsberg bridges. Introduces the reader to the branches of topology which deal with networks. Introduced by a note from James R. Newman.

A classic, valuable for the insight it gives to the history of mathematics.

FEHR, HOWARD F. "On Teaching Angle and Angle Measure." *Mathematics Teacher* 50: 551–56; December 1957.

Although the title implies topics other than algebra, the article contains many useful algebraic topics. The discussion on rectangular coordinates and graphing of inequalities is especially good.

Good reading for a second-year algebra class and for all teachers.

FEHR, HOWARD. "On Teaching Dihedral Angle and Steradian." *Mathematics Teacher* 51: 272–75; April 1958.

A good article for teachers who have not done any solid geometry teaching in their plane geometry class. It discusses dihedral angles and how they can be fitted in, along with steradian measure, to a revised geometry curriculum.

Useful for a geometry class.

FEHR, HOWARD. "On Teaching Trihedral Angle and Solid Angle." *Mathematics Teacher* 51: 358–61; May 1958.

This is another article for the teacher who has been teaching plane but not solid geometry in the classroom. It has nice diagrams and a clear discussion of some of the facts on trihedral angles and their relation to angles of a sphere.

Good material for a geometry class.

GARDNER, MARTIN. "Flexagons." *Scientific American* 195: 162–66; December 1956.

A lightly diverting article, taking us into geometry with some paper folding.

For geometry classes.

GRAY, NELSON S. "Right Triangle Construction." *Mathematics Teacher* 53: 533–36; November 1960.

This includes the familiar formulas for finding a, b, and c—the sides of the right triangle. It also includes a large table of Pythagorean triples and some interesting relationships dealing with these numbers.

Good material for a geometry class.

GRUHN, E. W. "Parabolas and the Pythagorean Triples." *Mathematics Teacher* 52: 614–15; December 1959.

This short article shows a relationship between parabolas and the Pythagorean theorem. It studies the intersection of conic sections and planes, and tells of interesting relationships of the two sets: (3, 4, 5) and (7, 24, 25).

Mostly geometry, but could be used in algebra or trigonometry classes.

GUGGENBUHL, LAURA. "An Unusual Application of a Simple Geometric Principle." *Mathematics Teacher* 50; 322–24; May 1957.

An amazing article, showing a method used for grafting skin and making incisions which is based on a simple geometric figure. The discussion as to size of angle requires some trigonometry.

Good plane geometry class could handle this.

HAHN, HANS. "Geometry and Intuition." *Scientific American* 190: 84–91; April 1954.

Translation and condensation of a lecture, includes clearly drawn figures. "It is not true, as Kant urged, that 'intuition is a pure *a priori* means of knowledge.' Rather it is force of habit rooted in psychological inertia." This is an interesting thesis, ably explored by the author.

For geometry classes.

HALLERBERG, ARTHUR E. "The Geometry of the Fixed Compass." *Mathematics Teacher* 52: 230–44; April 1957.

This is a variation of the Mascheroni constructions, in that a "rusty compass" is used. The article is complete with a large number of problems and worked-out diagrams of the problems.

Excellent for geometry classes.

HALLEY, ROBERT R. "Prove As Much As You Can." *Mathematics Teacher* 49: 491–92; October 1956.

Presents an excellent idea: let the student go on from certain theorems in geometry and prove several theorems from the one start. There are several opportunities for this procedure in a standard course.

Good material for a standard geometry class.

HART, PHILIP J. "Pythagorean Numbers." *Mathematics Teacher* 47: 16–21; January 1954.

An interesting tabular arrangement of the well known Pythagorean numbers. Gives the history, various formulas for computing them, the tabulated results, and formulas for use with more dimensions.

Recommended for a geometry class.

HLAVATY, JULIUS H. "Mascheroni Constructions." *Mathematics Teacher* 50: 482–87; November 1957.

This is one of several articles on this subject in the *Mathematics Teacher*. Classes are always excited to try a few of these after being shown the way.

Good material for better classes in geometry.

HOOD, RODNEY T. "A Chain of Circles." *Mathematics Teacher* 54: 134–37; March 1961.

This presents an interesting problem of inversion of circles, and could be well used to expand the scope of a plane geometry class to include some projective geometry.

Worthwhile for a geometry class or a math club.

JONES, BURTON W. "Miniature Geometries." *Mathematics Teacher* 52: 66–71; February 1959.

A fine topic to include in a unit for the better students. It includes a discussion of the most common finite geometry, the seven point geometry and some of its applications.

Excellent for better classes in geometry.

JONES, PHILLIP S., editor. "Cevians, Nedians and Radians." *Mathematics Teacher* 44: 496–97; November 1951.

Some interesting theorems and discussions about Ceva's theorem and Menelaus' theorem to bring out a topic from "Modern Geometry."

Students in geometry could handle some of this with the teacher's help.

JONES, PHILLIP S., editor. "More about Cevians, Nedians and Radians." *Mathematics Teacher* 45: 380–82; May 1952.

This second article on the subject continues the discussion in a previous issue and gives a few new ideas and different approaches.

For plane geometry students.

KENNER, MORTON R. "Helmholtz and the Nature of Geometrical Axioms: A Segment in the History of Mathematics." *Mathematics Teacher* 50: 98–104; February 1957.

An excellent study of the works of Helmholtz on the axioms of Euclid, making a comparison of their adaptations to planes, spheres, and pseudospheres.

For good classes in geometry. All geometry teachers should read this.

KEARSLEY, J. "Curves of Constant Diameter." *Mathematical Gazette* 36: 176–79;

Some notes on the curves of constant diameter or width.

Upper-grade level material.

KLINE, MORRIS. "Projective Geometry." *Scientific American* 192: 80–86; January 1955.

A beautifully done introduction to a geometry that is basic to all geometry. The illustrations are good. The approach is light rather than pedagogical, but the information is clear, complete, correct.

Good reading for beginning geometry students.

KLINE, MORRIS. "The Straight Line." *Scientific American* 194: 104–14; March 1956.

"A brief survey," the author calls it. It is beautifully done, leading us from Euclid to non-Euclid and back again, just suggesting the "darkness at the interior" of a straight line. (This, we would have liked more fully developed.) Geodesics and light rays are mentioned in their place in this expert discussion of the line itself and the length of segments.

LINDGREN, H. "Going One Better in Geometric Dissections." *Mathematical Gazette* 45: 94–97; May 1961.

Methods are described which facilitate the discovery of economical dissections of various geometric figures.

For lower-grade level.

LOKRE, SHRI M. R. "Construction of a Teaching Aid for 'Nine-Points Circle'." *School Science and Mathematics* 56: 459–65; June 1956.

Detailed instructions for the construction of a mechanical model of the many exciting properties of the Euler line and the nine-points circle.

For plane geometry classes.

McCARTHY, J. P. "A Difficult Converse." *Mathematical Gazette* 22: 365–71; October 1938.

A history of and different proofs for the theorem: "If the bisectors of the base angles of a triangle are equal, the triangle is isosceles."

For middle-grade level.

MESERVE, BRUCE E., and PINGRY, ROBERT E. "Some Notes on the Prismoidal Formula." *Mathematics Teacher* 45: 257–63; April 1952.
An interesting article showing the use of one general formula to find the volume of any common solid and of some others not so common.
To be studied by students taking solid geometry.

MESERVE, BRUCE E. "Topology for Secondary Schools." *Mathematics Teacher* 46: 465–74; November 1953.
This article should be read by every geometry teacher, and many of the topics incorporated into the class work. There are numerous popular puzzles and tricks discussed, and proved, or proved impossible.
Good material for a geometry class or math club. A "must" for teachers.

MILLER, WILLIAM G. "Tangent Circles and Conic Sections." *Mathematics Teacher* 46: 78–81; February 1953.
An interesting article describing eight combinations of tangency of circles and lines. Complete with diagrams to show the various conic sections generated by the centers of the tangent circles.
Good study for plane geometry, algebra, or analytic geometry classes.

MIRSKY, L. "Problems of Arithmetical Geometry." *Mathematical Gazette* 44: 182–91; October 1960.
A discussion of four interesting and easy-to-understand problems, but the solutions are not easy.
For upper-grade level.

MOSHAN, BEN. "Primitive Pythagorean Triples." *Mathematics Teacher* 52: 541–45; November 1959.
This continues on, from the familiar formulas that are used to calculate the Pythagorean triples, to other formulas that may be used to group the sets of numbers.
Worth reading by geometry teachers.

ORE, OYSTEIN. "An Excursion into Labyrinths." *Mathematics Teacher* 52: 367–70; May 1959.
Gives an historical background of labyrinths or mazes, and lists some of the more famous ones. Then describes several methods for solving the problems.
Interesting for the better students in geometry or for a math club.

PERISHO, CLARENCE R. "Colored Polyhedra: A Permutation Problem." *Mathematics Teacher* 53: 253–55; April 1960.
This offers an excellent opportunity to tie solid geometry into algebra by permutations. Presents the problem of the number of different ways the various faces of polyhedra can be painted.
Excellent for a solid geometry or advanced algebra class.

PICKETT, HALE. "Trisecting an Angle." *Mathematics Teacher* 51: 12–13; January 1958.
A simple little article any student could read to get the answer to the question: "Why can't I trisect an angle?"
For geometry students.

RANSOM, WILLIAM R. "A Six-Sided Hexagon." *School Science and Mathematics* 52: 94; February 1952.
The construction of an interesting little model—a hexagon which has six sides, not just two (front and back).
Doable by any student of mathematics, from the grades through high school.

RANUCCI, E. R. "The Weequahic Configuration." *Mathematics Teacher* 53: 124–26; February 1960.

Presents an old problem which has made the rounds of the drafting rooms for years. Two views of a solid are given, and the student has to supply the third view. There are many solutions, and it gives the student an excellent chance to practice visualization of three dimensions.

Excellent for a plane or a solid geometry class.

ROBUSTO, C. CARL. "Trisecting an Angle." *Mathematics Teacher* 52: 358–60; May 1959.

A discussion of the fact that the trisection is not possible; includes a proof of the fact that it is impossible.

Recommended for some of the better students in geometry.

ROSSKOPF, M. F., and EXNER, R. M. "Some Concepts of Logic and Their Applications in Elementary Mathematics." *Mathematics Teacher* 48: 290–98; May 1955.

Includes some elementary concepts of logic. Gives some definitions and some symbols that are similar to but different from the symbols of set notation. Works with truth tables, the deduction principle, and the generalization principle. The author illustrates by means of geometric proofs.

Good reading for geometry teachers.

SATTERLY, JOHN. "The Trisection of the Area of a Circle." *School Science and Mathematics* 53: 124–30; February 1953.

An interesting discussion of the problem and generalizations of it (division into 4, 5, 6, 7, 8, n).

A student of trigonometry could read this article with appreciation.

SATTERLY, JOHN. "The Morley Triangle and Other Triangles." *School Science and Mathematics* 55: 685–701; December 1955.

Proofs of the famous theorem using trigonometric functions, and also extensions of the theorem.

Very fine for enthusiasts in plane geometry who know some trigonometry.

SATTERLY, JOHN. "Equilateral and Other Triangles Associated with Triangles." *School Science and Mathematics* 56: 11–20; January 1956.

Equilateral triangles are constructed on the sides of any triangle, externally. This article is a study of the many special cases that arise from studying equilateral triangles formed by the incenters of these triangles.

Plane geometry and a little trigonometry needed for comprehension.

SATTERLY, JOHN. "The Circumcircle, the Incircle, the Nine-Point Circle, and the Circle through the Feet of the Bisectors of the Angles of a Triangle." *School Science and Mathematics* 56: 517–28; October 1956.

Interesting review of the Euler line, the nine-point circle. Extensions. Proof that the incircle, the nine-point circle, and the new circle described in the title all pass through the Feuerbach point.

Plane geometry required; the proofs require some knowledge of coordinate geometry and of a second course in algebra.

SATTERLY, JOHN. "Another Approach to the Nine-Point Circle." *Mathematics Teacher* 50: 53–54; January 1957.

Discusses the construction and proof of the nine-point circle in a not-so-lengthy form; and shows other facts that are related to this famous construction.

For the better classes in geometry.

SATTERLY, JOHN. "Meet Mr. Tau." *School Science and Mathematics* 56: 731–41; December 1956.

An interesting collection of much miscellaneous data on the number *tau* = $(\sqrt{5} + 1)/2$; its occurrence in algebra, geometry, and trigonometry; its relation to the most esthetic rectangle, the regular pentagon, the pentagram, the icosahedron, the dodecahedron, and Fibonacci numbers.

Plane geometry and some solid geometry required.

SATTERLY, JOHN. "Meet Mr. Tau Again." *School Science and Mathematics* 57: 150–51; February 1957.

Further relations between *tau* and the regular pentagon, and another related pentagon.

Plane geometry required.

SATTERLY, JOHN. "A Theorem of Related Triangles, the Centers of Gravity of a Full and a Skeleton Triangle." *School Science and Mathematics* 57: 367–74; May 1957.

Comparison of various geometric theorems arising from considering a triangle as a disk, and as a figure composed of three thin rods.

For plane geometry classes.

SATTERLY, JOHN. "A Note on Transversals in Triangles." *School Science and Mathematics* 57: 479–84; June 1957.

A proof of Ceva's theorem, using areas. Extensions of the theorem to take up various related theorems, including some due to Leonardo da Vinci.

For plane geometry classes.

SATTERLY, JOHN. "A Problem with Touching Circles." *Mathematics Teacher* 53: 90–95; February 1960.

An interesting set of construction problems. Follows along the idea of filling the plane with circles. The circles so constructed are all tangent to other circles. Methods of construction are included.

Fine for geometry classes.

SCHAAF, WILLIAM L. "The Theorem of Pythagoras." *Mathematics Teacher* 44: 585–88; December 1951.

Excellent reference material, for teachers or students, on the proofs, special cases, etc. of the Pythagorean theorem.

For plane geometry students.

SCHEID, FRANCIS. "Square Circles." *Mathematics Teacher* 54: 307–12; May 1961.

Describes circles other than the conventional geometric ones that we are familiar with. With suitable definitions you can have square circles. The radii are equal in taxicab geometry, or the kings walk of the game of chess.

Excellent for geometry class or a math club.

SISTER MARIE STEPHEN, OP. "The Mysterious Number PHI." *Mathematics Teacher* 49: 200–04; March 1956.

These pages in the reviewer's issue are just about worn out from referring to them and having children use them at math fairs. This article seems to tie the golden section, Fibonacci's series, and all the rest together very nicely.

Excellent for geometry classes.

SMITH, CYRIL STANLEY. "The Shape of Things." *Scientific American* 190: 58–64; January 1954.

This well written and nicely illustrated article sets forth convincingly the thesis that "the structure of any aggregate of matter—whether organic or inorganic, biological, natural or artificial—depends not only on the interplay

of forces but also on the very simple but inescapable mathematical requirements of space-filling."
Excellent for geometry students.

SMITH, DAN. "Vectors—An Aid to Mathematical Understanding." *Mathematics Teacher* 52: 608–13; December 1959.
An argument for more use of vectors in the mathematics curriculum. Includes a test which has 45 good problems on vectors.
Could be used by students in many different classes in mathematics.

SORNITO, JUAN E. "Involutions Operated Geometrically." *Mathematics Teacher* 48: 243–44; April 1955.
A short article describing two methods of construction and their proof of finding a power of a number, given a line equal in length to the number.
Fine for geometry classes.

SORNITO, JUAN E. "Two Cube Root Curves." *Mathematics Teacher* 48: 560–62; December 1955.
Contains two different curves, one for the construction of a line equal to $\sqrt[3]{2}$, and the other construction more generalized for extracting any number greater than one.
Might best be used in an algebra class.

STRUYK, ADRIAN, and CLIFFORD, PAUL C., editors. "Theme Paper, a Ruler and the Parabola." *Mathematics Teacher* 46: 588–90; December 1953.
The first of three articles on interesting methods of constructing the various conic sections by means of lined composition paper and a ruler.
Should be interesting for all algebra groups discussing the conics.

STRUYK, ADRIAN, and CLIFFORD, PAUL C., editors. "Theme Paper, a Ruler and the Hyperbola." *Mathematics Teacher* 47: 29–30; January 1954.
The second of three articles showing how the hyperbola can be constructed graphically on composition paper.
Of interest for any group studying conic sections.

STRUYK, ADRIAN, and CLIFFORD, PAUL C., editors. "Theme Paper, a Ruler and the Central Conics." *Mathematics Teacher* 47: 189–93; March 1954.
The last of a series of three articles showing how the hyperbola or the ellipse can be constructed graphically on composition paper.
Of interest for any group studying the conic sections.

SYNGE, J. L. "The Geometry of Many Dimensions." *Mathematical Gazette* 33: 249–63; December 1949.
An interesting introduction to geometry of many dimensions.
For upper-grade level.

TRIGG, CHARLES W. "Unorthodox Ways to Trisect a Line Segment." *School Science and Mathematics* 54: 525–28; October 1954.
Five unusual techniques for trisecting a line segment. Generalization to dividing a line segment into more parts.
For elementary geometry students.

TRIGG, CHARLES W. "Geometry of Paper Folding: I. Properties of a Drinking Cup." *School Science and Mathematics* 54: 453–55; June 1954.
Discusses the forming of a drinking cup, the pattern of the creases, the lengths of the creases, the relationships of the areas, some trigonometric functions, and the incommensurability of $\sqrt{2}$.
Elementary geometry and a little trigonometry helpful for background.

TRIGG, CHARLES W. "Geometry of Paper Folding: II. Tetrahedral Models." *School Science and Mathematics* 54: 683–89; December 1954.

Deals with surfaces derivable from an equilateral triangular pattern: collapsible patterns with closed edges, collapsible patterns with one open edge, patterns for non-regular tetrahedra, tetrahedra from lateral cylindrical surfaces, tetrahedron from tape, tetrahedra from envelopes, and equilateral triangle into tetrahedron with closed edges.

For plane and solid geometry classes.

TUCKER, ALBERT W., and BAILEY, HERBERT S., JR. "Topology." *Scientific American* 182: 18–24; January 1950.

This article has become a classic in the popularization of the then-new field of topology: not the point-set topology of modern algebraists, but the (often intuitive) geometry of properties remaining invariant under drastic transformations of figures.

MATHEMATICS STAFF OF THE COLLEGE, University of Chicago. "Three Algebraic Questions Connected with Pythagoras' Theorem." *Mathematics Teacher* 49: 250–59; April 1956.

This includes the questions: are there natural numbers that satisfy (a) $x^2 + y^2 = h^2$, (b) $x^2 + y^2 = u^4$ and $y - x = v^2$, and (c) $x^2 + y = u^2$ and $x + y = v^2$. The article includes guesses at the answers and proofs.

Recommended for use by a good geometry class when refreshing its algebra.

MATHEMATICS STAFF OF THE COLLEGE, University of Chicago. "Coloring Maps." *Mathematics Teacher* 50: 546–50; December 1957.

A discussion of the famous four-color map problem of topology. The article includes other problems related to this map, but unfortunately leaves out the coloring of the torus.

Fine for a geometry class.

VON BARAVALLE, H. "The Number π." *Mathematics Teacher* 55: 340–48; May 1952.

An excellent article showing uses of π and series and other methods of computing π. There is a geometric method of finding π. There is a method of squaring the circle based on an approximation type construction by Kochansky.

Most of this is suitable for geometry students. The discussion of series is suitable for advanced students.

WHEELER, R. F. "The Flexagon Family." *Mathematical Gazette* 42: 1–6; February 1958.

A discussion of the making and coloring of an infinite family of flexagons.

For middle-grade level.

WISEMAN, JOHN D., JR. "Some Related Theorems on Triangles and Circles." *Mathematics Teacher* 54: 14–16; January 1961.

Brings out the fact that the words are different but the configurations of lines and points are about the same. It gives a few examples, but paves the way for the teacher to be alert for other possibilities.

For geometry classes.

WISEMAN, JOHN D., JR. "Introducing Proof with a Finite System." *Mathematics Teacher* 54: 351–52; May 1961.

Gives two examples of a finite geometry which could be used to show the true reason for studying geometry. These examples could be followed by another geometry with its undefined and defined terms, its postulates, and theorems. Helpful for a geometry class.

WRENCH, J. W., JR., and EVES, HOWARD, editors. "The Evolution of Extended Decimal Approximation to π." *Mathematics Teacher* 53: 644–50; December 1960.

Contains several series and formulas for computing π. Gives the historical background of many of the series and the results of their calculations. The most extensive approximation (16167 places) has been tabulated to the distribution of the digits. A very extensive bibliography accompanies the article.

TRIGONOMETRY

AMI-MOEZ, ALI R. "Teaching Trigonometry through Vectors." *Mathematics Magazine* 32: 19–23; September–October 1958.

Considers only vectors with beginning at the origin O of a rectangular Cartesian coordinate system. (Denote $\overrightarrow{OA}$ by $\overrightarrow{A}$.) Includes: addition of vectors, inner product of two vectors, projection.

An interesting article, well written, with excellent presentation. Can be used in any level from 9–12.

BRISTOL, JAMES D. "Construction and Evaluation of Trigonometric Functions of Some Special Angles." *Mathematics Teacher* 54: 4–7; January 1961.

Presents some interesting ideas relating geometry, trigonometry, and algebra. Includes a proof of the construction of a pentagon, and shows construction of angles of 18°, 36°, 72°, 54°, and 108°.

Good reference material for the teacher, but a student would have to be familiar with trigonometry to fully appreciate this article.

CLIFFORD, EDWARD L. "An Application of the Law of Sines: How Far Must You Lead a Bird to Shoot It on the Wing?" *Mathematics Teacher* 54: 346–50; May 1961.

As the title implies, this is an application—though approximate or impractical —of the Law of Sines. Gives several suggestions on the general solving of word problems.

Best used for a regular trigonometry class.

CORNELL, ALVIN P. "Theorems and Formulas Concerning Fundamental Right-Angle Triangles." *School Science and Mathematics* 54: 533–38; October 1954.

An interesting excursion into Pythagorean triples. FRAT's (fundamental right-angle triangles), DRAT's (derivative right-angle triangles), and others are discussed; also, how to generate an infinite set of certain kinds of Pythagorean triples.

For elementary geometry students.

GREENBERG, BENJAMIN. "A Geometric Proof of Half-Angle Formulas for Triangle *ABC* by Means of an Escribed Circle." *School Science and Mathematics* 59: 682–85; December 1959.

An elegant, elementary geometric proof of the half-angle formulas.

For plane geometry and trigonometry students.

MAZKEWITSCH, D. "Geometric Derivations of Some Trigonometric Formulae." *School Science and Mathematics* 58: 213–16; March 1958.

Geometric derivations of the double-angle formulae, of tan $(A + B)$, and some others.

For a plane trigonometry class.

PEDLEY, ARTHUR H. "Complex Numbers and Vectors in High School Mathematics." *Mathematics Teacher* 53: 198–201; March 1960.

It is argued in this article that after studying the idea of complex numbers and how they are manipulated the matter should not be dropped; complex numbers should be carried on into vector applications in the solution of trigonometric problems.

Fine for a trigonometry class or an analytic geometry class.

RISING, GERALD R. "A Geometric Approach to Field Goal Kicking." *Mathematics Teacher* 47: 463–66; November 1954.

An interesting article on an exciting sport. In general the team may consider this for setting up a kick, but it will be more beneficial to the geometry class. There is one drawback to the article—trigonometry is needed to work out the solution.

Good material for geometry or trigonometry classes.

SATTERLY, JOHN. "Trigonometry and the Practical Descriptive Geometry of Oblique Planes." *School Science and Mathematics* 54: 539–55; October 1954.

Presents a very fine use of numerical and analytical trigonometry to extend it to the related field of descriptive geometry.

For elementary geometry and trigonometry classes.

SEEBECK, C. L., JR. "The Logarithm Function to the Base *e*—A Development for High School Seniors." *Mathematics Teacher* 54: 2–3; January 1961.

An approach to the logarithm function to the base *e* that emphasizes some important properties of the function. This is a follow-up of an article by the same author, which appeared in the October 1959 issue of this magazine.

Recommended for use with seniors.

WEGNER, KENNETH W. "Trigonometric Values That Are Algebraic Numbers." *Mathematics Teacher* 50: 557–61; December 1957.

A nice discussion of trigonometric values that are arrived at by equations from algebra. Discusses linear, quadratic, cubic, quartic, and equations of the 5th, 6th, 7th, 8th, and even a 12th degree equation.

Excellent for a trigonometry class.

TIERNEY, JOHN A. "Trigonometric Functions of Real Numbers." *Mathematics Teacher* 50: 38–39; January 1957.

An argument for the change from trigonometric functions of angles to the trigonometric functions of real numbers in the high school curriculum.

Worth considering for the trigonometry class.

YOUNG, FREDERICK H. "The Addition Formulas." *Mathematics Teacher* 5: 45–48; January 1957.

An interesting article on the number of different ways that the addition formulas in trigonometry can be derived. The article is complete with the derivations.

Stimulating to a trigonometry class.

APPLICATIONS

BELLMAN, R., and BLACKWELL, D. "Red Dog, Blackjack, Poker." *Scientific American* 184: 44–47; January 1951.

An introduction to mathematical theory of games, using progressively more complicated coin and card games to illustrate points of the theory.

Perhaps not mathematical enough for the serious student of mathematics.

BERGER, EMIL J. "A Model for Explaining How Latitude May Be Determined by Making Observations on Polaris." *Mathematics Teacher* 47: 405–06; October 1954.
The model is shown and its use is discussed. Applies the plane geometry theorem: "If two angles have their sides perpendicular right side to right side and left side to left side they are equal."
Good for use in plane geometry classes to show a practical application.

BERKELEY, EDMUND C. "Simple Simon." *Scientific American* 183: 40–43; November 1950.
Good article, introducing us to big computers through the characteristics of a small and easily understood one named Simon. Simon can transfer information automatically from one register to another, and can perform reasoning operations of indefinite length. Shows how we can build cheap, simple machines to teach us the fundamentals of computer operation, programming, and design.

BERNAL, J. D. "The Structure of Liquids." *Scientific American* 203: 124–34; August 1960.
The author presents this article as a progress report on his efforts to produce a theory of the geometry of liquid structure. Some surprising polyhedra arise.

BERNSTEIN, ALEX, and ROBERTS, M. DEV. "Computer vs. Chess-Player." *Scientific American* 198: 96–105; June 1958.
Good fun and interesting for the student interested in computer programming and in the question of the possibility of building machines which think. Describes an IBM 704, programmed to play a game of chess considerably above that played by a novice.

BUSHELL, W. F. "Calendar Reform." *Mathematical Gazette* 45: 117–24; May 1961.
An interesting discussion of the history of making a good calendar.
For lower-grade level.

CARNAP, RUDOLF. "What Is Probability?" *Scientific American* 189: 128–38; September 1953.
The final article in an entire issue devoted to "fundamental questions in science," science the "interrogation of nature." Carnap sets forth arguments for the reinstatement of inductive probability alongside statistical probability as a tool for investigating the hypotheses of science.

COHEN, JOHN. "Subjective Probability." *Scientific American* 197: 128–38: November 1957.
Good article bringing probability into the realm of human behavior. Experiments are described and the results mathematized in an understandable way.

COOPER, WILLIAM W., and CHARNES, ABRAHAM. "Linear Programming." *Scientific American* 191: 21–23; August 1954.
Linear programming is a method of pure mathematics which can be applied to human affairs, e.g., to calculate the best possible solution to a problem involving several variables. This article considers a hypothetical manufacturing situation to illustrate the method.

CUNDY, H. MARTYN. "A Demonstration Binary Adder." *Mathematical Gazette* 42: 272–74; December 1958.
Utilizes the design of an electrical circuit to demonstrate binary addition.
For middle-grade level.

DAVIS, HARRY M. "Mathematical Machines." *Scientific American* 180: 28–39; April 1949.
A study of the progress made before April 1949 in the construction and use of computers. The article, being a progress report, is now definitely dated.

GAMOW, GEORGE. "The Principle of Uncertainty." *Scientific American* 198: 51–57; January 1958.
Fine article which states, in terms which the layman can just understand, the dilemma in which modern physics finds itself and the solution proposed by Werner Heisenberg in his uncertainty principle. There is no particular tie-up made with mathematics, but the history is interesting and particularly the history of reaction by Einstein and others to the principle of uncertainty.

GARDNER, MARTIN. "Mathematical Games." *Scientific American* 199: 136–42; November 1958.
A regular monthly feature, which in this issue is devoted to a consideration of the problem of constructing perfect squares and rectangles, i.e., those which can be subdivided into squares no two of which have sides the same length. Some surprising electrical network theory is seen to be applicable to the problem.

HORTON, GEORGE W. "Hobladic." *Mathematics Teacher* 54: 212–16; April 1961.
An interesting report on how students can make out cards to show the working of the punched cards of a computer. It stresses the fact that the computer can not think, but will do only what it is told to do.
Good material for any class.

HURWICZ, LEONID. "Game Theory and Decisions." *Scientific American* 192: 78–83; February 1955.
Some simple examples are used to give the reader an idea of the logic underlying the theory of games, rather than to give him a comprehensive picture with involved and heavily mathematical overtones.

INGALLS, ALBERT G., editor. "The Amateur Scientist." *Scientific American* 188: 104–10; May 1953.
This installment of a regular monthly feature is devoted to a detailed discussion and description of some simple mathematical machines built by amateurs. The drawings are particularly well done, and the descriptions probably more than adequate for most readers.

JONES, PHILLIP S. "Napier's and Genaille's Rods." *Mathematics Teacher* 47: 482–87; November 1954.
An interesting article, complete with diagrams and reproductions of old prints. This is certainly a good topic to discuss as an introduction to the slide rule or use of logarithms. It takes you from Napier's bones, to the Genaille rods, to Bollee's Arithmografo—which links the old devices to the new computing machines.
Good material for a second-year algebra class.

KANE, ROBERT B. "Linear Programming, an Aid to Decision Making." *Mathematics Teacher* 52: 177–79; March 1960.
Presents an interesting use of two- and three-dimensional graphs in which the student has to use both equations and inequalities. It shows which operations on a machine could be more profitable, and takes some of the guesswork out of decision making.
For an algebra class.

KIRKPATRICK, PAUL. "Probability Theory of a Simple Card Game." *Mathematics Teacher* 47: 245–48; April 1954.
A practical application of the theory of probability to the card game *Concentration*. Works out a scheme for discovering the number of cards it is necessary to remember in order to be successful at the game.
For an algebra or advanced mathematics class.

LICHTENBERG, D., and ZWANG, M. "Linear Programming Problems for First-Year Algebra." *Mathematics Teacher* 52: 171–76; March 1960.
This description of a unit on linear programming is complete with graphs. It uses modern, or set, terminology. It might suggest other problems that could be considered for study by this method.
Good material for first-year algebra class.

LITTLEFIELD, DARREL G. "Computer Programming for High Schools and Junior Colleges." *Mathematics Teacher* 54: 220–23; April 1961.
This is actually a proposed outline for a course in computer programming. One must have at least "visitation rights" to a computer to make the course successful.
Recommended as a second, senior course or a project for a math club.

LITTLEWOOD, J. E. "Large Numbers." *Mathematical Gazette* 32: 163–71; July 1948.
An elementary, interesting discussion of physical problems which lead to very large numbers.
For middle-grade level.

MAUNSELL, F. G. "Why Does a Bicycle Keep Upright?" *Mathematical Gazette* 30: 195–99; October 1946.
An interesting discussion of the theory of the stable operation of a bicycle.
For upper-grade level.

MORGENSTERN, OSKAR. "The Theory of Games." *Scientific American* 180: 22–25; May 1949.
Tells of the work of the "Princeton Group," (von Neumann and others), who are forging a new tool for use in the study of economic and social behavior. Much has been written more recently which perhaps dates, but definitely does not invalidate, this article. A good primer in the field.

NADIR, NADI. "Abū al-Wafā on the Solar Altitude." *Mathematics Teacher* 53: 460–63; October 1960.
Based on an ancient treatise on the method of determining the time of day by the solar altitude. Shows the use of trigonometry in spherical astronomy, and also brings in versines to show one of their applications.
Good material for a trigonometry class.

NEMITZ, WILLIAM, and REEVES, ROY. "A Mathematical Theory of Switching Circuits." *Mathematics Magazine* 33: 1–6; September–October 1949.
Definitions and examples of circuits are discussed first; then, (a) study of binary functions of binary n-tuples; (b) study of functions including postulates and operations which are explained in the article; (c) definition of $f + g$, $f \cdot g$, and $f \times g$; (d) "Christmas tree" circuit. It is assumed that the reader has some familiarity with this.
Can be used with students who have some knowledge of electricity and mathematics (12th grade?). Some knowledge of basic ideas of switching circuit con-

struction and mathematical induction is necessary. Also knowing something about sums $\sum\limits_{k=1}^{v}$.

PFEIFFER, JOHN E. "Symbolic Logic." *Scientific American* 183: 22–24; December 1950.

Introductory non-technical, emphasis on practical applications in business and engineering worlds rather than on mathematics. "These applications merely suggest the fruitful future. . . ."

PHILLIPS, JO MCKEEBY. "Motivating the Study of Solid Geometry Through the Use of Mineral Crystals." *School Science and Mathematics* 42: 743–48; December 1952. 53: 134–38; February 1953.

Two articles on the six crystal systems and their relation to solid geometry.

Readable by students in plane and in plane-and-solid geometry.

PINKERTON, RICHARD C. "Information Theory and Melody." *Scientific American* 194: 77–86; February 1956.

Considers an exercise in the analysis of melody as an entertaining way to shed light on "the elusive creative processes of the human mind." Some simple nursery tunes are investigated to uncover the properties which make them tuneful. Harmony as well as melody is briefly considered. Perhaps, "as we begin to understand more about the property of creativeness, our enjoyment of the arts should increase a thousandfold."

SHANNON, CLAUDE E. "A Chess-Playing Machine." *Scientific American* 182: 48–51; February 1950.

Valuable for its introduction to the basic notions of computer operation. Does an electronic computer think? "The limits within which thought is really necessary need to be better defined . . . the automaton can do many things that are properly classed as thought." So said the inventor of an end-game machine in 1914.

STONG, C. L., editor. "The Amateur Scientist." *Scientific American* 199: 107–42; August 1958.

A regular monthly feature, devoted in this issue to "an excursion into the problem of measuring irregular areas." The methods investigated run all the way from intuition to the calculus, from pocket knives to polar planimeters.

VAN TASSEL, LOWELL T. "Notes on a 'Spider' Nomograph." *Mathematics Teacher* 52: 557–59; November 1959.

Nomographs are sadly lacking in the curriculum of the mathematics and science department—as far as their use and design go. This nomograph is a device for solving reciprocal type equations.

Good reading for second-year algebra and trigonometry classes.

VAN TASSEL, LOWELL T. "Digital Computer Programming in High School Classes." *Mathematics Teacher* 54: 217–19; April 1961.

Shows the workings of the programmer and how it is possible to set up programs. The only drawback is that of studying and becoming familiar with the code.

Could be used in any class that is working on word problems.

WALKER, G. T. "The Physics of Sport." *Mathematical Gazette* 20: 172–77; July 1936.

An interesting non-mathematical discussion of examples of dynamics from the world of sport.

For middle-grade level.

WEAVER, WARREN. "Probability." *Scientific American* 183: 44–47; October 1950.

Excellent article, clearly and completely exploring three or four facets of the history, growth, fields of useful application, and possible future uses of this branch of mathematics. A probabilistic view of scientific phenomena and of discoveries in nature is plausibly broached.

WOODBY, LAUREN G. "Computational Nomographs." *Mathematics Teacher* 48: 554–56; December 1955.

Discusses nomographs—devices which are not used or discussed enough in the high school mathematics or science departments. This particular nomograph may be used for products, quotients, and square roots.

Any algebra class would be interested in this.

MISCELLANEOUS

BAKST, AARON. "Hyperbolic Functions." *Mathematics Teacher* 46: 71–77; February 1953.

Discusses the value of study of hyperbolic functions and argues for its inclusion in the half-year course in trigonometry. Worth considering, and should be available for extra reading by the better student.

Good material for trigonometry classes.

CARNAHAN, WALTER H. "*Pi* and Probability." *Mathematics Teacher* 46: 65–66; February 1953.

Gives several methods of approximating *pi*, including Buffon's method with the stripes of the flag. Includes a proof that this is a reasonable approximation of the constant.

Good reading for a geometry or a probability and statistics class.

COHEN, I. BERNARD. "An Interview with Einstein." *Scientific American* 193: 68–73; July 1955.

Delightful interview in which Einstein gives estimates of the lives and works of other scientists.

Of particular interest to the student of the history of science.

COHEN, I. BERNARD. "Isaac Newton." *Scientific American* 193: 73–80; December 1955.

A good short biography, setting forth the scope of Newton's work and some of the flavor of his personal life.

Of particular value to the student of the history of science.

CROMBIE, A. C. "Descartes." *Scientific American* 201: 160–73; October 1959.

In this adequate short biography, the author shows that Descartes, while remembered for his publications in analytic geometry, attempted far more: "to reduce nature to mathematical law."

Of special interest for the student of the history of mathematics and science.

DIMMICK, EDGAR L. "Introducing Symbolic Logic." *Mathematics Magazine* 30: 18–24; September–October 1956.

Mentions Aristotle in a good introduction of the subject. Discusses: "If - then" statements; disjunction, negation X; alternation V; conjunction; truth functions; use of tables—many examples are introduced. A very elementary and clear article. A list of references includes: *Mits, Wits and Logic* by Lieber; *Methods of Logic* by Quine; *An Introduction to Symbolic Logic* by Langer; *Fundamentals of Symbolic Logic* by Ambrose.

Could be used at any level from 9–12.

DUREN, W. L., JR. "The Maneuvers in Set Thinking." *Mathematics Teacher* 51: 322–35; May 1958.

Presents the argument that sets are the basic mathematics; from there the author goes on to develop the ideas of algebra. Includes set notation, subsets, matching, induction, counting, Cartesian product, and methods of classifying the information.

Worthwhile for algebra students and teachers.

EINSTEIN, ALBERT. "On the Generalized Theory of Gravitation." *Scientific American* 182: 13–17; April 1950.

Excellent article, particularly good in its presentation of the philosophy of scientific inquiry and history of developments leading to the publication of the theory.

The mathematics mentioned is small in amount and "relatively" easy to follow.

FLETCHER, T. J. "The Solution of Inferential Problems by Boole Algebra." *Mathematical Gazette* 36: 183–88; September 1952.

A discussion of Boolean algebra with applications to solving inferential problems.

For upper-grade level.

GARDNER, MARTIN. "Mathematical Games." *Scientific American* 201: 166–77; December 1959.

In this, one of his monthly columns, Gardner explores diversions which clarify group theory. A network-tracing game and braid-weaving are used to get transformations.

GEORGES, J. S. "Parabolic Functions." *School Science and Mathematics* 58: 138–47; February 1958.

Definition and a short study of functions which are analogous to circular and hyperbolic functions.

Beginnings of the calculus needed.

GIBBINS, N. M. "Chess in Three and Four Dimensions." *Mathematical Gazette* 28: 46–50; May 1944.

An account of chess in three and four dimensions.

For middle-grade level.

GOODSTEIN, R. L. "Proof by *Reductio Ad Absurdum*." *Mathematical Gazette* 32: 198–204; July 1948.

An interesting lecture on the logic of proof by *reductio ad absurdum*.

For upper-grade level.

HALMOS, PAUL R. "Innovation in Mathematics." *Scientific American* 199: 66–73; September 1958.

Lead article in an entire issue devoted to innovation in science. Well done, it tells of the lives of many innovators in mathematics. Well illustrated. "The mathematician seeks a new logical relationship, a new proof of an old relationship, or a new synthesis of many relationships."

HALMOS, PAUL R. "Nicolas Bourbaki." *Scientific American* 196: 88–99; May 1957.

A progress report on the mathematical treatise of the age (*Eléments de Mathématique* or the "Bourbaki Treatise"—20 volumes so far) and the authors of this treatise which was started in 1939. This report is done in a light style, as befits the members of the Bourbaki group and the stories they have allowed (caused?) to grow up concerning them.

HOFFMAN, A. A. J., and OSBORN, ROGER. "Concerning Simultaneous Solutions of Polar Equations by the Graphical Method." *Mathematics Teacher* 53: 133–34; February 1960.

Brings out the difference between the number of graphical intersections of two curves in polar coordinates and the number of simultaneous solutions of two equations in polar coordinates.

Suitable for all analytic geometry groups.

KARST, OTTO J. "The Limit." *Mathematics Teacher* 51: 443–49; October 1958.

Uses a conversation between a student and his teacher as a device to clarify the idea of a limit. Brings out many good points, and many hazy ideas held by students are revealed by the questions.

Good reading for the teacher who has not taught or had a course in calculus for a few years.

LARSEN, H. D. "Mathematics on Stamps." *Mathematics Teacher* 48: 477–80; November 1955.

Presents a topic often overlooked by mathematics teachers: the number of countries that have honored mathematicians on stamps. Gives a list of about 40 mathematicians and the countries, date of issue, and denomination of the stamps.

Excellent for a math club or for historical background.

LE CORBEILLER, P. "The Curvature of Space." *Scientific American* 191: 80–86; November 1954.

An account of the material contained in Riemann's paper on the foundations of geometry given at Gottingen in 1854. The author beautifully weaves three threads—Gauss, Riemann, Einstein—together. The drawings are well done. A delightfully interesting article.

NAGEL, ERNEST, and NEWMAN, JAMES R. "Gödel's Proof." *Scientific American* 194: 71–86; June 1956.

Gives the background of the problem and the substance of Gödel's findings. Gödel showed, by reasoning not to be followed by the untrained mind, that it was impossible for the entirety of mathematics to be systematized by the axiomatic method.

Readable, even by the layman, and will be of interest to every serious student of mathematics.

NAGEL, ERNEST. "Automatic Control." *Scientific American* 187: 44–47; September 1952.

Introductory article in an issue which is entirely devoted to the subject of self-regulatory machines. *Note:* All eight articles in this issue plus other materials have been compiled by the editors of the *Scientific American* in the paperback: *Automatic Control*, New York: Simon and Schuster, 1955.

NEWMAN, JAMES R. "The Rhind Papyrus." *Scientific American* 187: 24–27; August 1952.

A fine article for the student of the history of mathematics, the written matter being complete and greatly enhanced by a carefully detailed diagram and accompanying key. "It would be misleading to describe the Rhind as a treatise (though it is the principal source of what we know of Egyptian mathematics)—it is a collection of mathematical exercises and practical examples worked out in a syncopated, cryptic style."

NEWMAN, JAMES R. "Laplace." *Scientific American* 190: 76–81; June 1954.
Short biography of Pierre Simon de Laplace, mathematician, physicist, astronomer . . . and "not an entirely admirable man."

PARKER, JEAN. "The Use of Puzzles in Teaching Mathematics." *Mathematics Teacher* 48: 218–27; April 1955.
Contains many ideas from paradoxes to crossword puzzles. Tells about the problems, gives the solutions, and discusses the errors of reasoning in the paradoxes. Contains a very extensive bibliography on puzzles.
Good reading for an algebra class or as reference for the teacher.

REEVE, J. E., and TYRRELL, J. A. "Maestro Puzzles." *Mathematical Gazette* 45: 97–99; May 1961.
A short discussion of puzzles which are concerned with packing a given set of figures to form a certain figure.
For lower-grade level.

RIPPEY, ROBERT M. "Rotation of Axes with Complex Numbers." *Mathematics Teacher* 53: 197; March 1960.
A very short article which states the methods or formulas used for rotating the axes to find the new coordinates of a point by means of complex numbers.
Useful for a group studying complex numbers or analytic geometry.

RIPPEY, ROBERT M. "Some Challenging Series for Gifted High School Students." *Mathematics Teacher* 54: 8–11; January 1961.
A good topic to fit in after arithmetic and geometric progressions. Includes Fibonacci series, sum of the squares of positive integers from 1 to n, and a generalization of figurate numbers.
Fine for second-year algebra or above.

ROBBINS, CHARLES K. "Analytic Geometry—The Framework of Mathematics." *Mathematics Magazine* 22: 201–10; March–April 1949.
A systematic development of coordinate geometry from the ideas of graphing the temperature of a thermometer through plane analytic geometry, the graphing of linear and quadratic functions, cycloids, trigonometric functions, and polar coordinates. Not a revolutionary article but a good development.
Some background in trigonometry is helpful. Could be used in 10th, 11th, 12th grade.

SANFORD, VERA. "September Hath XIX Days." *Mathematics Teacher* 45: 336–39; May 1952.
An interesting article on the various forms that the calendar has taken in the past. It might be a good introduction to the methods used today in calculating the calendar, and it could be tied in with spherical trigonometry or spherical geometry.
Any one can read and enjoy this, but it might be best used in trigonometry or solid geometry.

SCORER, R. S.; GRUNDY, P. M.; and SMITH, C. A. B. "Some Binary Games." *Mathematical Gazette* 30: 96–103; July 1944.
A discussion and generalization of the two games: Tower of Hanoi, and Chinese Rings.
For middle-grade level.

SMITH, C. A. B. "The Counterfeit Coin Problem." *Mathematical Gazette* 31: 31–39; February 1947.
A rather extensive treatment of the counterfeit coin problem and its various versions.
For upper-grade level.

STRUIK, D. J. "Omar Khayyam, Mathematician." *Mathematics Teacher* 51: 280–84; April 1958.
This brings out the "other" side of Omar Khayyam and his works in algebra. The work includes the cubic equation, the binomial theorem, and a discussion of the parallel postulate.
Good background reading for algebra teachers.

WEAVER, WARREN. "Lewis Carroll: Mathematician." *Scientific American* 194: 16–28; April 1956.
Short biography of interest to the student of personalities in mathematics.

WEINER, L. M. "Take a Number." *Mathematics Teacher* 48: 203–04; April 1955.
A short but very interesting article about a mind–reading trick to simplify algebraic expressions and solve linear equations.
Good reading for all teachers and some algebra classes.

WHITTAKER, SIR EDMUND. "Mathematics." *Scientific American* 183: 40–42; September 1950.
Survey article on mathematics in an issue devoted entirely to surveys of developments in various fields of scientific endeavor from 1900 to 1950.
Good article for history and commentary on the more significant mathematical developments of this half-century.

WIRSZUP, IZAAK. "The Seventh Mathematical Olympiad for Secondary School Students in Poland." *Mathematics Teacher* 51: 585–89; December 1958.
This will give the teacher an idea of what is expected of the Soviet Union's gifted students in mathematics. The article includes the tests and an appraisal of the grades on the tests.
Good reading for teachers and stimulating for students.

YATES, ROBERT C. "The Cardiod." *Mathematics Teacher* 52: 10–15; January 1959.
A very interesting study of the cardiod. Includes the method of constructing it, some linkages for generating it, curve stitching, and applications as cams.
Very good material for a senior mathematics class or math club.